Charting the Next Pandemic

Ana Pastore y Piontti · Nicola Perra · Luca Rossi
Nicole Samay · Alessandro Vespignani

Charting the Next Pandemic

Modeling Infectious Disease Spreading in the Data Science Age

With contributions by Corrado Gioannini, Marcelo F. C. Gomes, and Bruno Gonçalves

 Springer

Ana Pastore y Piontti
Northeastern University
Boston, MA
USA

Nicola Perra
University of Greenwich
London
UK

Luca Rossi
Institute for Scientific Interchange
Torino
Italy

Nicole Samay
Northeastern University
Boston, MA
USA

Alessandro Vespignani
Northeastern University
Boston, MA
USA

Collaborators
Corrado Gioannini
Institute of Scientific Interchange
Torino
Italy

Marcelo F. C. Gomes
Oswaldo Cruz Foundation
Rio de Janeiro
Brazil

Bruno Gonçalves
Center for Data Science
New York, NY
USA

ISBN 978-3-319-93289-7   ISBN 978-3-319-93290-3   (eBook)
https://doi.org/10.1007/978-3-319-93290-3

This book is in memory of Duygu Balçan, who has been a fantastic scientist and friend. She was suddenly taken away from us and from science, but she will always be with us in our dearest memories and through her outstanding scientific contributions.

. . . .

ACKNOWLEDGMENTS

It is the work on many collaborative projects that led to the tools and methodologies presented in the pages of this volume. We benefited from more than 10 years of collaborations and scientific exchanges with a long list of colleagues who helped us in shaping our view and knowledge of computational epidemiology through invaluable scientific discussions.

First of all, we want to thank Vittoria Colizza for her invaluable scientific collaborations. She has also been one of the engines behind the development of the GLEAM framework. We can never thank her enough for the work and contributions to many of the ideas and concepts illustrated in this book.

We are also grateful to Alain Barrat, Marc Barthelemy, and Romualdo Pastor-Satorras, who have been early companions in the analysis of global mobility networks and global epidemics.

Paolo Bajardi, Hao Hu, José J. Ramasco, Daniela Paolotti, Chiara Poletto, Michele Tizzoni, and Qian Zhang have been key collaborators of ours for many years and have greatly contributed in many tangible and intangible ways to this volume. We have grown with and learned so much from them.

Marco Ajelli and Stefano Merler contributed to the development of computational models and data-driven agent-based models that have literally defined the state of the art in the field. We are very fortunate to have the privilege of collaborating and working with them through several common projects.

We are simply honored and humbled by the possibility of collaborating with pioneers in the field of the caliber of M. Elizabeth Halloran and Ira M. Longini. We are thankful to them for the many projects and ideas that have truly shaped our way of looking at epidemic modeling.

We started to think about global epidemics as a data visualization problem because of Wouter Van den Broeck. He is the one who translated the GLEAM framework into images in the first place. Marco Quaggiotto is the one who, for many years, has created the GLEAMviz visual appearance. We are thankful to them for their constant inspiration and lessons in working with data visualization as an art.

We thank Dina Mistry for her invaluable comments and careful proofreading of the book manuscript.

A special thanks goes to all colleagues, students, and coauthors who have shared our journey and excitement in the science of epidemic modeling: Luca Cappa, Claudio Castellano, Ciro Cattuto, Dennis Chao, Matteo Chinazzi, Gerardo Chowell, Fabio Ciulla, Natalie E. Dean, Laura Fumanelli, Maria Litvinova, Paolo Milano, Delia Mocanu, Yamir Moreno, Evangelia (Evelyn) Panagakou, Daniela Perrotta, Piero Poletti, Diana Rojas, Michele Roncaglione, Mauricio Santillana, Mohamed Selim, Lone Simonsen, Kaiyuan Sun, Alain Jaques Valleron, Piet Van Mieghen, and Cécile Viboud. We are infinitely grateful for their outstanding support, ideas, and work.

Vibrant research environments and occasions of discussion with colleagues of all ages and expertise are the key to good science. We have been very lucky to be part of the Center for Inference and Dynamics of Infectious Diseases (CIDID), a NIH-funded Center of Excellence, the Summer Institute in Statistics and Modeling in Infectious Diseases (SISMID), and the Network Science Institute at Northeastern University and their wonderful scientific communities.

We also want to thank all those institutions that have supported our work and the completion of this book. Northeastern University in Boston and the Institute for Scientific Interchange in Turin have provided the authors of this book with great environments for their research and work. We are incredibly grateful to the Sternberg Family for the outstanding support of the MOBS Laboratory at Northeastern University. Finally we acknowledge the generous funding support at various stages of our research in the area of predictive epidemic modeling from the Models of Infectious Disease Agent Study, National Institute of General Medical Sciences Grant U54GM111274, and the Fondazione-CRT Lagrange laboratory.

....

Boston, USA   April 2018
Torino, Italy
London, UK

About the authors

Ana Pastore y Piontti is an Associate Research Scientist in the Laboratory for the Modeling of Biological and Socio-technical Systems (MOBS Lab) at Northeastern University in Boston, USA. Her research focuses on the characterization and modeling of the spread of infectious diseases, by integrating methods of complex systems with statistical physics approaches.

Nicola Perra is a Senior Lecturer in network science at the University of Greenwich, UK. His research interests revolve around the study of human dynamics, digital epidemiology, and network science.

Luca Rossi is a Senior Researcher at the Institute for Scientific Interchange in Torino, Italy. His research focuses on the mathematical and computational modeling of contagion processes in structured populations, in particular human transmittable infectious diseases.

Nicole Samay is a Senior Graphic Designer in the Network Science Institute at Northeastern University in Boston. She works with researchers to develop and adapt data visualizations, with a particular focus on information design and spreading processes.

Alessandro Vespignani is the Sternberg Family Distinguished Professor at Northeastern University in Boston. He is a fellow of the American Physical Society, member of the Academy of Europe, and fellow of the Institute for Quantitative Social Sciences at Harvard University. Vespignani is focusing his research activity in modeling diffusion phenomena in complex systems, including data-driven computational approaches to infectious disease spread.

This book is also a result of the special support and work at various stages of the editorial process from three fantastic collaborators:

Corrado Gioannini worked for several years in the private sector, in IT companies, developing his skills in software development and management. He is now a research leader at the Complex Systems and Networks group at ISI Foundation, where he coordinates the development of the GLEAMviz Simulator software framework.

Marcelo F. C. Gomes is a researcher on infectious disease modeling and surveillance at Fundação Oswaldo Cruz's Scientific Computing Program (Fiocruz, PROCC). His main research focus is on combining reported cases from public health agencies and human mobility for the development of risk analysis and nowcasting tools.

Bruno Gonçalves is a Moore-Sloan Fellow at NYU's Center for Data Science. With a background in Physics and Computer Science, his career has revolved around the use of datasets from sources as diverse as Apache web logs, Wikipedia edits, Twitter posts, Epidemiological reports, and census data to analyze and model human behavior and mobility. More recently he has focused on the application of machine learning and neural network techniques to analyze large geolocated datasets.

CONTENTS

05 THE NUMERICAL FORECAST OF PANDEMIC SPREADING: THE CASE STUDY
OF THE 2009 A/H1N1 PDM   55

 PANDEMIC INFLUENZA   73

 CORONAVIRUS   167

 EBOLA VIRUS   183

06 COMPUTATIONAL MODELING OF "DISEASE X"   197

PART II: PANDEMIC CHARTS

INTRODUCTION

. . . .

OUR INTENT IS TO INTRODUCE NON TECHNICAL READERS TO THE PROCESS THAT, STARTING FROM REAL-WORLD DATA, MAKES IT POSSIBLE TO DEVELOP SIMULATED SCENARIOS AND REAL-TIME FORECASTS OF THE GLOBAL SPREADING OF INFECTIOUS DISEASES.

THE FIRST SUCCESSFUL WEATHER FORECAST performed by a digital computer can be traced back to 1950. Five years later, a joint project by the US Air Force, Navy, and Weather Bureau began operational numerical weather prediction in the United States. Nowadays, computer-generated weather forecasts are at the fingertips of billions of individuals through government or commercial services, accessible by any mobile platform on the Internet. Weather forecasts can be very specific, focusing on local area models, or global. They can span a few hours or several days. They are part of our daily life and have helped popularize concepts like the "butterfly effect."[1]

A non-expert reader might reasonably think that the use of computer simulations for infectious disease outbreak forecast is as developed as the one

---

1 James Gleick, *Chaos: Making a new science* (Viking Press, 1987).

regarding weather forecast; unfortunately this is far from the case. Although the birth of mathematical epidemiology dates back to the eighteenth century, public health scientists and policy-makers started only in the last 20 years or so to increasingly rely on simulated models to understand epidemiological patterns and guide control measures in real time. It may come as a surprise, but while the weather forecast research community has built thousands of weather stations around the world, put satellites in orbit, and connected large supercomputing infrastructures, the modeling of epidemic outbreaks has long suffered from the lack of near-real-time, high-quality data on populations, human and animal mobility/behavior patterns, and pathogens' biology. For instance, the 1985 pioneering work of Ravchev and Longini[2] on a mathematical model for the global spread of influenza had to wait on a shelf for almost 20 years before a full-fledged implementation integrating the complete International Airlines database[3] saw the light. Similarly, the first cross-comparison of three different data-driven, individual-based, stochastic models of pandemic flu examining the consequences of intervention strategies in the United States had to wait until 2008.[4]

In the last 10 years, we have seen dramatic advances in data collection and availability in a number of areas, ranging from pathogen genetic sequences to human mobility patterns and social media data. These advances, often dubbed as the "big data" revolution, have finally lifted many of the limitations affecting epidemic predictive modeling. The big data paradigm is generally associated with "inductivist" approaches such as statistical modeling, phenomenological models, or machine learning-based methodologies. However, data availability also allows the development of detailed mechanistic models based on the construction of synthetic populations that statistically mimic the real world. They explicitly account for the dynamics of epidemics by calculating the future state of the system from its initial state, through time- and space-dependent equations, as well as the stochastic simulation of individual disease transmission processes.

Mechanistic approaches are clearly data hungry, and the amount of data integration depends on the scale of the model—global, regional, and local—as well as the level of detail in the population description. The latter can go down to the level of single households and specifically consider multiple transmission settings, such as schools or workplaces. It is worth stressing, however, that mechanistic models contain assumptions and approximations too. The theory and equations used to describe the system dynamics are often based

2 Leonid A. Rvachev and Ira M.Longini, Jr., "A mathematical model for the global spread of influenza," Mathematical Biosciences, **75**:3 22 (1985).

3 Lars Hufnagel et al., "Forecast and control of epidemics in a globalized world," Proceedings of the National Academy of Sciences of the United States of America **101**, 15124–15129 (2004).

4 M. Elizabeth Halloran et al., "Modeling targeted layered containment of an influenza pandemic in the United States," PNAS **105**, 4639–4644 (2008).

on effective or coarse-grained integration of degrees of freedom that are informed by the questions the model is set up to answer. No model fits all diseases or spans all scales and geographical resolutions.

In this context, predictive epidemic modeling is emerging as an interdisciplinary field that promises to advance the capabilities of projecting the course of an epidemic already underway or to anticipate the effectiveness of possible interventions or clinical trials. Indeed, computational modeling has been used to support responses to recent outbreaks such as the 2009 H1N1 pandemic, the 2014 West Africa Ebola outbreak, the Zika epidemic in the Americas in 2016, and the highly pathogenic avian influenza A(H7N9) in 2014 and 2015. Although predictive epidemic modeling is not yet as developed as weather forecasting, and many practical and foundational challenges still need to be addressed, the promise of computer simulations to improve epidemic preparedness and response is now recognized.[5]

The many types of descriptive and predictive models that were used during recent, large-scale outbreaks are often hidden behind the veil of technical jargon, mathematical and statistical language, and computational implementation. In order to reach out beyond the circle of practitioners and convey the transformative potential of computer simulations for public health preparedness and response, this book aims to provide a visual journey through the data and model integration process at the core of large-scale computational approaches. This overview is mostly done by using a storyboard that exemplifies data and algorithms through concrete examples and illustrations. Our intent is to introduce non technical readers to the process that, starting from real-world data, makes it possible to develop simulated scenarios and real-time forecasts of the global spreading of infectious diseases.

THE PROMISE OF DATA-DRIVEN MODELING APPROACHES TO IMPROVE EPIDEMIC PREPAREDNESS AND RESPONSE IS NOW RECOGNIZED.

In its first part, this book guides the reader in the construction of the modern frameworks used to project and analyze the global spread of epidemics and pandemics. The results of these modeling activities are intuitively communicated by powerful infographics; in particular, we present examples of results obtained from numerical simulations concerning the international spreading of potentially pandemic pathogens. The second part of this book is focused on a set of pandemic charts that illustrate, through the infographic tools described in the first part, the possible scenarios of future pandemics. This atlas is meant to show commonalities and patterns in emerging health threats, as well as explore the wide range of possible scenarios that can be used by policy-makers to anticipate trends, evaluate risks, and eventually manage future events. In a nutshell, the second part of this book is a visual catalog that captures the possible evolution of future pandemics and introduces the reader to a vast range of interventions characterizing the fight against infectious diseases.

---

5 National Science and Technology Council Report, "Towards Epidemic Prediction: Federal Efforts and Opportunities in Outbreak Modeling" (2016).

THE RIGOROUS ANALYSIS AND DISCUSSION OF PANDEMIC RISK IS NOW ONE OF THE RESEARCH FRONTIERS OF COMPUTATIONAL EPIDEMIC MODELING.

In order to exemplify numerical epidemic modeling, throughout the book we used GLEAM, the global epidemic and mobility framework,[6] developed and supported by a team of researchers and institutions around the world. This framework is by no means to be considered prototypical; however, it integrates many of the data and concepts common to the many types of descriptive and predictive models available to the scientific community. We feel this framework conveys to a general audience the kind of work and results that can be achieved in computer simulated epidemic models. We also refer to several other models and approaches that attain the same level of complexity and results that we present here; we apologize in advance to all the colleagues that have done extraordinary work in this area if their contributions are not referenced or explained in detail. Indeed, this book is not meant as a technical review of the field, but rather as an introduction for non-practitioners to the richness of this approach, showing the potential of looking to the future of global epidemic modeling through data-driven numerical approaches.

It is also important to stress that the pandemic charts are not to be considered an exhaustive catalog of epidemic events: a full exploration of all the possible scenarios, as well as the risks associated to pandemic events, is far beyond the scope of this book and necessarily involves a large effort from the entire scientific community.

The rigorous analysis and discussion of pandemic risk is now one of the research frontiers of computational epidemic modeling, where a number of major scientific challenges still need to be addressed in the coming years, as acknowledged in the book's final outlook chapter. Indeed, we hope that this book will contribute to fueling the interest in solving these challenges and advancing numerical epidemic modeling to the point of an operational framework analogous to the one used for numerical weather forecasting.

---

6 The Global Epidemic and Mobility model, www.gleamviz.org

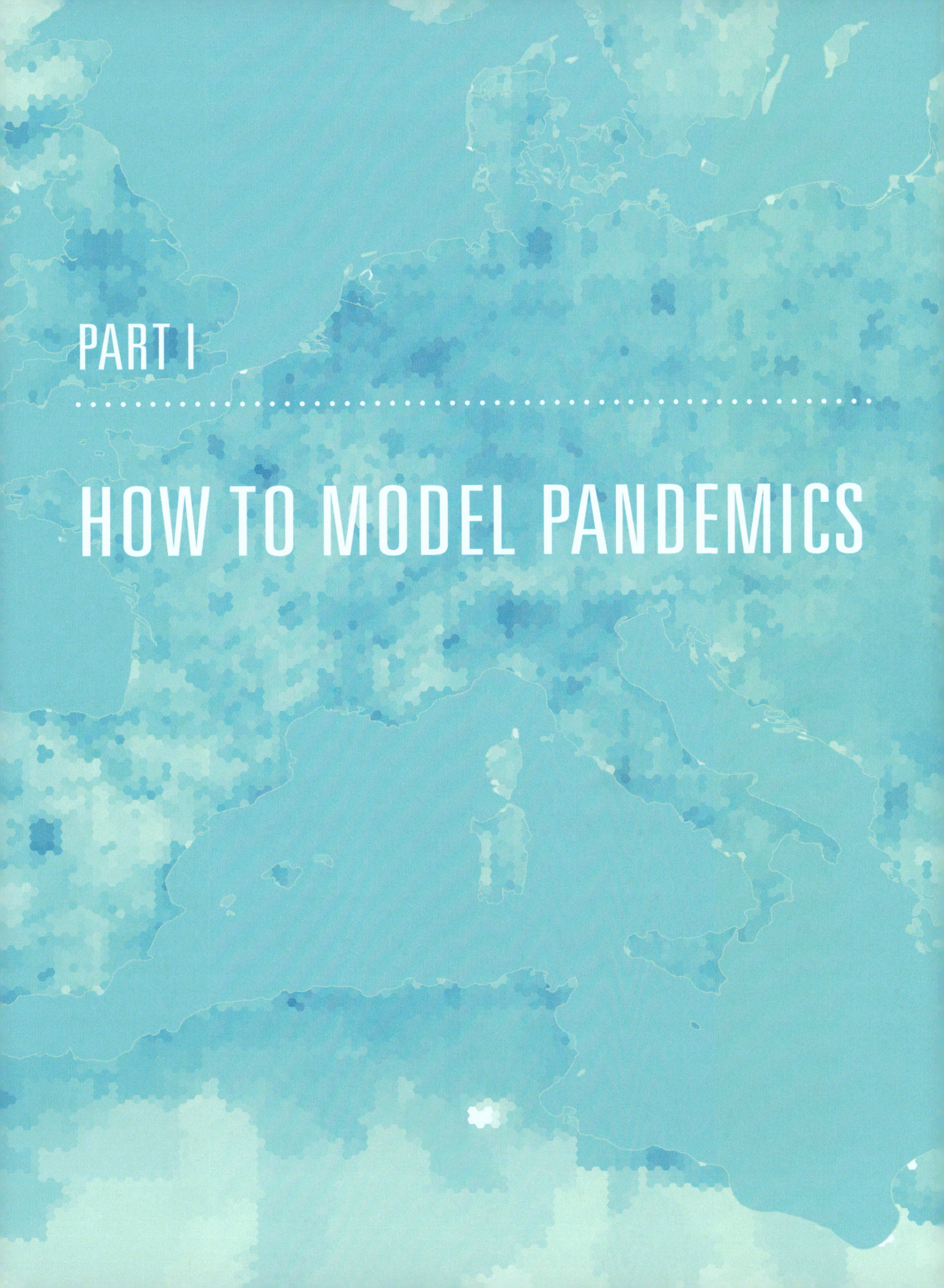

PART I
....

HOW TO MODEL PANDEMICS

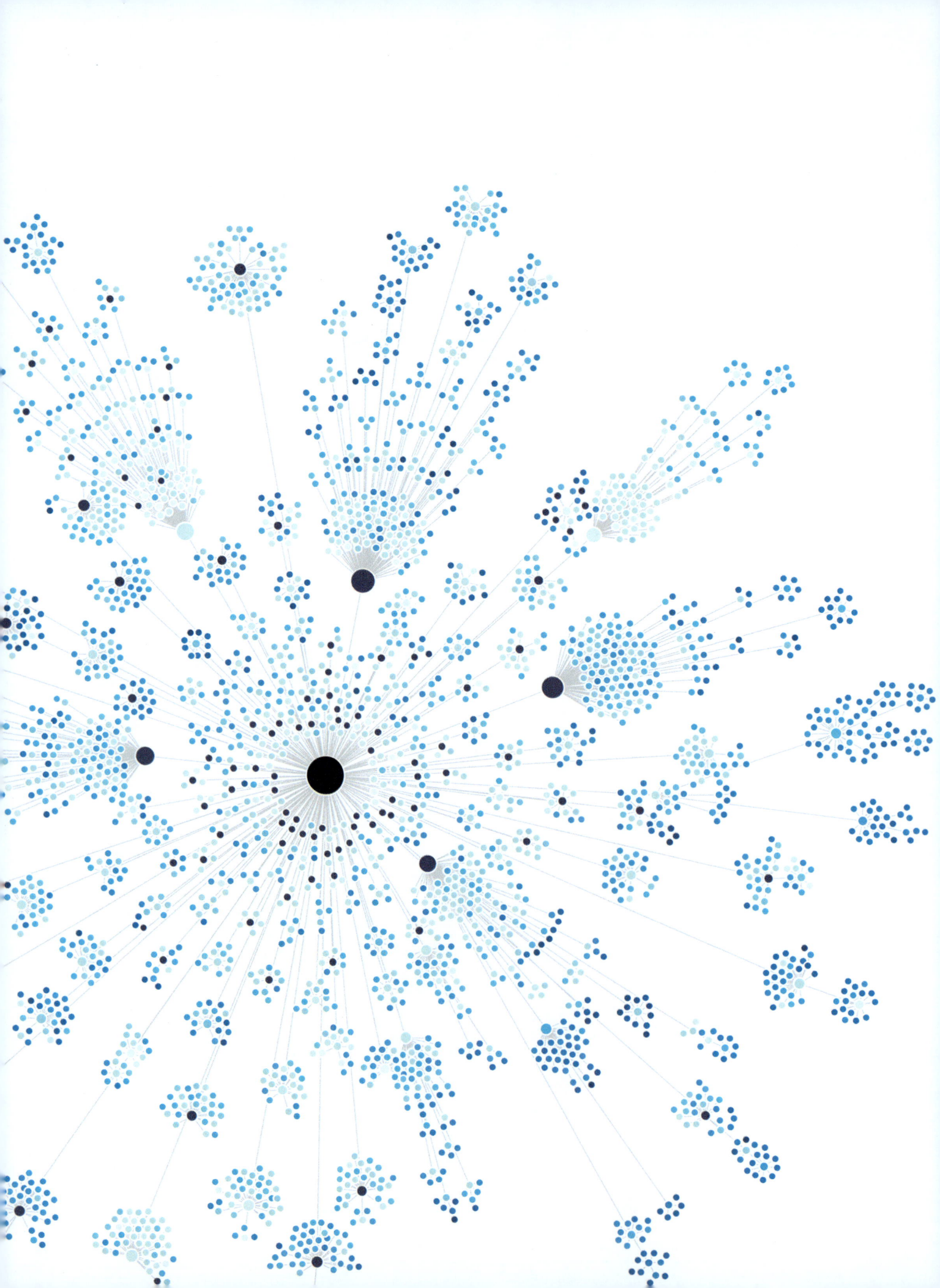

CHAPTER 1

···

INFECTIOUS DISEASE SPREADING: FROM DATA TO MODELS

WE LIVE IN AN INCREASINGLY INTERCONNECTED WORLD where every day one billion cars take the road and more than two billion people travel each year by plane. Urbanization, growing populations, and global migrations are creating a new and complex battlefield in the fight against new and old diseases. As a result, we demand ever-increasing predictive power to anticipate future epidemic outbreaks and evaluate associated risks. In scientific terms, this power corresponds to the mathematical description of patterns found in real-world data needed to develop models that can be used to predict future events.

In the natural sciences, we are used to predicting the complex properties of new materials through precise measurements of physical quantities and the implementation of numerical models or studying the performance of a new airplane by means of computers before we even assemble one of its parts. One of the most successful examples of predictive power is that of weather forecasts, which are generated using sophisticated simulations on supercomputer infrastructures which integrate present data and huge collections of previous meteorological patterns into large systems of non-linear equations. Although weather forecasts may help us plan for our next barbecue, more importantly the science behind this allows us to prepare for extreme and disruptive natural events like tornadoes, storms, and hurricanes, saving countless lives every year.

In sharp contrast, for a long time, achieving predictive power in areas such as epidemic spreading meant confronting the insurmountable obstacle of a lack of high-quality data on human behavior and mobility at all scales. In fact, although mathematical models have been important tools in analyzing the spread and control of infectious diseases for more than a century, in most cases it was impossible to gather detailed data concerning individual and collective behavior and turn those models into real-time predictive computational tools. In the last 20 years, this foundational limitation has started to lift: indeed, every 1.2 years, more human-driven socioeconomic data is produced than during all preceding human history.[1] This data revolution has started the quest of a

---

1 James Manyika, Michael Chui, Brad Brown, Jacques Bughin, Richard Dobbs, Charles Roxburgh, and Angela H Byers, "Big data: The next frontier for innovation, competition, and productivity," (2011).

A MULTIDISCIPLINARY APPROACH TO EPIDEMIC ANALYSIS

01
MODELING
Elaborate stochastic infectious disease models to support a wide range of epidemiological studies, covering different types of infections and intervention scenarios.

02
REAL-WORLD DATA
Real-world data on population and mobility networks are integrated in structured spatial epidemic models to generate data-driven simulations of the worldwide spread of infectious diseases.

03
COMPUTATIONAL THINKING
The computer is the laboratory. Models run on high-performance computers to create in silico experiments that would be hardly feasible in real systems, to guide our understanding of typical non-linear behavior and tipping points of epidemic phenomena.

04
TOOLS DEVELOPMENT
Computational tools help in modeling the spread of a disease, understanding observed epidemic patterns, and studying the effectiveness of different intervention strategies. These tools are available to researchers, healthcare professionals, and policy-makers.

new mathematical and data-driven understanding of human networks and dynamics and finally ignited a transformative science cycle based on the following components:

- Collection, acquisition, and integration of human dynamics data from the individual to the societal scale
- Development of data infrastructures, collaborative information platforms, enabling the production of knowledge from data
- Identification of general principles and laws that characterize social complexity and capture the essence of human dynamics (data analysis and integration)
- Development of mathematical and data-driven models endowed with a high level of realism, able to offer novel quantitative understanding and predictive power in the area of socio-technical systems

THE COMPUTER IS NOT JUST A NUMBER CRUNCHING TOOL: IT ALLOWS RESEARCHERS TO SIMULATE AND STUDY SYSTEMS THAT DO NOT FIT IN A LABORATORY.

The spreading of infectious diseases has been one of the first scientific areas where this research program has been put in motion. The scientific community is finally in the position to envision the development of large-scale models that, by combining theory, data, and computational thinking, can deliver real-time or near-real-time situational awareness during infectious disease outbreaks.

COMPUTATIONAL MODELING OF INFECTIOUS DISEASE SPREADING

The development of data-driven models is rooted in the combination of large-scale data mining techniques, computational approaches, and the mathematical modeling of infectious diseases. Data-driven computational approaches have solid foundations on the wealth of data that can be integrated at different scales, from the biology of the pathogen and the behavior of the single individual to the "social aggregate," where the "social aggregate" is a large-scale social system consisting of millions of individuals whose dynamics can be characterized in space (geographic and social) and time. Those data are integrated in mathematical epidemic models, which are adapted to deal with the multi-scale and complex nature of the real world. In this context, the mathematical and statistical modeling framework has evolved from simple compartmental models to structured approaches in which the heterogeneities and details of the population and system under study are becoming increasingly important features (FIGURE 1.1). Modeling approaches explicitly include spatial structures consisting of multiple sub populations coupled by traveling fluxes, while the epidemic within each sub population is described according to approximations depending on the specific case studied. This patch, or meta population modeling framework, has then grown into a multi-scale framework in which the various possible granularities of the system (country,

| HOMOGENEOUS MIXING | SOCIAL STRUCTURE | CONTACT-NETWORK MODELS | MULTI-SCALE MODELS | AGENT-BASED MODELS |

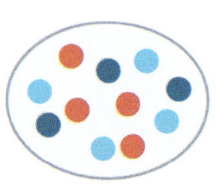 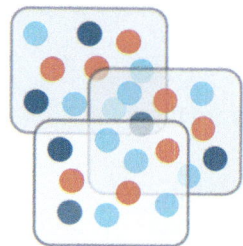 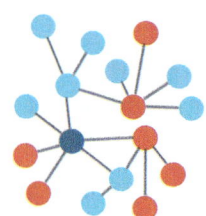 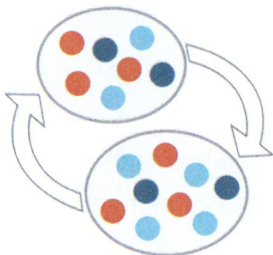 

Figure 1.1 | Different scale structures used in epidemic modeling

Circles represent individuals and each color corresponds to a different stage of the disease. From left to right: homogeneous mixing, in which individuals are assumed to homogeneously interact with each other at random; social structure, where people are classified according to demographic information (age, gender, etc.); contact-network models, in which the detailed network of social interactions between individuals provides the possible disease propagation paths; multi-scale models, which consider sub population coupled by movements of individuals, while homogeneous mixing is assumed on the lower scale; agent-based models, which recreate the characteristic movements and interactions of any single individual on a very detailed scale (a schematic representation of a city is shown).

Adapted from V. Colizza et al., C. R. Biologies 330 (2007)

inter city, and intra city) are considered through different approximations and combined with interaction networks describing the flows of people and/or animals. At the most detailed level, the introduction of agent-based models (ABMs) has enabled the usual modeling perspective to stretch even further by achieving a full description of the society and a complete characterization (household, workplace, etc.) of each individual.

The advances in data-model integration have highlighted complex properties and heterogeneities, which cannot be neglected in the description of epidemics. Although these characteristics have long been acknowledged as relevant factors in determining the properties of dynamical processes, it is clear now that the large-scale epidemics often elude the straightforward linear thinking we are used to and surprise us with tipping points, emergent behaviors, and unexpected shifts in dynamical regime that characterize complex phenomena. In this context, the computer plays a fundamental role; on the one hand, it allows the creation of in silico experiments hardly feasible in real systems, and on the other hand, it can access and compare quantities and observables across many different models. This computational approach is the guide for understanding typical non-linear behaviors and tipping points not accessible by analytical means. In this way, the computer is not just a number crunching tool: it allows researchers to simulate and study systems that do not fit in a laboratory, such as the traffic patterns in a big city or the spreading of a new pandemic. In this perspective, it has been crucial to introduce computational algorithms that rely on the repeated computation of stochastic processes. These techniques, generally named Monte Carlo methods or stochastic algorithms, are especially useful in studying systems with a large number of coupled degrees of freedom and where stochastic effects are a fundamental component of the system. While these algorithms have been first used in physical sciences, their use in the area of human systems, economics, and social sciences has become increasingly popular. The possibility of handling significant uncertainty in inputs and occurrences allows us to finally study large-scale systems where stochasticity is an intrinsic element of the system.

DATA-INTENSIVE COMPUTATIONAL TOOL FOR THE ANALYSIS OF INFECTIOUS DISEASES

The Global Epidemic and Mobility (GLEAM) model is a publicly available computational framework based on a metapopulation approach in which the population of the world is spatially structured into geolocalized patches or subpopulations (e.g., cities) where individuals mix. These patches are connected by the mobility patterns of individuals. GLEAM is capable of generating stochastic simulations of epidemic spread worldwide, yielding (among other measures) the incidence and seeding events at a daily resolution for 3,253 subpopulations in 232 countries and dependent territories (www.gleamviz.org).

Other high-performance computational tools are also available to the public for the spatial analysis and modeling of epidemics. These tools differ in their underlying modeling approaches and in the implementation, flexibility, and accessibility of the software itself.

The Spatiotemporal Epidemiological Modeler (STEM) is a modeling system for simulating the spread of an infectious disease in a spatially structured population. Contrary to other approaches, STEM is based on an extensible software platform, which promotes the contribution of data and algorithms by users (www.eclipse.org/stem).

Agent-based models describe the stochastic propagation of a disease at the individual level, thus taking into account the explicit social and spatial structure of the population under consideration. In this respect, CommunityFlu is a software tool that simulates the spread of influenza in a structured population of approximately 1,000 households with 2,500 persons (www.cdc.gov/flu/tools/communityflu).

A larger population is considered in FluTe, a publicly available tool for the stochastic simulation of an epidemic in the United States at the individual level (www.cs.unm.edu/~dlchao/flute). The model is based on a synthetic population, structured in a hierarchy of mixing social groups, such as households, household clusters, neighborhoods, and nation-wide communities.

The Bruno Kessler Foundation hosts the most detailed European-wide agent-based model for infectious diseases (dpcs.fbk.eu). The model is informed by routine socio demographic data collected throughout all European countries (e.g., school and workplace attendance, household structure, etc.).

EpiFast involves a parallel algorithm implemented using a master-slave approach, which provides for scalability on distributed memory systems. This tool allows for the detailed representation and simulation of a disease in social contact networks among individuals that dynamically evolve over time and adapt to actions taken by individuals and public health interventions (www.vbi.vt.edu).

FRED (A Framework for Reconstructing Epidemiological Dynamics) is an open source modeling system developed by the University of Pittsburgh Public Health Dynamics Laboratory. The system uses agent-based modeling derived from census-based synthetic populations that capture the demographic and geographic distributions of the population, as well as detailed household, school, and workplace social networks (www.phdl.pitt.edu/index.php/research/software18).

Computational modeling has led to a qualitative change in the ways we model epidemic and social contagion processes. Visualization and analysis tools that are able to cope with multiple levels of representation are being developed along with computer simulations that provide experiments not feasible in the real world. For the first time, epidemic processes can be studied in a comprehensive fashion that addresses the complexity inherent to real-world health problems. Whereas data availability is pointing out the limits of our conceptual and modeling frameworks, it is also allowing the validation of results across different modeling approaches, mathematical techniques, and approximation schemes.

GLOBAL EPIDEMIC AND MOBILITY MODEL

In this book, we want to illustrate the methodology described earlier and convince you that this approach is mature enough to be encapsulated in actionable tools for the simulation of case studies, the analysis of risk through model scenarios, and the forecasts of newly emerging infectious diseases (INFOBOX 1.1).

In order to tell this story, we use the Global Epidemic and Mobility (GLEAM) framework. GLEAM combines real-world data on populations and human mobility with complex stochastic models of disease transmission. This computational model is the product of the multidisciplinary work of a 10-year collaboration among several teams across the world. The model has been integrated in a computational platform that can model the world wide spread of epidemics for human transmissible diseases, offering extensive flexibility in the design of the compartmental model and scenario setup, including computationally optimized numerical simulations based on high-resolution global demographic and mobility data.

In the following chapters, we show the reader the what, how, and why of the computational approaches to modeling infectious diseases. First, we present what data we can leverage in informing state-of-the-art modeling approaches. Next, we show how the data-model integration is made; in other words, we take a look under the hood of the computational modeling approach. Finally, and most importantly, we make the case for why computational modeling is important to support public health decisions and intervention plans. Indeed, computational modeling does more than forecast and provides rationales and quantitative analysis in a number of areas that we briefly summarize here.

Preparation and contingency planning

The decision-making process in the fight against diseases relies on a multitude of options such as vaccination, contact tracing, quarantine, administration of drugs, and social distancing, which must be implemented in a manner that addresses the complexity inherent to the biological, social, and behavioral

COMPUTER SIMULATIONS PROVIDE EXPERIMENTS NOT FEASIBLE IN THE REAL WORLD AND ADDRESS THE COMPLEXITY INHERENT TO REAL-WORLD HEALTH PROBLEMS.

aspects of health-related problems. Large-scale computational models provide scenario analysis tools that can be used to inform the policy-making process quantitatively and to explore the effectiveness of specific strategies and mitigation/containment plans.

Epidemiological explanations

Computational modeling allows the testing and validation of specific assumptions of the potential transmission mechanisms of a disease. It also offers the possibility of defining counterfactual numerical experiments aimed at disentangling the causal mechanisms and effects driving the disease dynamics. While no model is reality itself, models allow us to have a structured reasoning framework that through validation and falsification may lead to a coherent picture of the epidemic unfolding.

Threats and systemic risk

One of the main challenges in assessing the threats of deliberate/natural disease outbreaks is the choice of key epidemiological parameters and the natural history of infection. The variability of the parameters creates a wide range of scenarios for which it is crucial to develop epidemic models able to gauge the actual threat of diseases with pandemic potential.

Forecast

A more difficult challenge than scenario analysis is the use of mathematical and computational models for the real-time forecasting of an epidemic unfolding. Ultimately, forecasting amounts to providing the mathematical description of patterns found in real-world data in order to develop models that can be used to anticipate trends, evaluate risks, and eventually manage future events.

CHAPTER 2

DATA, DATA, AND MORE DATA

G IVEN THE DECADES OF SUCCESS achieved in weather forecasts and the phenomenal results in the computational modeling of new material properties, it is natural to wonder why we are at a very different stage in the quantitative forecasting of the next pandemic or the progression of seasonal influenza. The main difference is that spreading and contagion models start with assumptions about the behavior of humans and society instead of the physical laws governing fluid and gas masses. In other words, while it is possible to produce a complete temperature analysis of the sea surface and satellite images of atmospheric turbulence, we do not yet have large-scale knowledge of commuting patterns or precise maps of contact among individuals.

In recent years, however, the tremendous progress made in data collection and information technology is lifting the limits we have faced for decades in the gathering of quantitative social, demographic, and behavioral data. Improved techniques and methodologies support the inter linkage and integration of digitalized datasets with geo-coding information, economical, and transportation databases.

SOCIOECONOMIC DATA ARE THE BRICKS THAT BUILD THE PILLARS OF ANY COMPUTATIONAL APPROACH TO THE ANALYSIS OF INFECTIOUS DISEASES.

Data-driven epidemic modeling relies on a multitude of datasets, ranging from population movement records to social and behavioral data, as well as census and health-related information. These data are the bricks that build the pillars of any computational approach to the analysis of infectious diseases. This chapter is dedicated to the illustration of the extraordinary high-quality datasets that can be leveraged to build those pillars.

POPULATION

The geographical distribution of the population is the first key ingredient of any epidemiological model. In order to model the spreading of an infectious disease that is transmitted from person to person, it is crucial to have detailed data of where individuals are living. Creating high-quality maps of the human population at the worldwide scale is per se a scientific challenge. Different projects tackled this problem using advanced techniques that merge different datasets, ranging from routinely collected census data to satellite imagery analysis. NASA's Landscan[1], Columbia University's Gridded Population of the World[2], and the WorldPop[3] projects all make worldwide maps available at different granularities and resolutions. More precisely, in these works the surface of the world is divided into a grid of cells that can have different resolution levels. Each cell has an estimated population value assigned. Data from the Gridded Population of the World project is shown in **FIGURE 2.1**. The bright regions are associated with highly populated locations. Not surprisingly,

Figure 2.1 | Mapping the population of the world

Global population distribution from the Gridded Population of the World project.

Source: Columbia University's Gridded Population of the World

1 landscan.ornl.gov
2 sedac.ciesin.columbia.edu
3 www.worldpop.org.uk

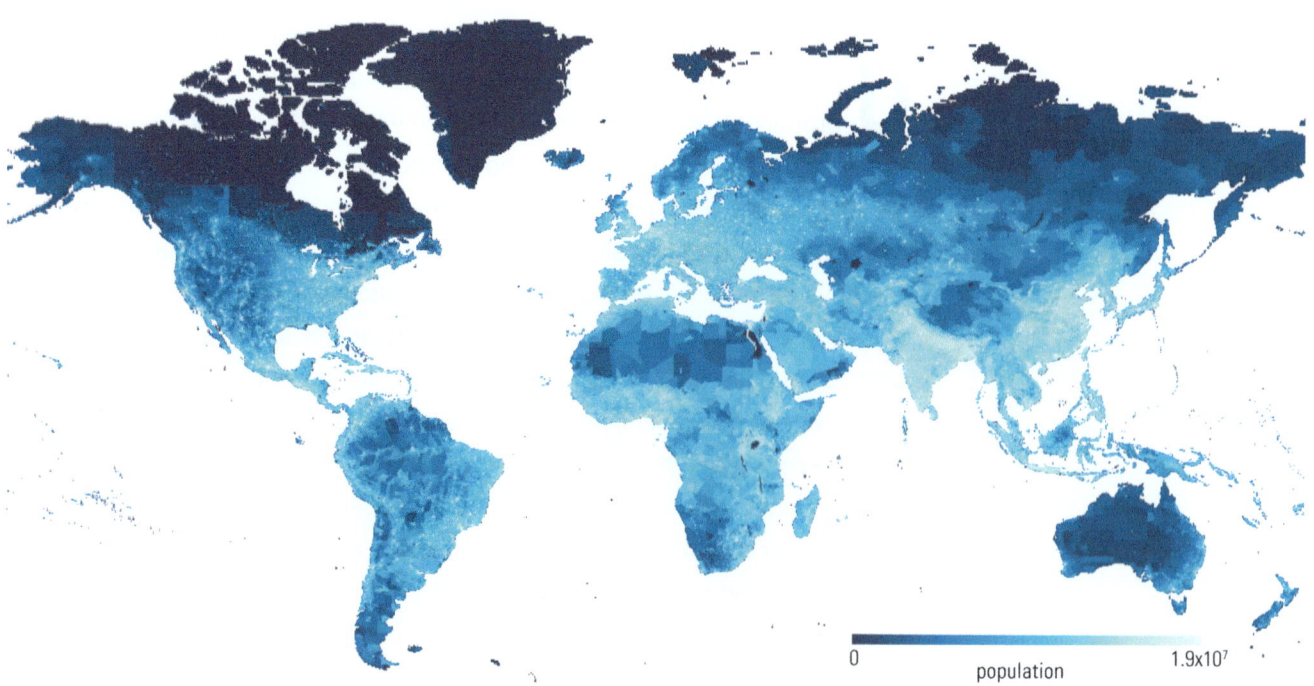

0   population   1.9x10⁷

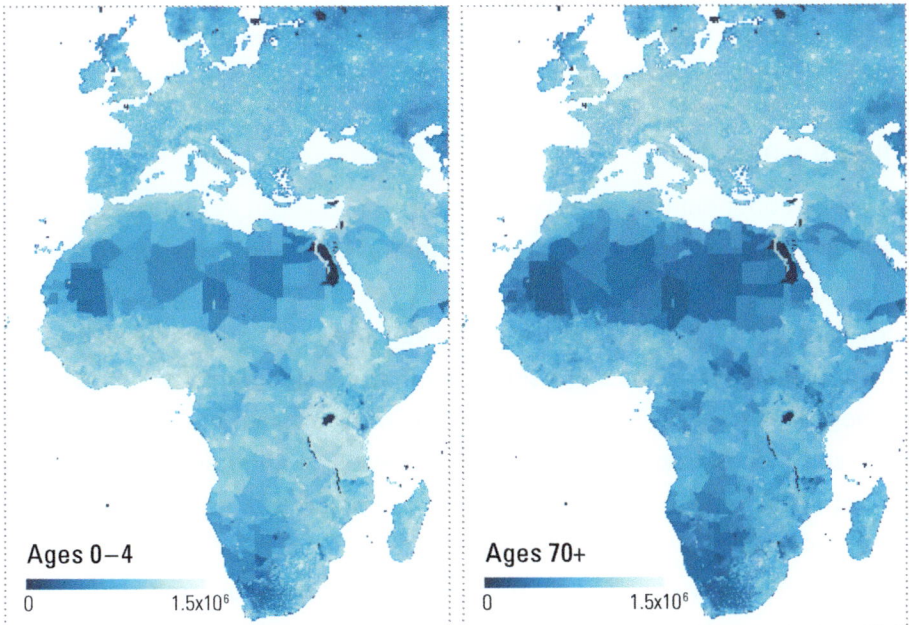

Ages 0–4

0   1.5x10⁶

Ages 70+

0   1.5x10⁶

China, India, and Mexico are extremely visible. These maps make clear how the population is heterogeneously distributed around the globe. Australia and China are two extreme cases. The first country has just a few highly populated regions. The second, instead, is the most populated country in the world, and its population is spread across a large part of the country. Historical, geographical, environmental, and political factors are at the root of these heterogeneities that play an important role in the geospatial diffusion of infectious diseases.

THE AGE STRUCTURE OF THE WORLD POPULATION

The accurate characterization of the structure of social contacts in mathematical and computational models of infectious disease transmission is another key element in the assessment of the impact of epidemic outbreaks and in the evaluation of effective control measures. For instance, the transmissibility potential of a disease and the final epidemic size depend on mixing patterns between individuals in the population, which in turn depend on sociodemographic parameters. A critical feature affecting disease spread is the age structure of the population of each country. Furthermore, individuals of different ages interact differently according to household size, fraction of workers and students in the population, etc. All of these indicators change consistently across the world. Here, we consider 10 age brackets: 0–4, 5–9, 10–14, 15–19, 20–29, 30–39, 40–49, 50–59, 60–69, and 70+ years old. More developed countries typically have larger numbers of the elderly and fewer numbers of children and young adults, while the picture reverses itself for developing countries where we observe a larger proportion of children than the elderly. This is clear from inspecting the two maps showing Europe and Africa (**FIGURE 2.2**). Indeed, the density of individuals over 70 is visibly much

Figure 2.2 | Age structure heterogeneity

Population distribution for the age bracket 0–4 (left) and 70+ (right), in Europe, Middle East, and Africa.

Source: Columbia University's Gridded Population of the World

THE TRANSMISSIBILITY POTENTIAL OF A DISEASE AND THE FINAL EPIDEMIC SIZE STRONGLY DEPEND ON MIXING PATTERNS BETWEEN INDIVIDUALS OF THE POPULATION, WHICH IN TURN DEPEND ON SOCIO DEMOGRAPHIC PARAMETERS.

higher in Europe than in Africa; alternatively, the density of children is evidently higher in Africa than Europe.

HUMAN MOBILITY

As we go about our daily lives, we come in contact with individuals who can carry viruses, bacteria, and other pathogens, several of which are capable of causing disease and infection. Whenever the conditions are favorable, they may be able to infect us, turning us into a new vehicle for the spreading of the disease. It is not surprising then that the geographical diffusion of pathogens is thus intrinsically intertwined with human mobility. Every time a person moves from home to work, or travels to another city or country, an opportunity arises for the pathogen to spread to a new population of potential hosts.

As technology evolves, so do our traveling habits: automobiles, trains, and airplanes help to shorten physical and temporal distances, for humans and for pathogens.

Long-Range Mobility

Over the course of 100 years, flying went from an occupation of the eccentric, to a luxury of the wealthy and finally to a necessity in the life of millions of passengers every year.

This shift was brought about by the increasing affordability and convenience of air transportation. Once distant cities are now just a few hours apart whether they are separated by hundreds or thousands of miles. This worldwide

THE GLOBAL AIRLINE NETWORK INCLUDES OVER 4,000 AIRPORTS CONNECTED BY TENS OF THOUSANDS OF DIRECT FLIGHTS.

Figure 2.3 | Global long-range mobility network

Each node represents a major transportation hub. Their size is proportional to the population served. Each link represents a direct flight connection. Their width is proportional to the number of available seats.

Source: Official Airline Guide

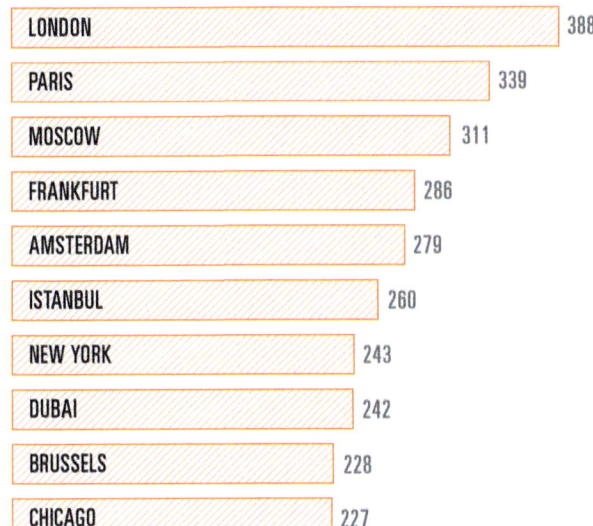

LONDON	388
PARIS	339
MOSCOW	311
FRANKFURT	286
AMSTERDAM	279
ISTANBUL	260
NEW YORK	243
DUBAI	242
BRUSSELS	228
CHICAGO	227

LONDON	86.59 M
NEW YORK	72.80 M
HONG KONG	64.17 M
PARIS	59 M
BEIJING	56.71 M
ATLANTA	56 M
CHICAGO	54.27 M
SHANGHAI	53.95 M
LOS ANGELES	53.26 M
TOKYO	49.11 M

network evolved organically during the last century through the decentralized decisions of hundreds of different companies and individuals. As a result, its global structure is extremely complex, with many non trivial properties (**FIGURE 2.3**). According to the Official Airline Guide (OAG)[4], the global airline network includes over 4,000 airports connected by tens of thousands of direct flights. Chaining different flights through a major hub creates connections between airports not directly connected. Cities like New York, Atlanta, Frankfurt, or London have large airports (or even multiple airports in close proximity) that serve as the hubs connecting destinations that do not have direct flights. The top 10 cities by airline traffic and number of connections in 2012 are listed in **FIGURE 2.4**.

Airline data also refers to the so-called origin-destination flows between airports. Origin-destination datasets report the actual number of travelers between airports, regardless of intermediate stops or connecting flights. The origin-destination flows network has many more links than the airline network, which only considers non stop connections. This network accounts for the number of individuals traveling from place to place, thus providing a more accurate picture of the dispersion of individuals potentially carrying diseases.

Short-Range Mobility

Trains, public transportation, and personal automobiles have impacted short-distance travel in much the same way that accessible airline transportation has affected long-distance spreading patterns. Now, more than ever, we can easily cover tens of miles just to go from home to the office or school

Figure 2.4 | Top cities by airline traffic and number of connections

List of the top 10 cities by the number of connections per day (left) and available seats per year (right); data refers to 2012.

Source: Official Airline Guide

---

4 www.oag.com

VARIATION ON THE GLOBAL AIR TRANSPORTATION INFRASTRUCTURE OVER TIME

In this infographic, we map the temporal variation on the number of available seats in all the airports from 2002 to 2012. The blue circles describe airports in 2002, while the orange circles represent the airports in 2012. Their size is proportional to the number of seats available. Interestingly, while in the United States and Europe we do not observe much variation, in Africa, Russia, India, and Asia we observe the emergence of many new airports as well as an increase in available seats in the major hubs.

EVALUATION OF THE NUMBER OF SEATS AVAILABLE BY REGION

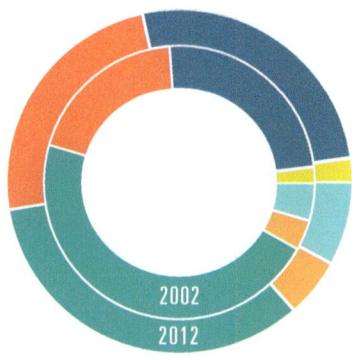

Regions as defined by the World Health Organization (WHO)

- Americas
- Africa
- E. Mediterranean
- Europe
- S.E. Asia
- W. Pacific

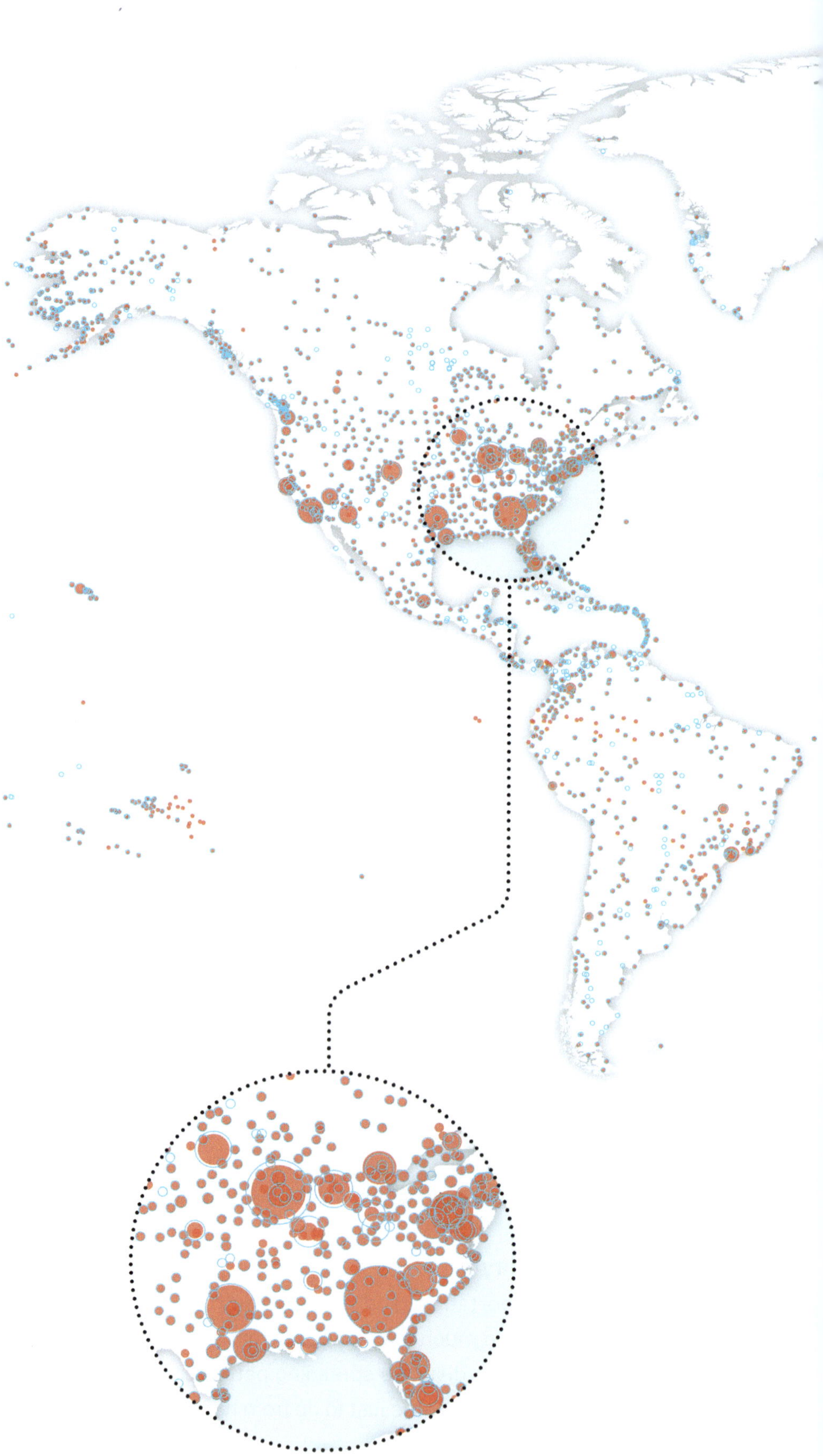

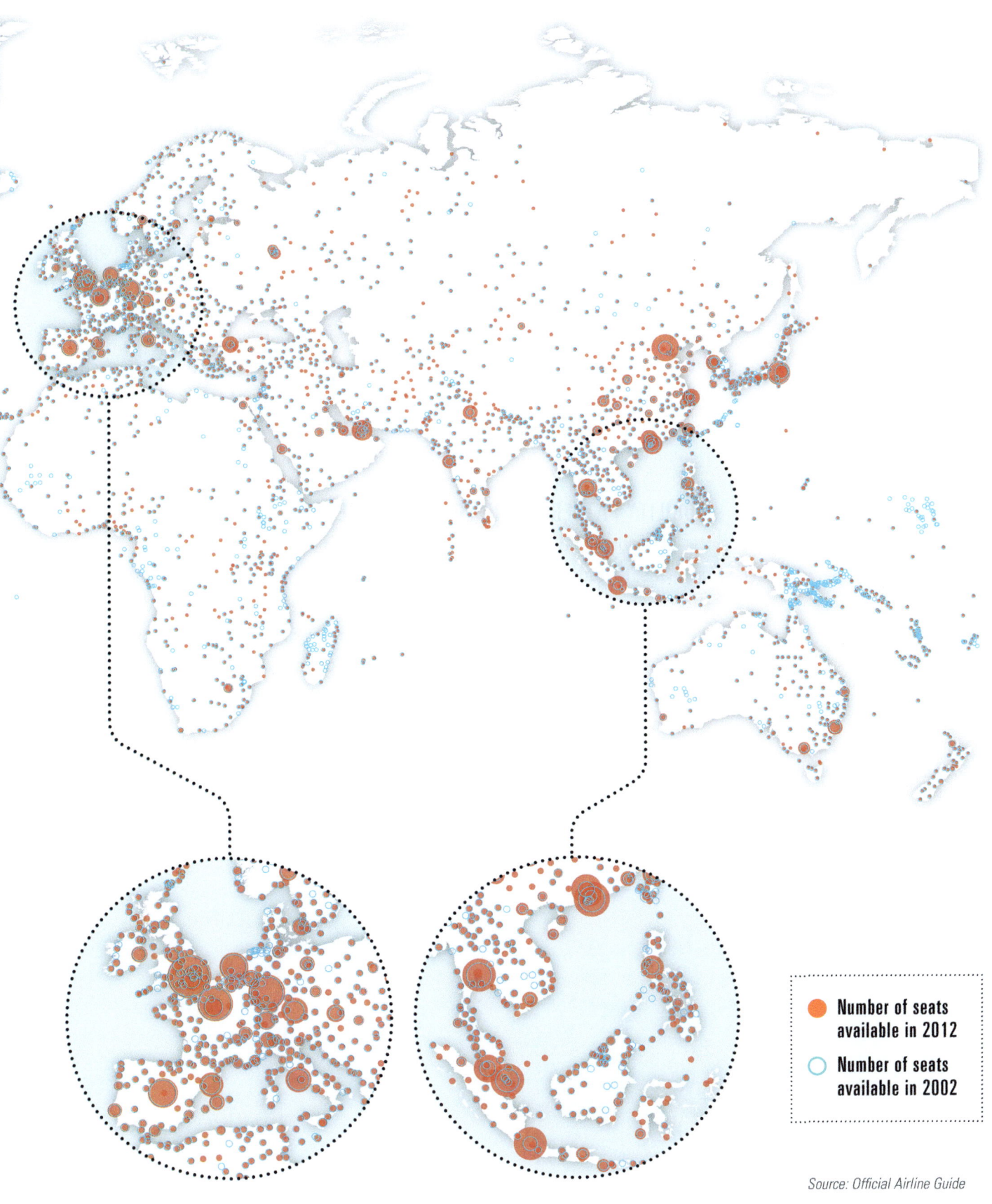

Number of seats
available in 2012

Number of seats
available in 2002

Source: Official Airline Guide

and return home at the end of the day. In doing so, we greatly increase the number of people we come in contact with daily and with it the opportunities for a disease to spread.

The cyclical nature of our commuting patterns tightly couples neighboring cities within a few hours. In this way, infections that first arrive in a city through airline connections are quickly diffused and spread locally. Such coupling is so evident that, by simply plotting the commuting patterns between neighboring cities, one is able to quickly identify the major metropolitan areas, even in the absence of any other information, as these naturally generate stronger flows. In **FIGURE 2.5**, we show two examples: the United States and Italy. In the maps, each node is a municipality and the weighted links between them are the flow of commuters. In the United States, commuting is more concentrated in the east coast where the density of municipalities is higher. Not surprisingly, large urban areas such New York City, Washington D.C., Atlanta, San Francisco, and Seattle are clearly visible and contain the majority of traffic. The same behavior is observed in Italy, where Milan and Turin (north) and Rome and Naples (center and south) are the most densely connected areas.

MOBILITY PATTERNS AND EPIDEMIC SPREADING

It is interesting, and extremely relevant in understanding the spread of infectious diseases, to notice the differences between the topologies of long-range and short-range mobility networks. Indeed, in the air transportation network, just the nodes are embedded in the earth's surface; links are not. They cross each other connecting a node with several others, possibly at any distance. In the commuting network rather, links are also embedded in the earth's surface: a node is only connected to geographically closest points. Understanding the differences between the two topologies is crucial for the characterization of the spreading of diseases. While through air transportation a disease can spread to very far distances directly, in the case of commuting, it is constrained

Figure 2.5 | Short-range mobility network in continental United States and Italy

Each node represents the center of a population cell. Each link represents a flow of commuters connecting two nodes. The width is proportional to the traffic per connection.

New York

to move in space going from one node to the next closest one, so that the epidemic spreads like a wave in water. Their differences explain the different behavior of pandemics in the modern times with respect to their unfolding before mass air transportation. **FIGURE 2.6** illustrates the differences in the case of New York: in orange are the connections to the nearest neighbors in the air transportation network and (inset) in the commuting network. The dissimilarities between the two are extremely clear. By using the flight network, an infected individual can move thousands of miles in less than one day, reaching, for example, Japan, while an individual is rather limited by commuting, to the order of a hundred miles.

Figure 2.6 | Direct flight connections to and from New York

The locations reachable from New York via one single flight. Inset: the cells within a circle of 100 miles from New York connected by short-range mobility.

DISEASES

Needless to say, the knowledge of the pathogenic agent itself is critical input in the analysis of infectious disease spreading. No single model fits all diseases. Their basic features and properties have critical effects in the spreading process, and each model must be tailored to the particular illness under study. In this book, we focus on a particular subset of infectious diseases, which are caused by viruses and are transmitted through human-to-human interactions. We explicitly consider influenza viruses, coronaviruses, and filoviruses. A different virus characterizes each of these diseases. Their biological features define the natural history, mortality, and transmissibility of the disease. In the following section, we provide a basic description of each of these diseases, summarizing their biological characteristics and historical spreading.

NO SINGLE MODEL FITS ALL DISEASES. THEIR BASIC FEATURES AND PROPERTIES HAVE CRITICAL EFFECTS IN THE SPREADING PROCESS, AND EACH MODEL MUST BE TAILORED TO THE PARTICULAR DISEASE UNDER STUDY.

DISEASES

 INFLUENZA

Every year, the flu spreads across countries during the winter season, usually in mild forms, killing nevertheless between 250,000 and 500,000 individuals worldwide. The biological structure of influenza viruses usually changes little from year to year, leaving the majority of people still immune from previous existing/circulating strains. Occasionally, however, the biological structure can change enough and a new strain emerges, against which the population has no immunity. This might happen through different mechanisms, often involving flu viruses circulating among birds or pigs. Pandemics occur when a new flu virus is able to spread from person to person in an efficient and sustained way, quickly affecting a large fraction of the population that has no immunity to the disease. On average, pandemics take place a few times every century, with reports of flu pandemics seeming to start in ancient Greece more than 2,500 years ago.[5] In **FIGURE 2.7**, we show the largest influenza pandemics since 1900. As is clear from the number of deaths, influenza viruses are a serious threat. The first influenza virus was isolated in the laboratory in the early 1930s. Since then, our knowledge of its biological structure and dynamics has greatly increased. **FIGURE 2.8** shows the geographical impact of the latest flu pandemic of 2009.

VIRUS STRUCTURE

There are three types of influenza viruses that affect humans: type A, B, and C. The first type is the most diffused and responsible for the majority of epidemics and pandemics in our collective history. Type B and C are rarer and associated with localized outbreaks. Influenza A viruses are further divided into subtypes based on their biological structure. They are made of eight single-stranded RNA segments with two major glycoproteins in the surface: HA hemagglutinin and NA neuraminidase. There are at least 16 different HA and 9 different NA. The predominant subtypes found in humans are H1, H2, H3, and N1, N2.

---

5   World Health Organization

Figure 2.7 | Largest flu pandemics

Death tolls of the largest influenza pandemics since 1900.

Source: Centers for Disease Control and Prevention

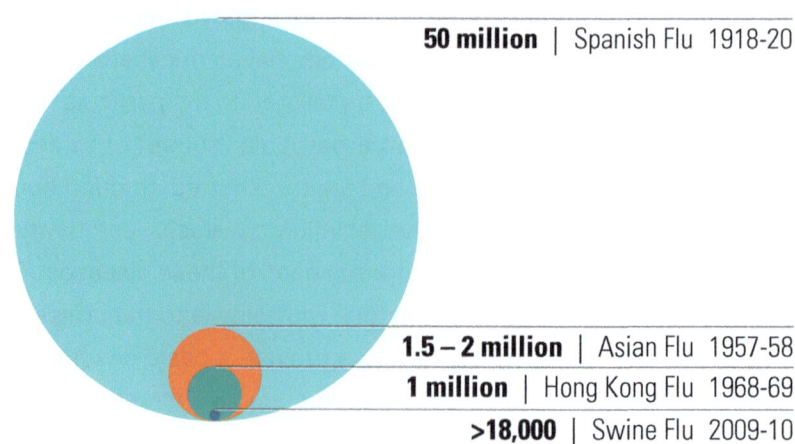

50 million | Spanish Flu 1918-20

1.5 – 2 million | Asian Flu 1957-58

1 million | Hong Kong Flu 1968-69

>18,000 | Swine Flu 2009-10

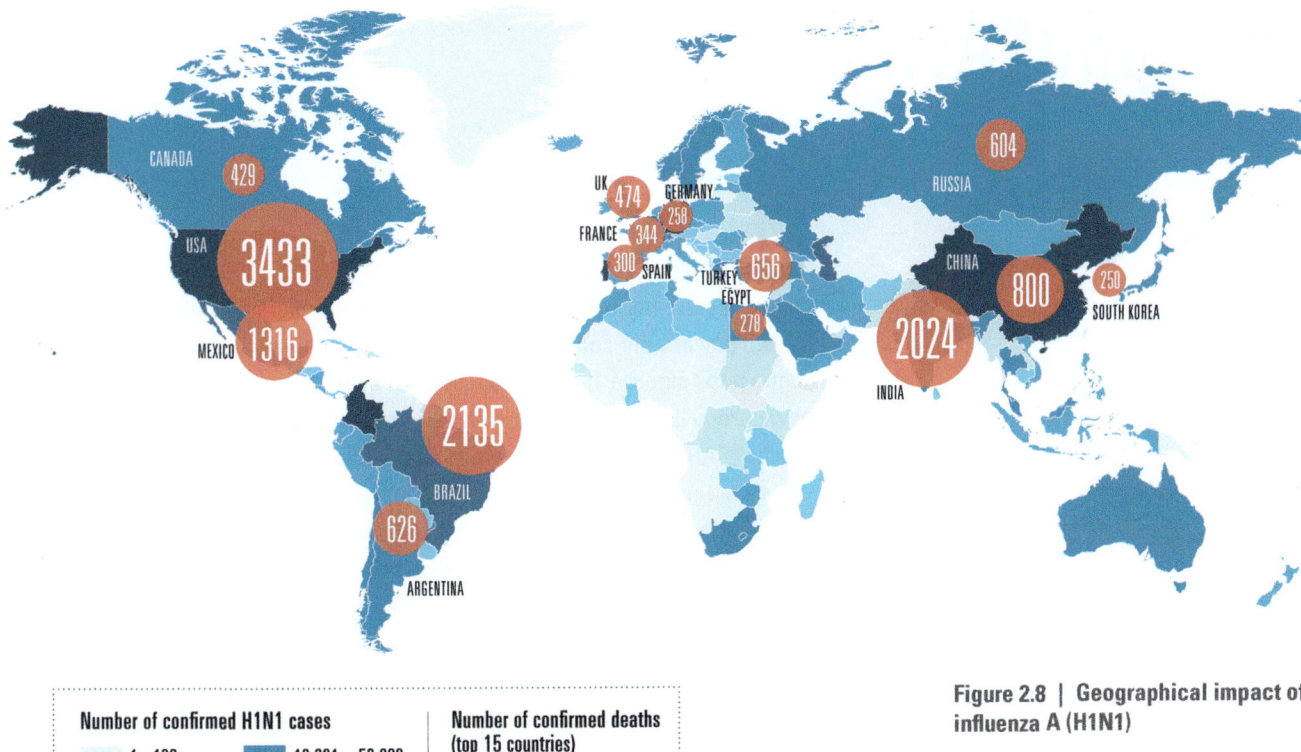

Figure 2.8 | Geographical impact of influenza A (H1N1)

The number of influenza cases in each country around the world with circles indicating the number of deaths due to it.

Source: FluNet, World Health Organization

Their combinations form the influenza virus. In recent history, we have been subjected to H1N1, H2N2, and H3N2 pandemics. Each of these subtypes can be further divided in different strains defined by the specific combination of genes and proteins.

VIRUS TRANSMISSION

The virulence and fatality rate change from strain to strain. Influenza viruses affect animals as well and can sometimes be transferred from animals to humans. In particular, swine (H1 and H3), ducks (H7), and chickens (H5 and H9) have been shown to infect humans. There are three mechanisms behind the diffusion of influenza viruses: droplet, airborne, and contact transmission. In the first class, an infected individual coughs or sneezes, diffusing large droplets that reach conductive or mucous membranes of a susceptible person. In the second case, there is no direct contact between the droplets and the susceptible person; they can be vaporized in the air and become breathable by susceptible individuals. In the last case, the virus can be transmitted from person to person through contact with the secretions of an infected individual or through physical contact between individuals. After being in contact with the virus, susceptible individuals might become infected. If this is the case, the virus starts reproducing inside the new host. Typically, the symptoms arise after the first day. The infectiousness increases, while the virus reproduces itself and reaches its peak after 2–3 days on average, after which the viral load then starts to decline thanks to the reaction of the immune system.

DISEASES

CORONAVIRUS

Coronaviruses are extremely diffused among humans and more in general among mammals and birds. They are the cause of a large percentage of all common colds in human adults. In contrast to influenza, just two human coronaviruses were known for many years. In 2003, a new coronavirus was identified as being responsible for the severe acute respiratory syndrome (SARS) outbreak that spread around the globe. SARS is one of the best examples of a new emerging disease. After AIDS in the early 1980s, this was the first new virus able to reach a global scale. Indeed, it spread to more than 30 countries across 5 continents in 2003. While the number of confirmed cases has been limited, around 8,000 in total, the associated mortality was much higher than a typical influenza strain. It ranged from 3% to 10%, creating large concerns that led to global efforts for its containment. The map of FIGURE 2.9 highlights the countries affected during this outbreak. In 2012, a new type of coronavirus was discovered in the Middle East, dubbed MERS-CoV. Then, in May 2015, an outbreak of MERS-CoV occurred in the Republic of Korea, causing one of the largest MERS-CoV outbreaks outside of the Middle East region.[6]

VIRUS STRUCTURE

Coronaviruses belong to the subfamily Coronavirinae in the family Coronaviridae. They are enveloped viruses with a positively stranded RNA genome. Several proteins contribute to the biological structure of coronaviruses. In particular, there are spike, envelope, membrane, and nucleocapsid proteins.

VIRUS TRANSMISSION

The transmission mechanisms of coronaviruses are typical of influenza-like illnesses (ILI). However, other characteristics are quite different. For example, in the case of SARS, the proportion of asymptomatic infections was relatively small. The maximal infectiousness registered occurred about 7 days after the onset of symptoms. The virus responsible for SARS is different from all other known coronaviruses. It appears to have originally been an animal virus that crossed to humans. Indeed, the virus has been isolated from civet cats in the Guangdong Province. In this region, there are many markets in which civet cats and other exotic animals are sold. A large fraction of workers in these markets were found to be seropositive for SARS. However, it is not clear if civet cats or other animals are the natural reservoir of the virus in the wild.

---

6   Centers for Disease Control and Prevention

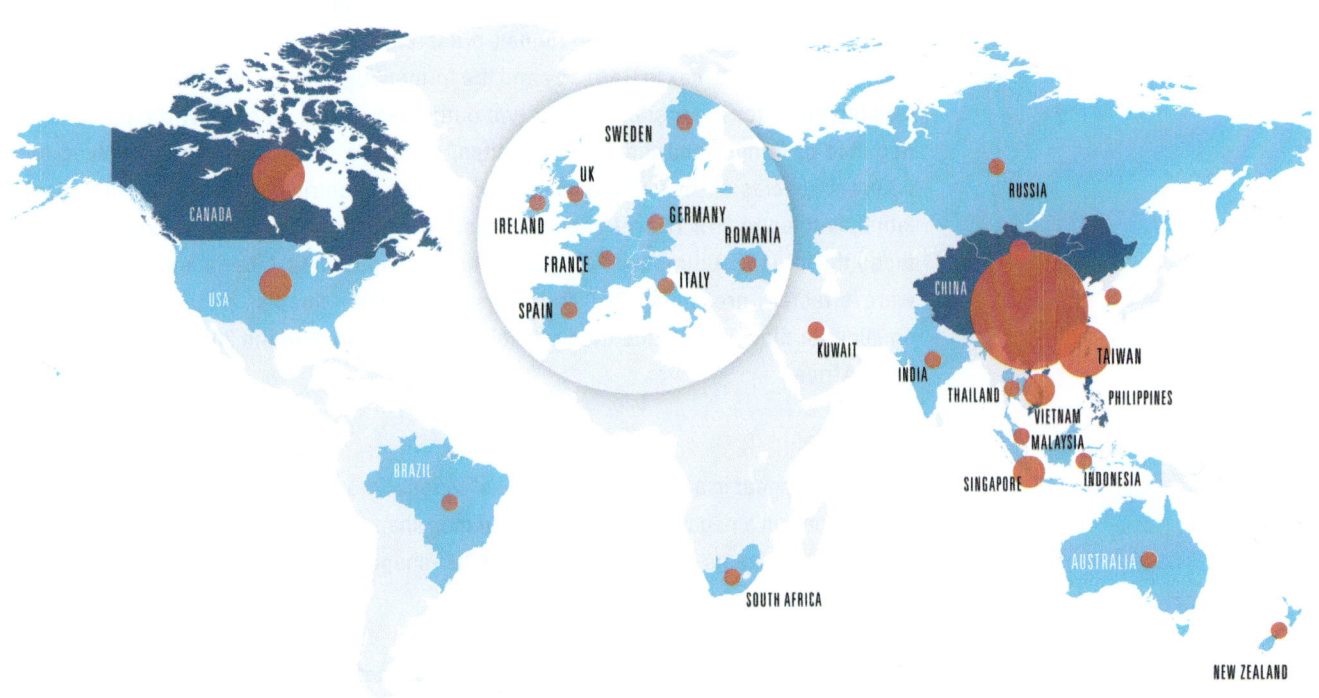

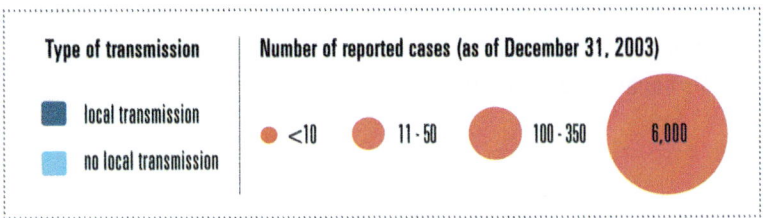

Type of transmission

- local transmission
- no local transmission

Number of reported cases (as of December 31, 2003)

<10   11 - 50   100 - 350   6,000

Figure 2.9 | Geographical impact of SARS

Total number of confirmed SARS cases in the 2003 outbreak.

Source: World Health Organization

FILOVIRUSES

Ebola and Marburg viruses belong to the family of filoviruses and cause severe hemor-rhagic fevers in human and non human primates. The first filovirus was identified in 1967, during an outbreak in Germany and the former Yugoslavia.[7] In 1976, the Ebola virus was identified for the first time when two outbreaks, one in Zaire (now the Democratic Republic of Congo) and the other in Sudan, occurred. These outbreaks involved two different species of the virus, named after the two countries. Ebola viruses are highly lethal, with up to 90% (Zaire) and 50% (Sudan) of cases being fatal. In 2014, the world faced the largest outbreak yet of Ebola (Zaire), when Guinea, Liberia, and Sierra Leone were affected, presenting 28,000+ cases and 11,000+ deaths. **FIGURE 2.10** shows a timeline for filoviruses since the first isolation of the virus, to the 2014 EVD outbreak in West Africa.

VIRUS STRUCTURE

Filoviruses appear in a variety of "threadlike" virions (infectious viral particles) and encode their genome in a negative-sense RNA. Common shapes include long branching filaments, and shorter filaments shaped like a "6," the "U"-shaped filament, and even circles.

VIRUS TRANSMISSION

It is unknown how the virus is transmitted from its natural reservoir to humans. Once a human is infected, ways of transmission include close personal contact with infected individuals or their bodily fluids. Caregivers are at a higher risk of becoming infected due to close contact with the infectious individual. During outbreaks, the isolation of patients and the use of protective clothing and disinfection procedures are crucial to interrupt the transmission of filoviruses. In 2015, the first vaccine was developed to treat the Ebola virus, and field trials began in West Africa shortly thereafter.

---

7 Centers for Disease Control and Prevention

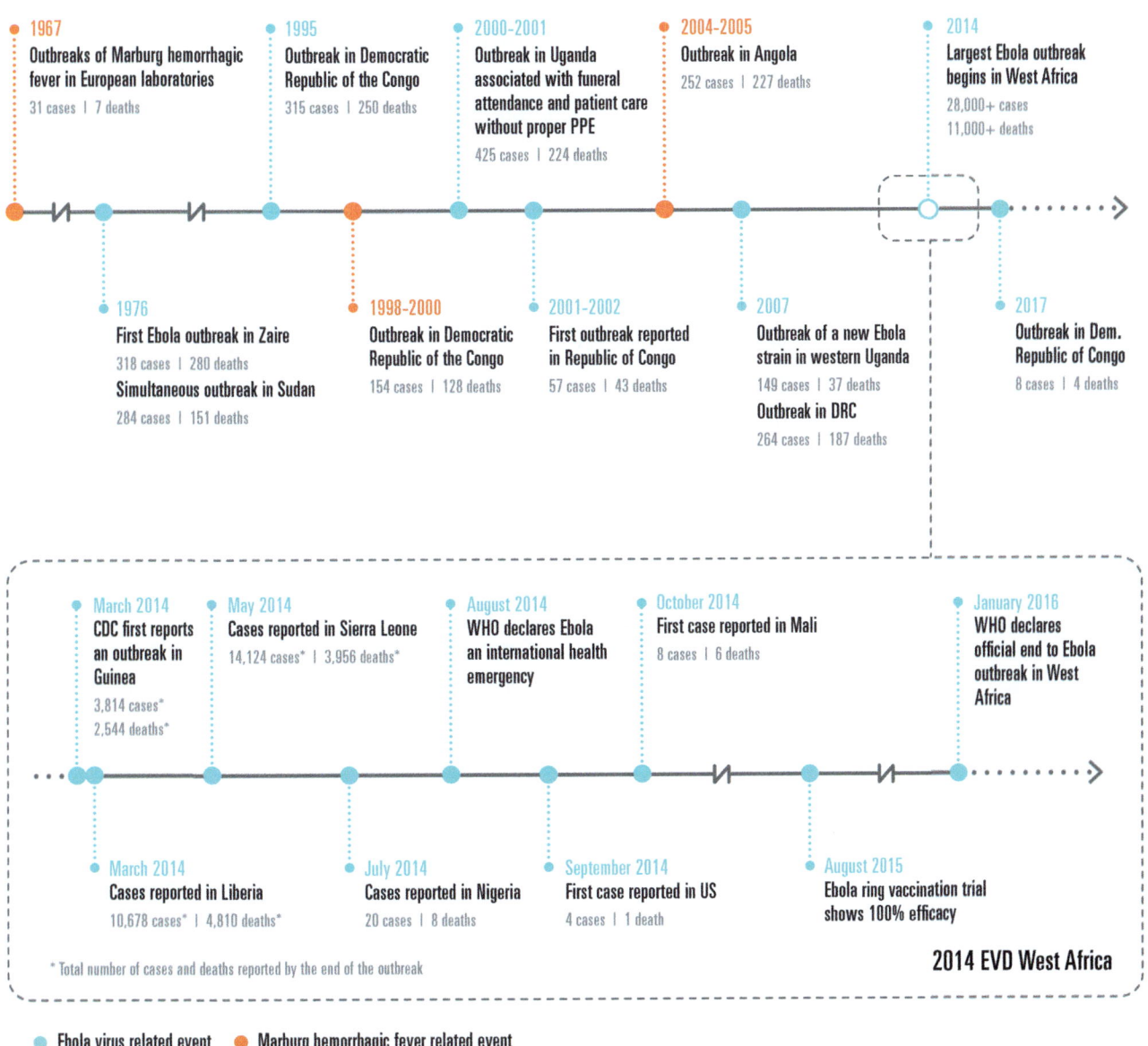

1967
Outbreaks of Marburg hemorrhagic fever in European laboratories
31 cases | 7 deaths

1995
Outbreak in Democratic Republic of the Congo
315 cases | 250 deaths

2000-2001
Outbreak in Uganda associated with funeral attendance and patient care without proper PPE
425 cases | 224 deaths

2004-2005
Outbreak in Angola
252 cases | 227 deaths

2014
Largest Ebola outbreak begins in West Africa
28,000+ cases
11,000+ deaths

1976
First Ebola outbreak in Zaire
318 cases | 280 deaths
Simultaneous outbreak in Sudan
284 cases | 151 deaths

1998-2000
Outbreak in Democratic Republic of the Congo
154 cases | 128 deaths

2001-2002
First outbreak reported in Republic of Congo
57 cases | 43 deaths

2007
Outbreak of a new Ebola strain in western Uganda
149 cases | 37 deaths
Outbreak in DRC
264 cases | 187 deaths

2017
Outbreak in Dem. Republic of Congo
8 cases | 4 deaths

March 2014
CDC first reports an outbreak in Guinea
3,814 cases*
2,544 deaths*

May 2014
Cases reported in Sierra Leone
14,124 cases* | 3,956 deaths*

August 2014
WHO declares Ebola an international health emergency

October 2014
First case reported in Mali
8 cases | 6 deaths

January 2016
WHO declares official end to Ebola outbreak in West Africa

March 2014
Cases reported in Liberia
10,678 cases* | 4,810 deaths*

July 2014
Cases reported in Nigeria
20 cases | 8 deaths

September 2014
First case reported in US
4 cases | 1 death

August 2015
Ebola ring vaccination trial shows 100% efficacy

* Total number of cases and deaths reported by the end of the outbreak

2014 EVD West Africa

● Ebola virus related event   ● Marburg hemorrhagic fever related event

Figure 2.10 | Ebola virus and Marburg hemorrhagic fever timeline

Source: Centers for Disease Control and Prevention, World Health Organization

HEALTH INFRASTRUCTURES

Some countries are more ready than others to deal with the risks associated with epidemic diseases. Different indicators are commonly used to evaluate each country's capacity to respond: for instance, the number of physicians per capita and the number of beds per 10,000 individuals. It is clear that countries with a good health infrastructure will be better at combating the spreading of an epidemic as well as reducing its impact on the population. Understanding and mapping health infrastructures and other socioeconomic indicators are extremely important in the building of models that can provide insight on the disease burden across different countries and measure the risk associated to emerging pathogens.

From the map of **FIGURE 2.11**, it is clear that more developed countries have a larger number of physicians per capita. **TABLE 2.1** reports the top 10 and bottom 10 countries ranked according to physicians per capita and hospital beds for 10,000 individuals. We consider 6 regions as defined by the World Health Organization (WHO): Africa, Americas, Eastern Mediterranean, Europe, South-East Asia, and Western Pacific. As shown in the rankings, European countries are in many of the top positions. The large majority of African countries are instead in the last positions. Indeed, concerning the density of physicians, 9 of the last 10 are in the Africa region.

The same pattern is observed for the number of beds (**FIGURE 2.12**). The precise ranking is a bit different, especially for the top 10, but the geographical distribution of these quantities is well correlated. Indeed both indicators are correlated with the Gross Domestic Products (GDP) of each country.

In the next chapter we discuss how all of this data can be incorporated within the framework of epidemic modeling.

Table 2.1 | Distribution of health infrastructures

List of top and last 10 countries per number of physicians and hospital beds per 10,000 people.

Source: Global Health Observatory data repository, World Health Organization, 2009

WHO Region	Country	Physicians
Americas	Cuba	75
Europe	Monaco	73
Europe	Greece	63
Europe	Austria	52
Europe	Georgia	48
Europe	Macedonia	46
Europe	Portugal	44
Europe	Norway	44
Europe	Lithuania	43
Europe	Germany	41
Africa	Liberia	<1
Africa	Malawi	<1
Africa	Niger	<1
Africa	Sierra Leone	<1
Africa	Ethiopia	<1
Africa	Burundi	<1
E. Mediterranean	Somalia	<1
Africa	Chad	<1
Africa	Lesotho	<1
Africa	Central African Republic	<1

WHO Region	Country	Hospital Beds
Western Pacific	Republic of Korea	143
Western Pacific	Japan	134
South-East Asia	Democratic Republic of Korea	115
Europe	Belarus	110
Europe	Ukraine	88
Europe	Germany	83
Europe	Russian Federation	82
Europe	Austria	76
Europe	Turkmenistan	74
Europe	Lithuania	73
Africa	Mali	1
Africa	Madagascar	2
Africa	Ethiopia	3
Africa	Guinea	3
Africa	Niger	3
South-East Asia	Nepal	3
Africa	Senegal	3
Africa	Burkina Faso	4
Africa	Côte d'Ivoire	4
Africa	Mauritania	4

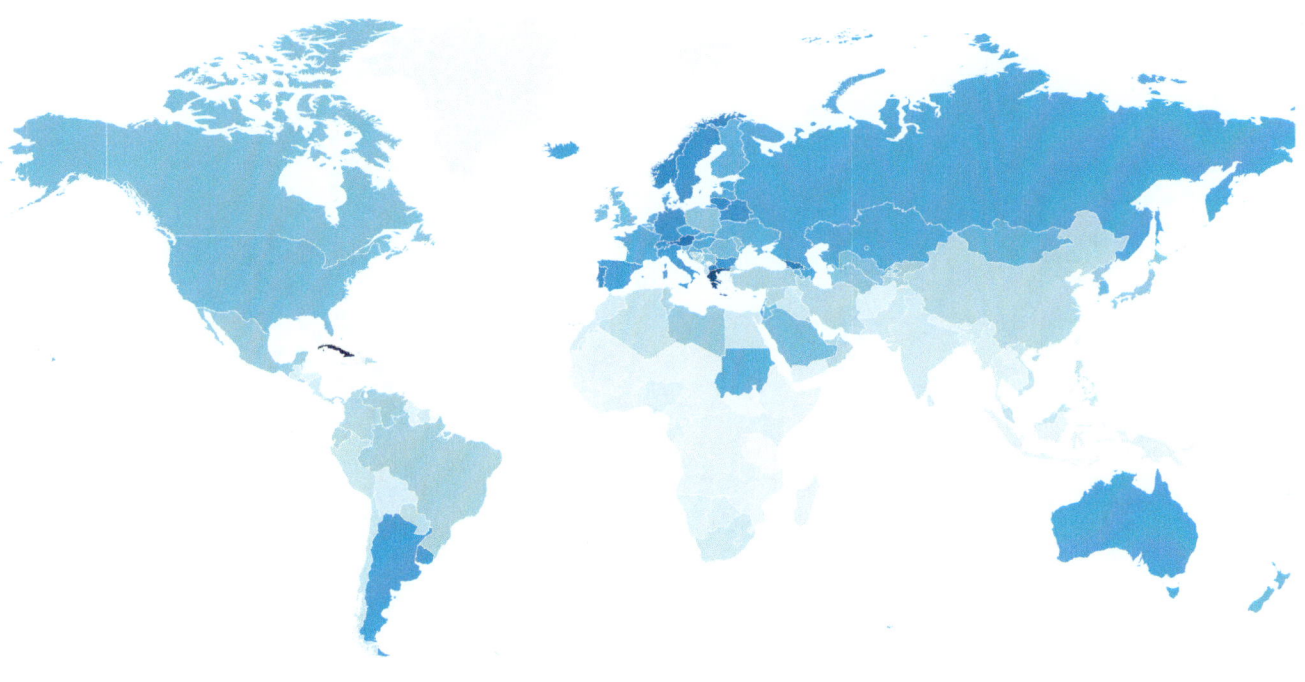

density of physicians (per 10,000)

0   37.5   75

Figure 2.11 | Mapping health infrastructures

Geographical distribution of the number of physicians per 10,000 people.

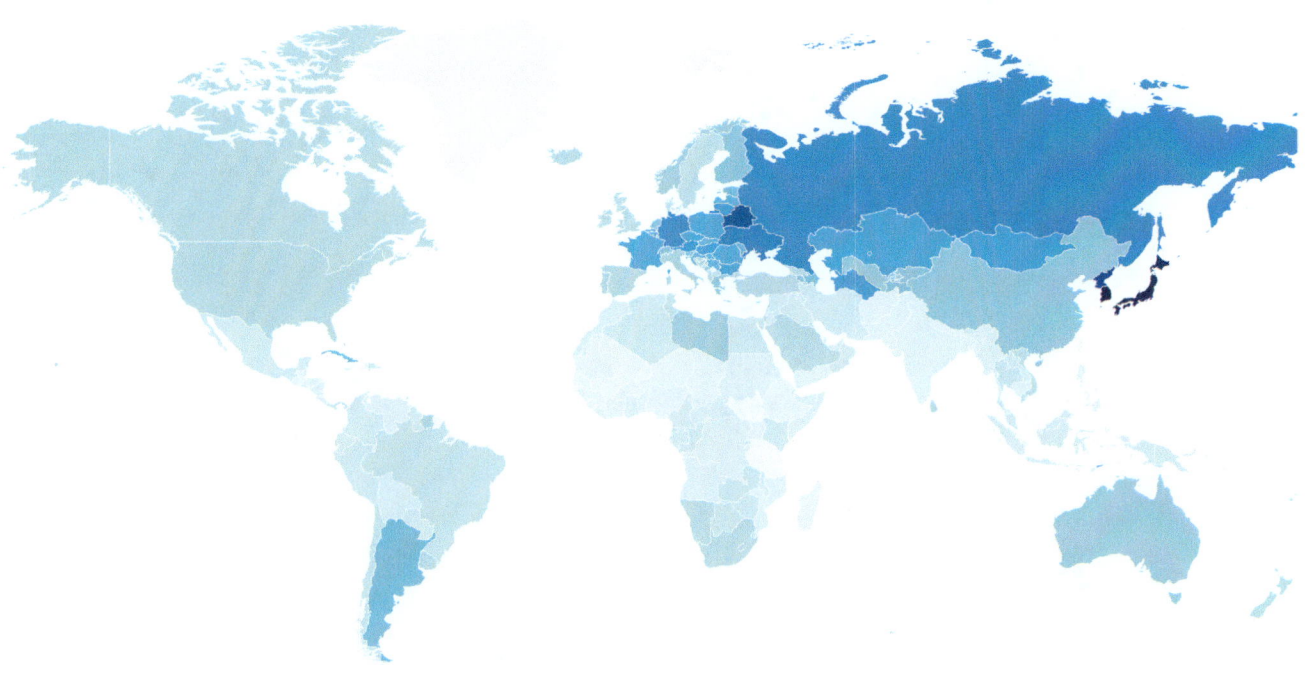

density of hospital beds (per 10,000)

0   71.5   143

Figure 2.12 | Mapping health infrastructures

Geographical distribution of the number of hospital beds per 10,000 people.

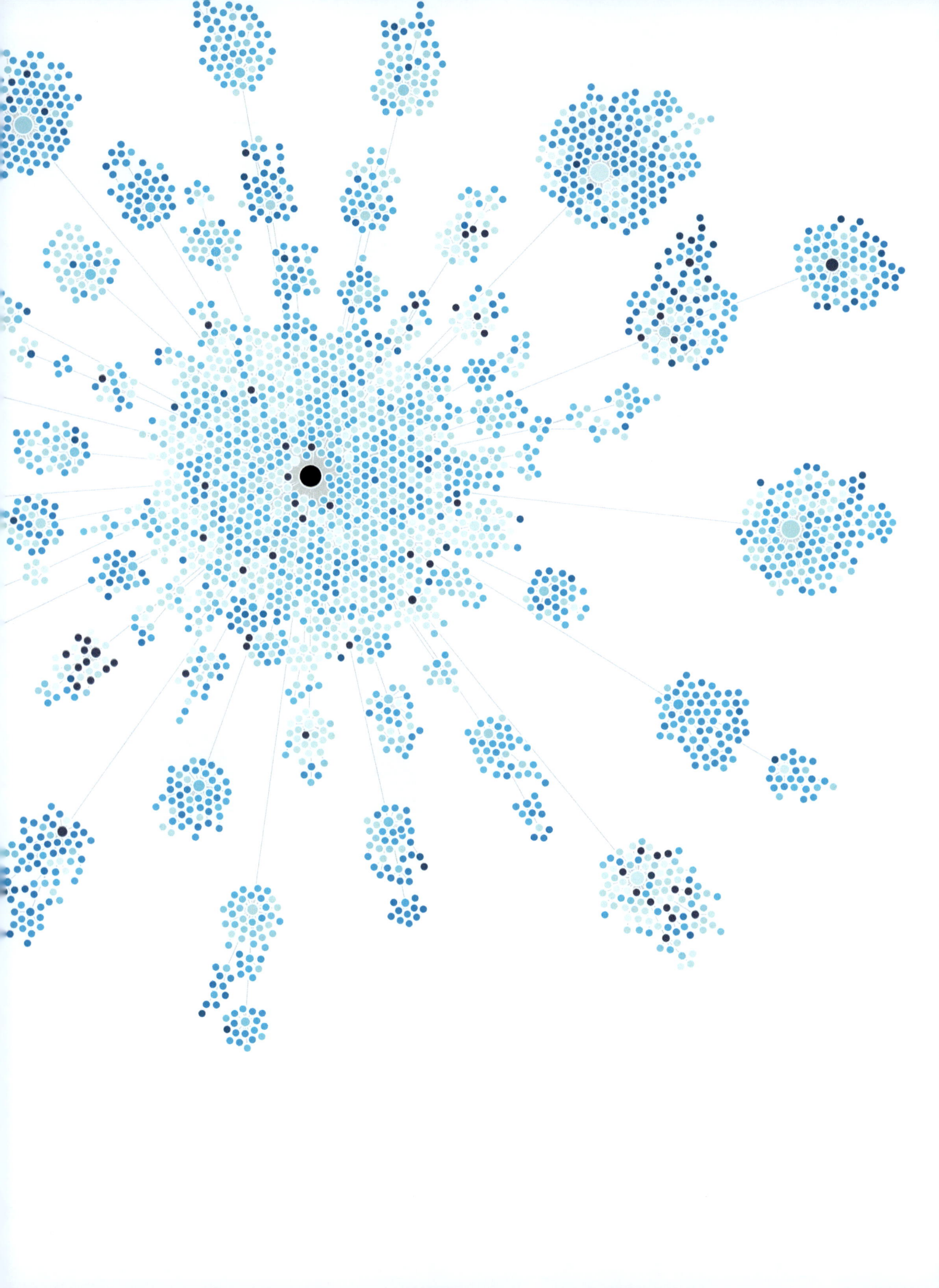

CHAPTER 3

DATA MODEL INTEGRATION: THE GLOBAL EPIDEMIC AND MOBILITY FRAMEWORK

GIVEN A SET OF INITIAL CONDITIONS for the local outbreak of a new potentially pandemic pathogen, the timeline of the arrival of the epidemic in each country is mainly determined by the human mobility network that couples different regions of the world. By looking at individual countries or a given continent in isolation, any estimate of the epidemic timeline would be based on assumptions about imported cases from the rest of the world. Human mobility patterns are hence key to consistently simulating the mobility of infectious individuals on the global scale, and thus providing ab initio estimates of the epidemic timeline in each country or urban area without assumptions on case importation.

The Global Epidemic and Mobility (GLEAM) framework produces realistic simulations of the global spread of infectious diseases with this in mind. It integrates (**FIGURE 3.1**) three data layers:

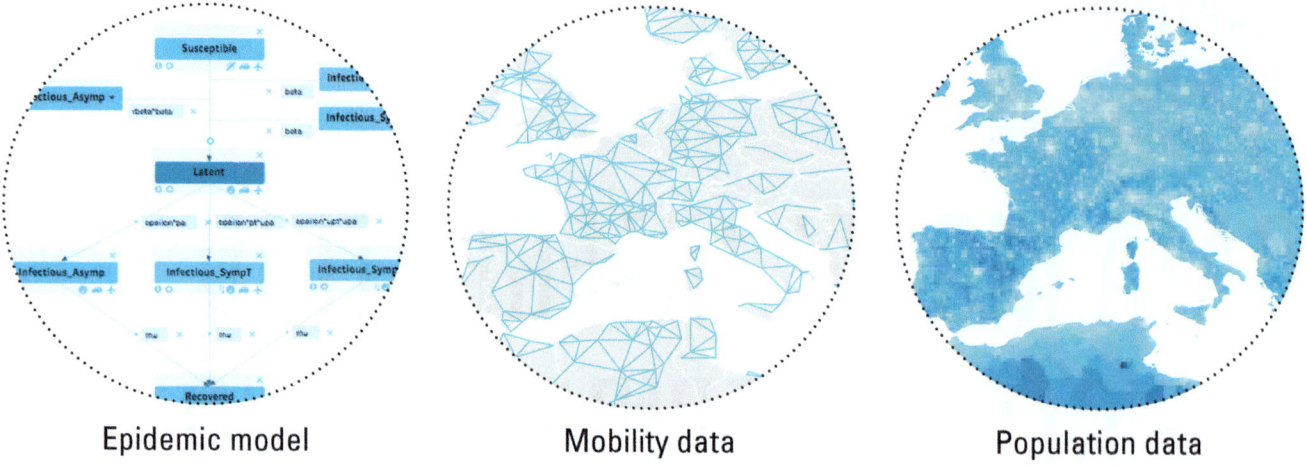

| Epidemic model | Mobility data | Population data |

Figure 3.1 | GLEAM data layers

The three data layers integrated into the GLEAM computational modeling framework.

- An individual-based stochastic mathematical model of the infection dynamics
- Real-world data on the mobility of this population
- Real-world data on the global population

The real-world population and mobility data discussed in the previous chapter are used to determine when and where people will interact and potentially transmit the infection. In order to do that, GLEAM divides the world into a grid of small square cells. Satellite and census sources are used to calculate the population density in each of these cells, which are then clustered into subpopulations centered on their nearest transportation hub.

GLEAM simulates human mobility and disease spreading in a sequence of time steps (representing full days). Within each population cluster, the spread of the infection among individuals is governed by the characteristics of the disease and the containment and mitigation responses specified in the epidemic model. The disease is transmitted between population clusters when people commute to work or school or travel longer distances on national and international flights. On high-performance computers, GLEAM executes millions of stochastic simulations, making it possible to generate for each population the statistical ensemble of possible epidemic evolutions and analytics for quantities such as newly generated cases, seeding events, time of arrival of the infection, and others.

BUILDING A SYNTHETIC WORLD

The GLEAM framework is based on a metapopulation approach in which the world is divided into geographical regions defining a subpopulation network where connections among subpopulations represent the individual fluxes due to the transportation and mobility infrastructure. The population layer is based on highly detailed data, with a granularity defined by a lattice of cells covering

HUMAN MOBILITY PATTERNS ARE THE KEY TO CONSISTENTLY SIMULATING THE MOBILITY OF INFECTIOUS INDIVIDUALS ON THE GLOBAL SCALE.

the whole planet at a resolution of 15 x 15 arc minutes (approximately 25 x 25 kilometers). In order to define the subpopulations that constitute the metapopulation structure of the model, a Voronoi tessellation of the Earth's surface is performed, defining census areas centered around the major transportation hubs obtained from the International Air Transport Association (IATA) and OAG database, as shown in **FIGURE 3.2**. By considering the distance between the cells and the transportation hubs, we assign each cell to a specific hub; this process generates over 3,200 subpopulations worldwide or more precisely census areas. In this tessellation, hubs generally correspond to major urban areas and airports. The cells belonging to a subpopulation allow us to determine the population of that census area. Other attributes, such as the age structure of the population, health infrastructures, etc., can be added according to available data.

MOVING PEOPLE AROUND

The spatio temporal patterns of the disease spreading are associated with the mobility flows that couple different subpopulations. These flows constitute the mobility data layer that is represented as a network of connections among subpopulations. This identifies the number of individuals going from one subpopulation to the others. The mobility network is made by different kinds of mobility processes, from short-range commuting to intercontinental flights with time scale and traffic volumes that span several orders of magnitude. The airline system layer integrates air travel mobility, containing the list of worldwide origin-destination flows between airport pairs on a daily schedule.

Figure 3.2 | GLEAM's tessellation of North America

The polygons define the census areas considered by the model. The circles represent the major transportation hub centers of each area. The colors of the census areas are proportional to the population of each cell.

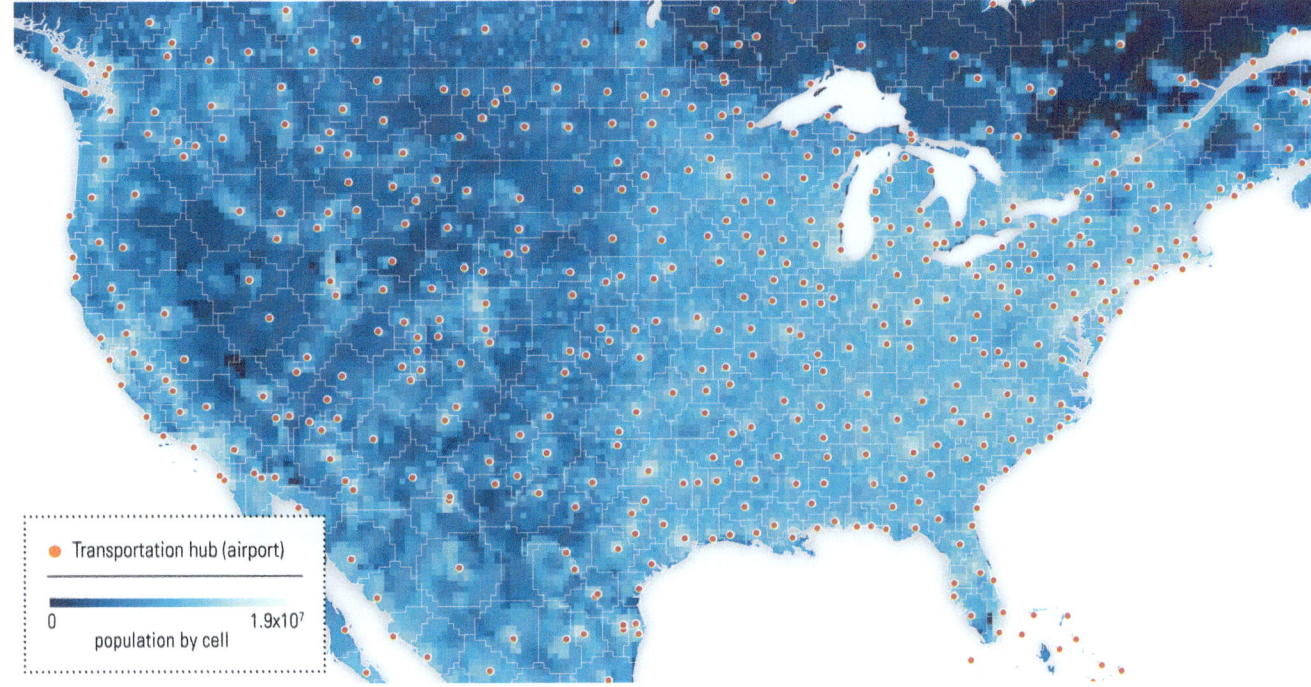

Transportation hub (airport)

0   1.9x10⁷
population by cell

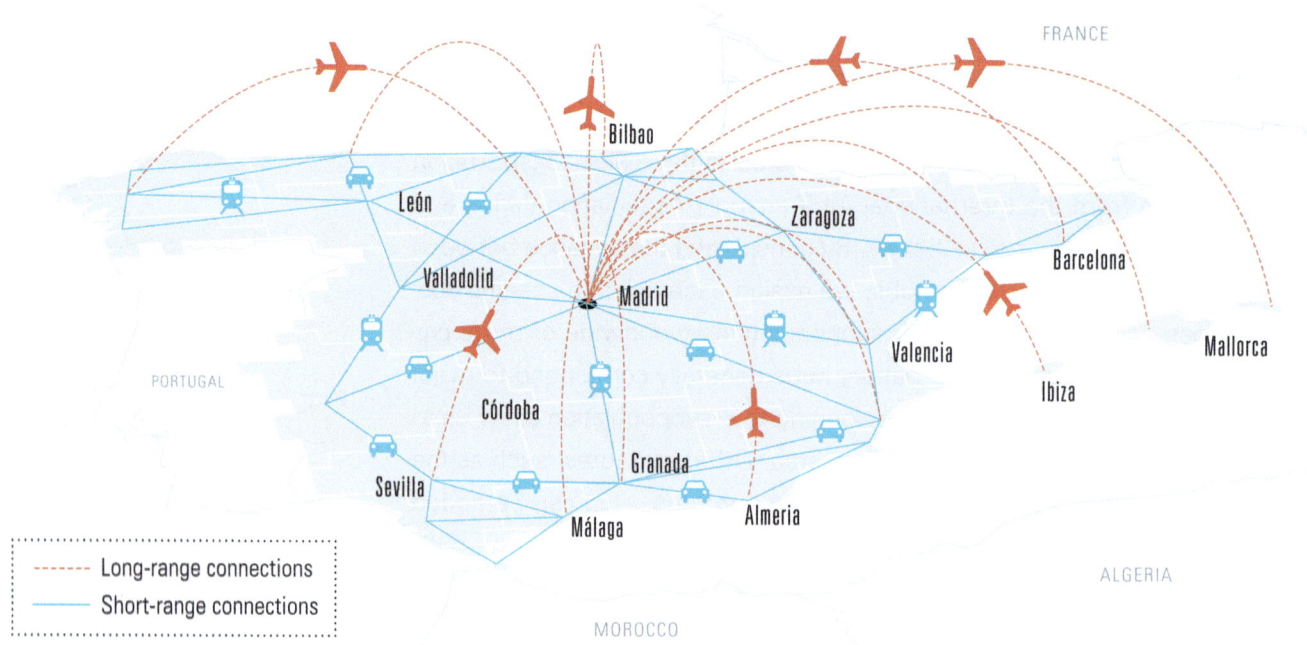

Figure 3.3 | Long- and short-range mobility implementation in GLEAM

Highlighting some of the air transportation connections (orange) and the short-range mobility network (blue) for Madrid, Spain.

Individuals travel on airplanes according to an explicit dynamic that considers the probability for each individual in the population to travel on a specific route.

For the short-range mobility, we rely on databases collected from the Offices of Statistics of 30 countries in five continents. The full dataset includes more than 80,000 administrative regions and over five million commuting flow connections between them. In order to overcome the differences in the spatial resolution of the commuting data across different countries, we define a worldwide homogeneous standard for GLEAM. We use the geographical census areas obtained from the Voronoi tessellation as the elementary units to define the centers of gravity for the process of commuting. This allows us to deal with similar units across the world with respect to mobility which emerge from the tessellation, and not country specific administrative boundaries. We map the different levels of commuting data into the geographical census areas. The mapped commuting flows can be seen as a second transportation network connecting subpopulations that are geographically close. Where data are not available, the short-range mobility layer can be generated synthetically by relying on the so-called gravity law [1,2] and the more recent approach dubbed the radiation law, both calibrated on the real data available.[3] The short-range mobility network can be overlaid on the airline system layer forming the mobility system of the GLEAM synthetic world (**FIGURE 3.3**).[4]

1 Duygu Balcan et al., "Multiscale mobility networks and the spatial spreading of infectious diseases," Proceedings of the National Academy of Sciences **106**, 21484–21489 (2009).
2 Cécile Viboud et al., "Synchrony, waves, and spatial hierarchies in the spread of influenza," Science **312**, 447–451 (2006).
3 Filippo Simini et al., "A universal model for mobility and migration patterns," Nature **484**, 96–100 (2012).
4 Duygu Balcan et al., "Modeling the spatial spread of infectious diseases: The global epidemic and mobility computational model," Journal of Computational Science **1**, 132–145 (2010).

The short-range commuting mobility of individuals is simulated by an effective approach that defines mixing subpopulations and which identifies the number of individuals $N_{ij}(t)$ of the subpopulation i effectively present in subpopulation j at time t (see **INFOBOX 3.1**). This methodology assumes the subpopulation i as having an effective number of individuals $N_{ij} \ll N_{ii}$ in contact with the individuals of the neighboring subpopulation j in a quasi-stationary

3.1 | INFOBOX
MIXING SUB POPULATIONS

In the case of commuting flows, we assume that individuals in the subpopulation i will visit anyone of the connected subpopulations with a per capita diffusion rate σ_i. As we aim at modeling commuting processes in which individuals have a memory of their location of origin, displaced individuals return to their original subpopulation with rate τ.

In order to model the commuting flows in the subpopulation network, we define mixing subpopulations. At any moment in time, each member of subpopulation i is either in their subpopulation of residence or outside and visiting one of the neighboring subpopulations j. By using the approach developed in Sattenspiel and Dietz (1995)[1] and Keeling and Rohani (2002)[2], we may group the members of i according to the location in which they are actually present at a given time t, $N_{ii}(t)$ and $N_{ij}(t)$ with $j \in v(i)$ where $v(i)$ are the subpopulations connected to i (**FIGURE 3.1.1**). The rate equations for the population sizes of different subgroups can be readily written by explicitly taking into account the mobility rates along the edges of the subpopulation network. This system of rate equations has a characteristic relaxation time that can be obtained by solving the appropriate differential equations. In particular, it is possible to show under the general assumption of $\sigma_i \ll \tau_i$ that the relaxation characteristic time is τ_i^{-1} and that the mixing subpopulations read as:

$$N_{ii} = \frac{N_i}{1 + \sigma_i/\tau_i} \quad \text{and} \quad N_{ij} = \frac{N_i}{1 + \sigma_i/\tau_i}\, \sigma_{ij}/\tau_i$$

This implies that in the regime $\sigma_i \ll \tau_i$, $N_j(t)$ represent a small perturbation to the overall subpopulation of size N_j. These expressions are used to obtain the effective force of infection taking into account the interactions generated by the commuting flows.

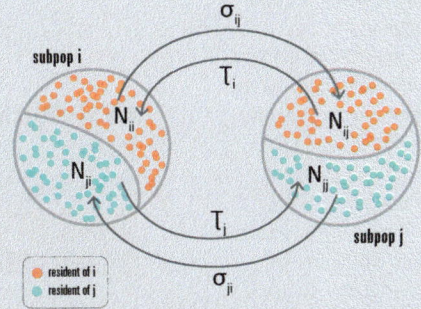

Figure 3.1.1 | Illustration of commuting and subdivision of population

At any time each subpopulation is occupied by its residents plus visitors from its neighbors. For instance, the population in subpopulation i is divided between individuals who reside and are present in the subpopulation (N_{ii}) and those who are residents in subpopulation j but present in subpopulation i (N_{ij}). Different classes of people move between connected subpopulations along the edges at the rates shown.

Adapted from D. Balcan and A. Vespignani, "Invasion threshold in structured populations with recurrent mobility patterns." J. Theor. Biol 293 (2012)

1 L. Sattenspiel and K. Dietz, "A structured epidemic model incorporating geographic mobility among regions," Mathematical Biosciences 128, 71–91 (1995).

2 Matt J Keeling and Pejman Rohani, "Estimating spatial coupling in epidemiological systems: a mechanistic approach," Ecology Letters 5, 20–29 (2002).

state, reached whenever the time scale of the epidemic spreading is larger than the commuting rate.

THE DISEASE DYNAMIC

Superimposed on the worldwide population and mobility layers is the epidemic model that defines the disease and population dynamics. Individuals move around and transmit the infection via the interactions with other people. The infection dynamics takes place within each subpopulation and assumes the classic compartmentalization scheme for the characterization of the disease (**FIGURE 3.4**). Each individual fits, at any given point in time, within a certain compartment that corresponds to a particular disease-related state (being, e.g., susceptible, symptomatic, or vaccinated). These compartments are connected by transitions that define how individuals may pass from one state to another (e.g., from susceptible to latent when being infected), while the associated parameters determine the likelihood that such transitions take place. GLEAM uses algorithms mathematically defined through individual-based stochastic processes to calculate the proportion of the population within each compartment for each subpopulation and how these proportions change

Figure 3.4 | Summary of basic definitions of the stages of a disease

As a function of time, the disease's evolution in two patients. Individuals who have not been in contact with the pathogen are generally labeled as susceptible. The pre infectious period, also called the latent period, defines the time from infection to when the host is on her turn able to transmit the infection. The incubation period is the time from infection to the onset of clinical symptoms. The infectious period is the time period in which the host can transmit the infection to other hosts. The clinical disease time refers to the duration of the clinical symptoms. It is important to stress that the time of each period can change in the case of pharmaceutical interventions specific to the disease considered. Some diseases may also require more detailed classification of the states characterizing the natural history of the disease.

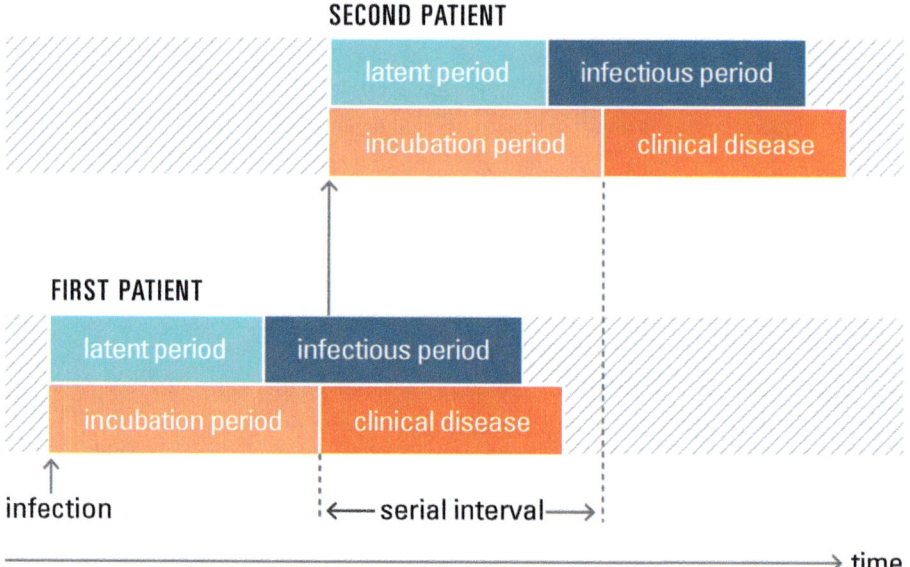

over time as individuals transition from one compartment to the next (**INFO-BOX 3.2**). The progression of the disease is then simulated at the individual level. GLEAM can also include the age structure of individuals in defining the transitions. It is clear that no model fits all diseases, and GLEAM needs the detailed process describing the evolution of the illness within each individual and the transmission process. In general, this is specified by the so-called natural history of the disease that maps the time-periods of the disease

3.2 INFOBOX
MODELING THE DISEASE TRANSMISSION

Although the realistic modeling of infectious diseases generally implements much more complicated compartmental structures, let's use as an example the basic compartmental structure where individuals can only ever be in one of three states: susceptible, infectious, and recovered. This simple three-state compartmentalization defines the classic susceptible-infected-recovered (SIR) model. In this case we have only two possible transitions. The first, denoted $S \rightarrow I$, is when a susceptible individual interacts with an infectious one and becomes infected. The second one, denoted $I \rightarrow R$, occurs when the infectious individual recovers from the disease and is assumed to have acquired permanent immunity from the disease. The two processes completely determine the epidemic evolution. The $I \rightarrow R$ transition is spontaneous and occurs after the individual has spent a certain time fighting the disease or taking a specific medical treatment; in other words the transition does not depend on any interaction with the other individuals in the population. The $S \rightarrow I$ transition instead occurs only because of the contact/interaction of the susceptible individual with an infectious individual. In this case the interaction dynamics between people is a specific feature of the transition and has to be taken into account. The conceptual abstraction of the SIR compartmental model is well represented by the flow diagram of **FIGURE 3.2.1** in which the different compartment transitions are schematized through arrows that indicate the possible change of state of the individuals.

The $I \rightarrow R$ transition is obviously the simplest one to model. For many type of diseases, the amount of time spent in the infectious class is distributed around a well-defined mean value. For the sake of realism, the probability that one person will move from the I class to the R class depends on

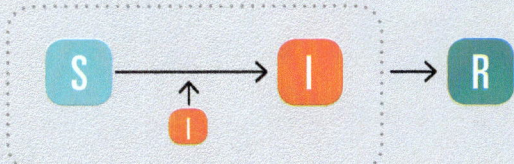

Figure 3.2.1 | Flow diagram of the SIR model

The model allows only the $S \rightarrow I$ and the $I \rightarrow R$ transitions. The transitions are denoted by arrows going from one compartment to the other. The $S \rightarrow I$ transition occurs only in the presence of a contact/interaction with infectious individuals. For this reason the transition arrow has a line callout to the infectious compartment.

how much time he/she has spent in the I class. The distribution of the "infectious period" and the transition probability can generally be estimated from clinical data; however, from a simplistic modeling assumption, the probability of transition is assumed constant. In this way it is possible to define a transition probability per unitary timestep μ, called the recovery probability. Since we are dealing with a probability per unit time, the time an individual will spend on average in the infectious compartment, the mean infectious period, is equal to μ^{-1}.

The definition of the transition probability in the case of the $S \rightarrow I$ transitions is more complicated than the recovery transition. The probability that a susceptible individual moves into the infectious compartment depends on the number of contacts with infectious people and the probability that in each contact with an infectious individual the disease is transmitted to the susceptible. The number of contacts with infectious individuals depends in turn on the per capita number of contacts per unit time with other individuals and the total number of infectious people present in the population. In the GLEAM framework, the transmission dynamics can be simulated at different levels of detail (see **FIGURE 3.2.2**). ▶

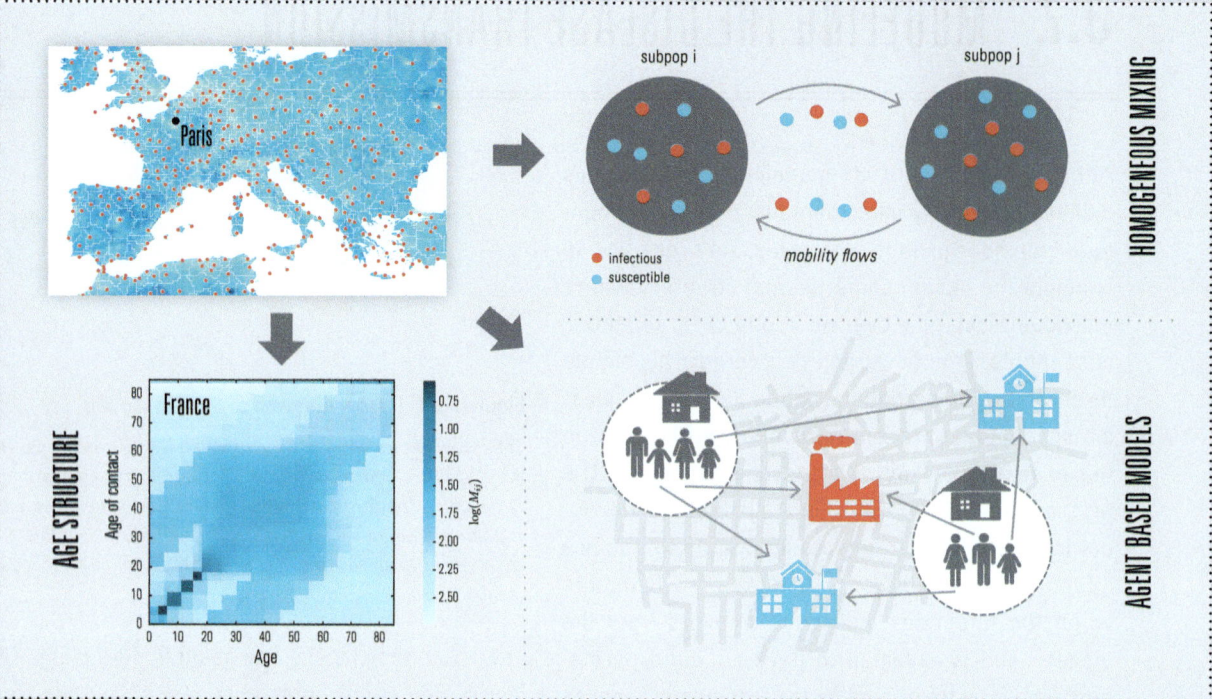

Figure 3.2.2 | Multiple schemes for the stochastic intra-population contagion dynamic

The simplest approaches consider chain binomial processes in which the discrete individuals are indistinguishable and characterized only by their compartmental state. These models can be made more realistic by including age structure or other features of the individuals. In this case the transmission of the disease is described by parameters that depend on those features. An example is provided by models implementing specific contact matrices that characterize the number of contacts among individuals in different age brackets.

At the finer level, synthetic population constructions are even more refined and consider a classification of location such as households, schools, offices, etc. The movements and time spent in each location can be used to generate individual-location bipartite networks whose unipartite projection defines the individual level, synthetic interaction network that governs the epidemic spreading. Also in this case, although the model underlying the computational approach is a network model, each individual is annotated with the residence place, age, as well as many other possible demographic information that can be exploited in the analysis of the epidemic outbreak. Detailed synthetic populations thus reconstruct a statistically equivalent picture of the actual population down to the level of the granularity of the data available.

For the sake of simplicity, let us consider here the example of a homogenous mixing approximation which assumes that individuals randomly interact among them. According to this minimal framework, the larger the number of sick and infectious people among one individual's contacts, the higher the probability of infection transmission. This readily translates in the definition of the force of infection λ, also called risk, that expresses the probability per unit time at which susceptible individuals may contract the infection. In the limit of small risk, it is possible to derive the explicit form $\lambda = \beta I_t / N$. Here β defines the transmissibility, average number of transmissions per unit time, that depends on the specific disease as well as the contact pattern of the population,

I_t the total number of infectious individuals at time t; I_t/N is therefore the density of infected individuals in the population. This form of the force of infection is called the *mass action law* and is used in many other reaction-diffusion problems in chemistry and physics. It is important to note that the force of infection is said to be *frequency dependent* as it assumes that the number of contacts is independent of the population size. Therefore, the force of infection depends only on the density of infectious individuals, and decreases for larger populations all the other factors being equal. This is indeed an assumption that fits with our intuition as the probability of getting infected by one single infectious individual in a city like Paris with about two million residents is necessarily much lower than the probability to be infected by the same infectious individual in Bloomington, Indiana, a campus town with only 80,000 residents.

In order to translate the above formal relation into an explicit equation, we can define the variables S_t, I_t and R_t denoting the number of individuals in the susceptible, infectious, and recovered compartment at time t, respectively. Given the assumption that μ and β are constant, we can easily define the associated stochastic processes that relate the stochastic variables at time t with the variables at time $t+1$ in the form of a simple binomial model of transmission for discrete contacts and discrete time. Each susceptible individual has a probability $\lambda_t = \beta I_t/N$ to contract the disease and transit to the infectious state.

As we assume to have S_t independent events occurring with the same probability, the number of new infected individuals I_+ generated at time $t+1$ is a random variable that will follow the binomial distribution $I_+ \sim Bin(S_t, \lambda_t)$. The binomial distribution provides the probability that among the S_t independent trials with probability λ_t, we have y positive events at time $t+1$. Analogously, the number of new recovered individuals at time $t+1$ is a random variable that will follow a binomial distribution $R_+ \sim Bin(I_t, \mu)$, where the number of independent trials is given by the number of infectious individuals I_t that might recover and the probability of

recovery in a timestep is given by the recovery probability μ. If we consider a specific value of the stochastic variables S_t, I_t and R_t, the stochastic equations regulating the behavior of the epidemic can be written as:

$$s_{t+1} = s_t - Bin(s_t, \lambda_t)$$
$$i_{t+1} = i_t + Bin(s_t, \lambda_t) - Bin(i_t, \mu)$$
$$r_{t+1} = r_t + Bin(i_t, \mu),$$

where $Bin(s_t, \lambda_t)$ and $Bin(i_t, \mu)$ are two random variables distributed according to the respective binomial distribution. It is worth remarking here that the unitary time step defines an actual time scale Δt and that the transition probability must be defined as a function of this time scale.

In the SIR model it is possible to readily calculate the basic reproduction number explicitly as $R_0 = \beta/\mu$. It is given simply by the transmissibility times the average duration of the infectiousness of the single individual; this provides the average secondary infections per infectious individual.

In such a computational approach, we deal with stochastic systems, and therefore we need to generate random variables according to the specified probability distributions defined in the model. In a stochastic simulation, each sequence of random values is generated through a random number generator. Each different random input therefore provides a single stochastic instance of the system's behavior. In the case of epidemic models, each stochastic realization will represent only one of the many possible epidemic outcomes that the same model with the same initial conditions and parameters can generate. A careful analysis of the quality of the random number generator used is advisable in all intensive large-scale computational applications.

The simple example discussed here has to be generalized to the more complicated compartmental structures used by GLEAM for the realistic modeling of infectious diseases. In many cases this implies the use of more advanced mathematical constructions and the use of multinomial stochastic processes.

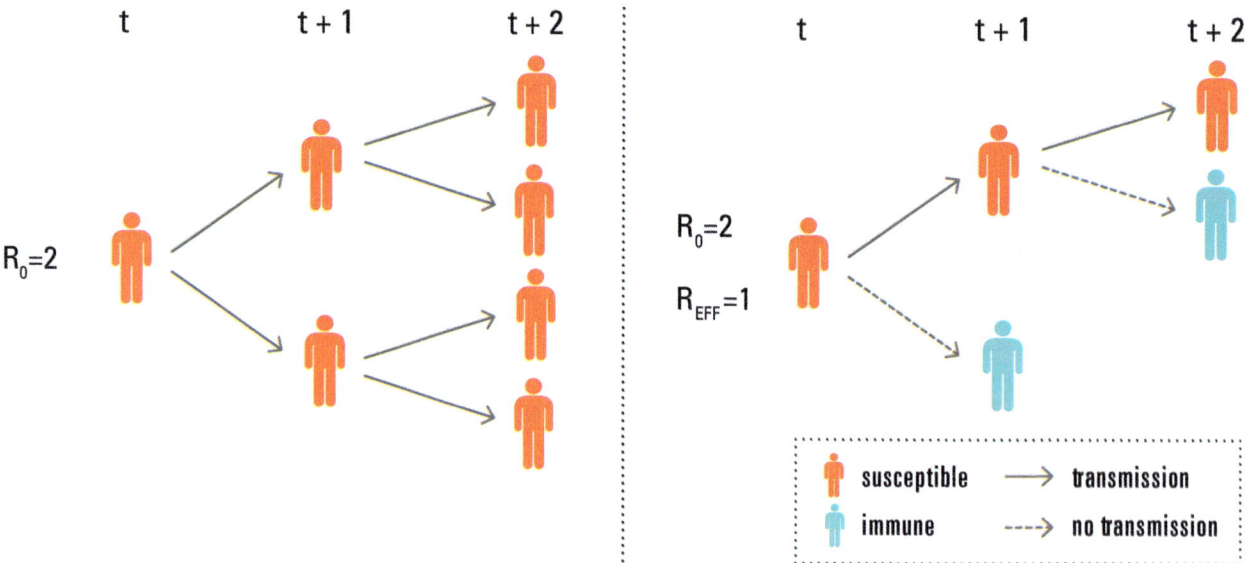

Figure 3.5 | Illustration of the chain of transmissions as a branching process

Each infectious individual generates a number of secondary cases according to the model transmission rate and the available susceptible individuals. The branching ratio of this process, defined as the average number of secondary generations in a fully susceptible population, defines the basic reproduction number or ratio, R_0. One of the main targets of public health intervention is the reduction of the transmissibility, for instance, by vaccinating a fraction of individuals, resulting in an effective reproduction number R_{eff} smaller than one.

Disease Type	R_0 Value
SARS	2 – 3
HIV	2 – 5
Smallpox	5 – 10
Pandemic Influenza	1.5 – 3.5
Ebola	1.5 – 3.5

Table 3.1 | Reproduction number

Ranges for the reproduction number of some infectious diseases.

progression. In **INFOBOX 3.3**, we show the basic compartmental structure of some of the diseases we will consider. It is important to note that it is possible to define more compartmental states by also considering the implementation of pharmaceutical and non-pharmaceutical interventions such as hospitalization, vaccination, quarantine, isolation, and so on. In addition, in GLEAM it is important to associate with each compartment the likelihood that the individual will travel long distance, commute, etc. In many cases, clinical symptoms are associated with reduced or no mobility of the sick individual. This is also true for compartments signaling the isolation or quarantine of individuals. Each compartment, therefore, carries additional information of the mobility and the potential interaction of the individual.

The disease progression is mostly defined by two quantities generally used to quantify the transmissibility potential and spreading time scale of the disease. The first quantity is the *basic reproduction number, R_0*, that is the average number of secondary cases produced by a primary case in a fully susceptible population. If each infectious individual does not generate a number of infectious individuals larger or equal to one, the number of infectious individuals will generally decrease, and the transmission chain will die out before an epidemic can take place. However, if each infectious individual generates more than one infectious individual in the transmission process, the number of infectious individuals will continue to increase in time. Intuitively, the larger the R_0, the more transmissible the disease is and the faster the number of cases will grow (**FIGURE 3.5**). Together with the basic reproduction number, the generation time G_t of the disease, defined as the time occurring from the infection of the host to the end of the infectious period, is also an important quantity that defines the time scale of the disease. For the same R_0, a smaller generation time indicates a much faster progression of the disease, as the same number of secondary cases are generated in a shorter time window. In general, a measure of the generation time is clinically offered by the *serial interval*: the time from the onset of a primary case to the onset of the secondary case. R_0 is a function of the parameters describing the natural history of the disease and can be calculated explicitly in different models (**INFOBOX 3.2**). Often, the level of threat of an infectious disease is measured as a function of the R_0.

SYNTHETIC EPIDEMICS

GLEAM defines a synthetic world in which we can simulate with the computer the unfolding of epidemics and pandemics. Each simulated time step represents a full day. The model needs the definition of the initial conditions that specify the number and location of individuals capable of transmitting the disease. At the start of the time step, we use the flight network to move travelers to their destination.

As a consequence, the arrival time for the infection is the day at which the first infected traveler arrives, and this seed individual is considered to have the chance of infecting others. The probability of traveling changes from day to day and can be generalized in order to consider the effects of location specific

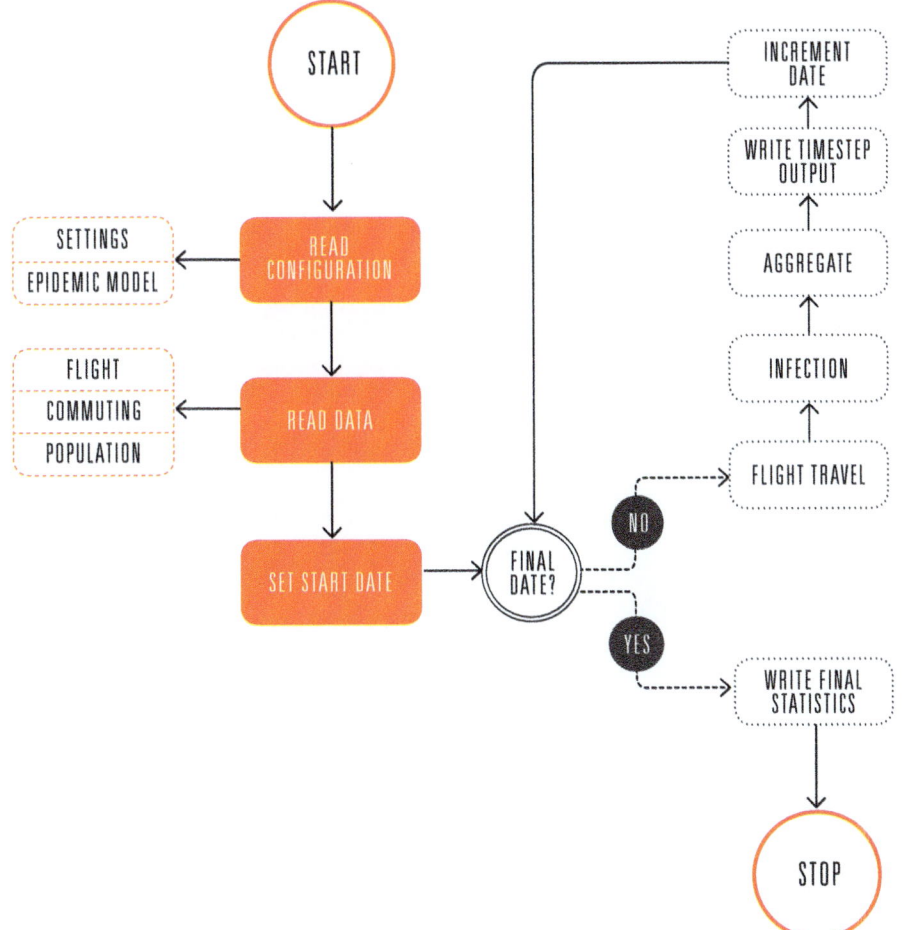

Figure 3.6 | GLEAM engine flow chart

The full procedure used by the GLEAM simulation engine. The left column represents input databases. The program flow occurs along the center.

DISEASE COMPARTMENTAL STRUCTURE

GLEAM labels individuals in each population according to the compartment describing the state of the disease and the possibility to travel, commute, etc.

Influenza

A susceptible individual in contact with a symptomatic or asymptomatic infectious person contracts the infection at rate β or $r_\beta \beta$, respectively, and enters the latent compartment where he is infected but not yet infectious.

At the end of the latency period ε^{-1}, each latent individual becomes infectious, entering the symptomatic compartments with probability $1-p_a$ or becoming asymptomatic with probability p_a. The symptomatic cases are further divided between those who are allowed to travel (with probability p_t) and those who would stop traveling when ill (with probability $1-p_t$). Infectious individuals recover permanently with rate μ.

Influenza with antiviral pharmaceutical interventions

A modified Susceptible-Latent-Infectious-Recovered model is considered to take into account the use of antiviral drugs as a pharmaceutical measure. In particular, infectious individuals are subdivided into: asymptomatic (Infectious$_a$), symptomatic individuals who travel while ill (Infectious$_{st}$), symptomatic individuals who restrict themselves from travel while ill (Infectious$_{snt}$), and symptomatic individuals who undergo the antiviral treatment (Infectious$_{AVT}$). A susceptible individual interacting with an infectious person may contract the illness with rate β and enter the latent compartment where he/she is infected but not yet infectious. The infection rate is rescaled by a factor r_a in case of asymptomatic infection and by a factor r_{AVT} in case of a treated infection. At the end of the latency period, of average duration equal to ε^{-1}, each latent individual becomes infectious, showing symptoms with probability $1-p_a$, and asymptomatic with

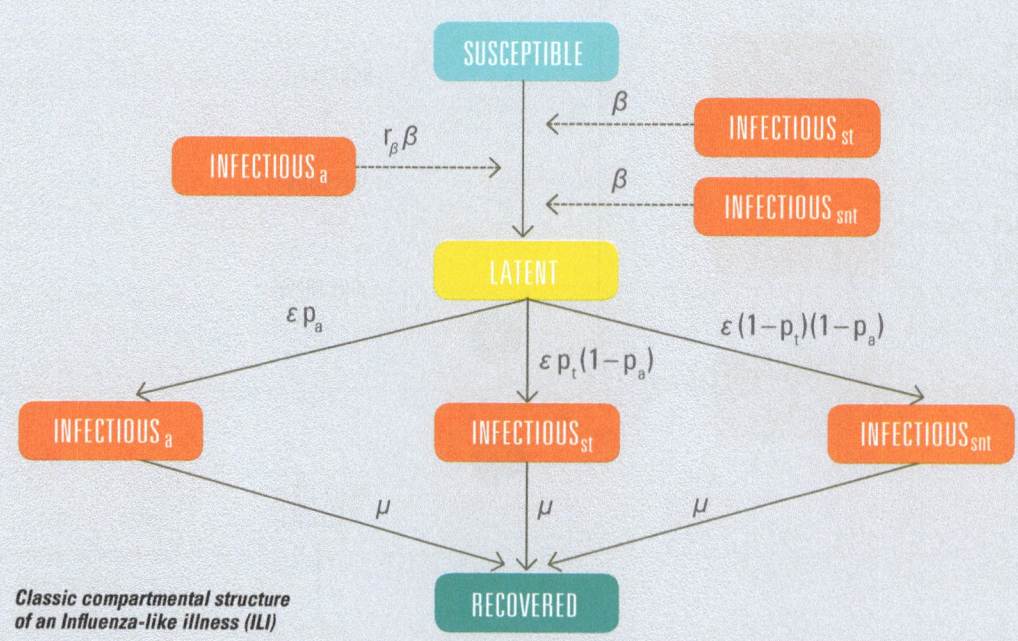

Classic compartmental structure of an Influenza-like illness (ILI)

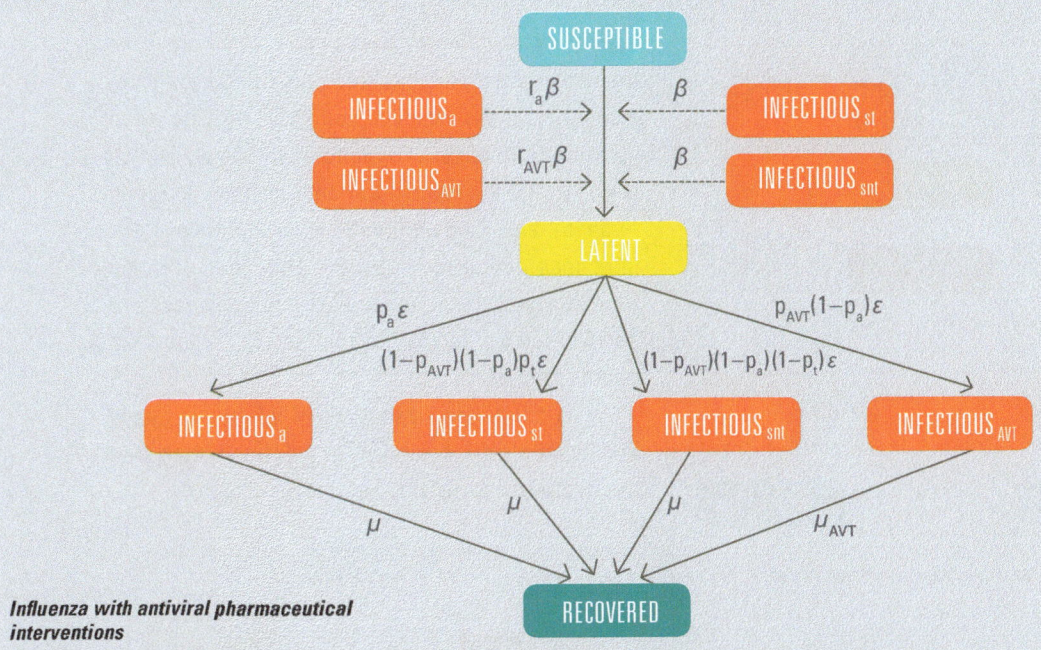

Influenza with antiviral pharmaceutical interventions

probability p_a. Change in traveling behavior after the onset of symptoms is modeled by p_t, which is the probability that individuals would continue traveling when ill. Infectious individuals recover permanently after an average infectious period μ^{-1}. We assume the antiviral treatment regimen is administered with p_{AVT} to the symptomatic infectious individuals within 1 day from the onset of symptoms, reducing the infectiousness and shortening the infectious period by one day.

SARS-like viruses and their non-pharmaceutical containment

The population of each city is classified into seven different compartments, namely, susceptible, latent, infectious, hospitalized who either recover or die, dead, and recovered individuals. We assume that hospitalized, as well as infectious individuals are able to transmit the infection, given the large percentage of the cases that were seen among healthcare workers. The actual efficiency of hospital isolation procedures is modeled through a reduction of the transmission rate β by a factor $r_\beta = 20\%$, as estimated for the early stage of the epidemic in Hong Kong. The infectiousness of patients in the hospitalized compartments HR and

HD are assumed to be equal (although this assumption can easily be changed in the model). Susceptible individuals exposed to SARS enter the latent class. Latents represent infected individuals who are not yet contagious and are assumed to be asymptomatic, as suggested by results based on epidemiological, clinical, and diagnostic data in Canada. They become infectious after an average time ε^{-1} (mean latency period). Individuals are classified as infectious during an average time equal to μ^{-1}, from the onset of clinical symptoms to their admission to the hospital where they eventually die or recover. Patients admitted to the hospital are not allowed to travel. The average periods spent in the hospital from admission to death or recovery are equal to μ_D^{-1} and μ_R^{-1}, respectively. The average death rate is denoted by d.

Viral hemorrhagic fever compartmental model

Legrand et al.[1] introduced a compartmental model for VHF virus where the individuals are classified in the following way: susceptible individuals S, who can acquire the disease after contact with infectious individuals, latent individuals L who are infected but do not transmit the disease and are asymptomatic, infectious ▶

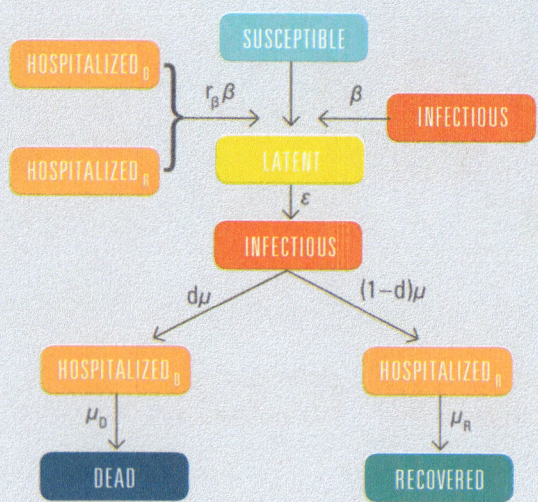

SARS-Like viruses and their non-pharmaceutical containment

according to different parameters. Similarly, from the compartment hospitalized and funeral, individuals can enter the removed compartment. The mean duration from onset of symptoms to hospitalization is γ_h^{-1}, γ_{dh}^{-1} is the mean duration from hospitalization to death, and γ_i^{-1} denotes the mean duration of the infectious period for survivors. The mean duration from hospitalization to end of infectiousness for survivors is γ_{ih}^{-1}, and γ_f^{-1} is the mean duration from death to burial.

θ_1 is computed so that $\theta\%$ of infectious cases are hospitalized. δ_1 and δ_2 are defined such that the overall case-fatality ratio is δ.

individuals I who can transmit the disease and are symptomatic, hospitalized infectious individuals H, dead individuals F that can infect through the burial ceremonies, and removed individuals R. The most distinctive feature of this model is that dead individuals can still transmit the disease.

Susceptible individuals, after contact with an infectious individual (I, H or F), enter the latent class at a rate β_I, β_H, or β_F. At the end of the latency period α^{-1}, each individual becomes infectious. Infectious individuals then can transition to the hospitalized, funeral, or removed compartments

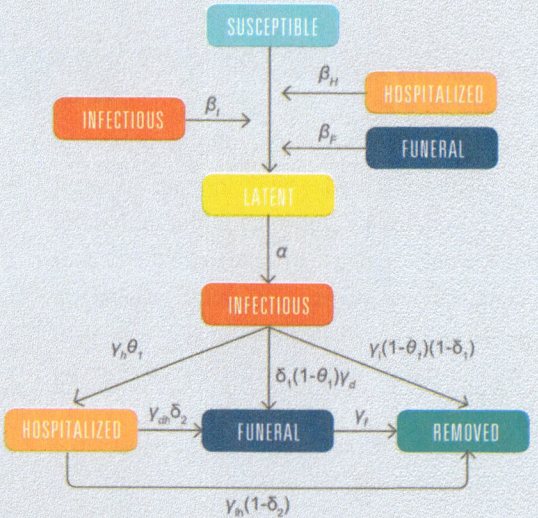

Viral hemorrhagic fever compartmental model

1 Judith Legrand, Rebecca Freeman Grais, Pierre-Yves Boëlle, Alain-Jacques Valleron, and Antoine Flahault, "Understanding the dynamics of ebola epidemics," Epidemiology & Infection 135, 610–621 (2007).

airline traffic reductions. The short-range mobility and the infection dynamics are modeled together by defining the probability of transition and risk of infection acting on each individual in each subpopulation. This process is repeated for every simulated day, keeping track of all the individuals and their traveling patterns as shown in the pseudo-code for the GLEAM algorithm (**FIGURE 3.6**).

GLEAM also allows the introduction of seasonal variations in the transmissibility of the disease, such as in the case of influenza. Seasonality effects are still an open problem in the transmission of ILI. In order to include the effect of seasonality on the observed patterns of ILI, a standard empirical approach can be used in which seasonality is modeled by a forcing that reduces the basic reproduction number by a factor ranging from 0.1 to 1 (no reduction). This forcing is described by a sinusoidal function over a 12-month period that reaches its peak during winter time and its minimum during summer time in each hemisphere, with the two hemispheres at opposite phases. The minimum rescaling of α_{min} of the reproduction number is a free parameter to be estimated from data. For scenario purposes it is possible to consider a mild seasonality and a strong seasonality scenario, with $\alpha_{min} \sim 0.5$ and $\alpha_{min} \sim 0.1$, respectively.[5]

Given the population and mobility data, infection dynamics parameters, and initial conditions, GLEAM performs the simulation of stochastic realizations of the worldwide unfolding of the epidemic. From these in silico epidemics, a variety of information can be gathered, such as the prevalence, morbidity, number of secondary cases, number of imported cases, hospitalized patients, amounts of drugs used, and other quantities for each subpopulation with a time resolution of 1 day. In the next chapter, we will see the results of the numerical simulations and why and how they can be useful to our analysis and understanding of infectious disease spreading.

IF EACH INFECTIOUS INDIVIDUAL GENERATES MORE THAN ONE INFECTIOUS INDIVIDUAL, THE NUMBER OF INFECTIOUS INDIVIDUALS WILL INCREASE IN TIME.

---

5 Ben S. Cooper et al., "Delaying the International Spread of Pandemic Influenza," PLoS Med **3**, e212 (2006).

....

CHAPTER 4

FROM DATA TO KNOWLEDGE: HOW MODELS CAN BE USED

BY INTEGRATING THE DATASETS we have explored in previous chapters and by performing the complex calculations necessary to numerically integrate the epidemic and mobility models, GLEAM can be used to produce microsimulations of infectious disease spreading.

Once all the parameters and initial conditions of an outbreak are plugged into the model, computers generate a large number of nominally identically initialized, numerical stochastic simulations of the global progression of the epidemic. The output provides, for each point in space and time allowed by the resolution of the model, the ensemble of possible epidemic evolutions. From this ensemble it is possible to define the median, mean, and confidence interval of epidemic observables, such as newly generated cases, seeding events, and the time of arrival of the infection, among others. It is important to note that the forecast ensemble and the statistical quantities depend on the key parameters of each specific disease and on the initial condition—date and place—of each outbreak. In order to provide understanding and intelligence on the unfolding of the epidemic, this stochastic forecast output (SFO) can be aggregated and explored in many different ways.

AFFECTED COUNTRIES

Computational models are able to capture the full stochasticity of epidemic spreading. Given a local outbreak and a number of index cases, the probability that the outbreak will take off and start affecting a large part of the community is not a given: indeed, the probability of having a major outbreak will depend on the transmissibility of the disease and other factors, such

Figure 4.1 | Countries affected by an epidemic as a function of time for the six WHO regions

The concentric circles refer to 90, 180, 270, and 365 days from the start of the spreading, moving outward, respectively. In each circle we report the number of countries affected by an outbreak. The circles are divided in colored sections proportional to the fraction of affected countries in each region.

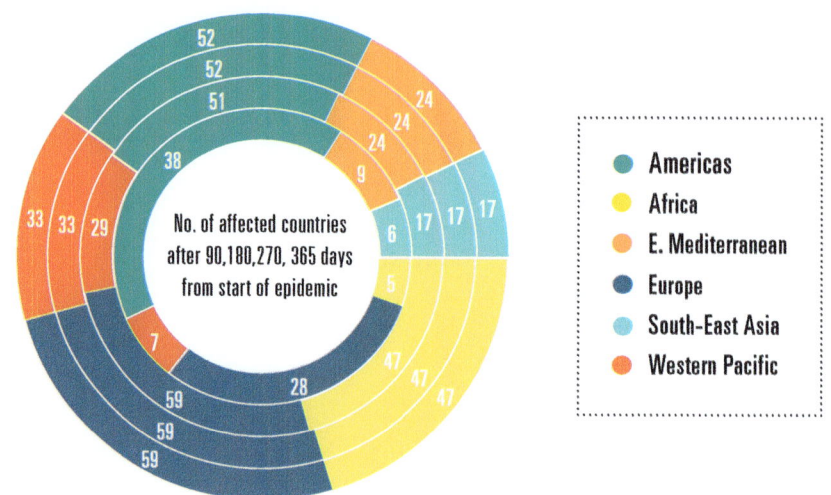

as the time of year of the outbreak. Furthermore, even if one population is affected by a major outbreak, it is not certain that the outbreak will affect other populations, countries, or regions of the world. In the case of a low basic reproduction number (e.g., $R_0 = 1.1$), approximately 80%–90% of the epidemics starting with a single infectious individual undergo a stochastic extinction. The number of countries affected by the epidemic as a function of time is, therefore, one of the main indicators that can be extracted from the stochastic numerical output of GLEAM.

For instance, **FIGURE 4.1** provides the number of countries with local transmission during the course of an epidemic for each WHO region after a period of 90, 180, 270, and 365 days.

INVASION TREE

One of the most interesting pieces of information we can gather from the numerical output is the spreading pathway of epidemics. This can be traced by tracking, across the world, the path along which the epidemic has moved from one population, the infector, to the other, the infectee. This amounts to recording the origin and date of importation of the first individual carrying the disease in each subpopulation of interest. This procedure allows the construction of the epidemic invasion tree, which represents the transmission of the infection from one subpopulation to the other during the epidemic's unfolding.

The stochastic nature of the epidemic process implies that each simulated scenario might produce an ensemble of different invasion trees. An overall epidemic invasion network can be obtained by defining weighted, directed links, T_{ij}, denoting the probability that subpopulation j is seeded by individuals belonging to the subpopulation i. This probability is given by the ratio between the number of realizations in which we have a seeding event $i \rightarrow j$ and the total number of realizations. Finally, to highlight only the most likely infection tree, we construct the directed minimum spanning tree using the Chu-Liu/Edmonds algorithm[1,2]. In our case, the cost of going from one subpopulation to another is given by the probability $(1-T_{ij})$. The method gives us a directed spanning tree where each subpopulation has only one incoming link, from its infector, which corresponds to the most likely path along which the outbreak has arrived. In **FIGURE 4.2**, we show an epidemic invasion tree originating in Barcelona, Spain. The nodes in this figure identify 3,200 populations worldwide, and the directed links indicate the path that the epidemic followed from one population to the other. The color map from red to blue

Figure 4.2 | Invasion tree

Each circle represents an urban area, and each link describes an infection path along which the epidemic has moved from one population to the other. The color of the circles refers to the time lag with respect to the start of the epidemic.

1 Yoeng-Jin Chu, "On the shortest arborescence of a directed graph," Science Sinica **14**, 1396–1400 (1965).
2 Jack Edmonds, "Optimum branchings," Journal of Research of the National Bureau of Standards B **71**, 233–240 (1967).

time (days)
32   64   128   256

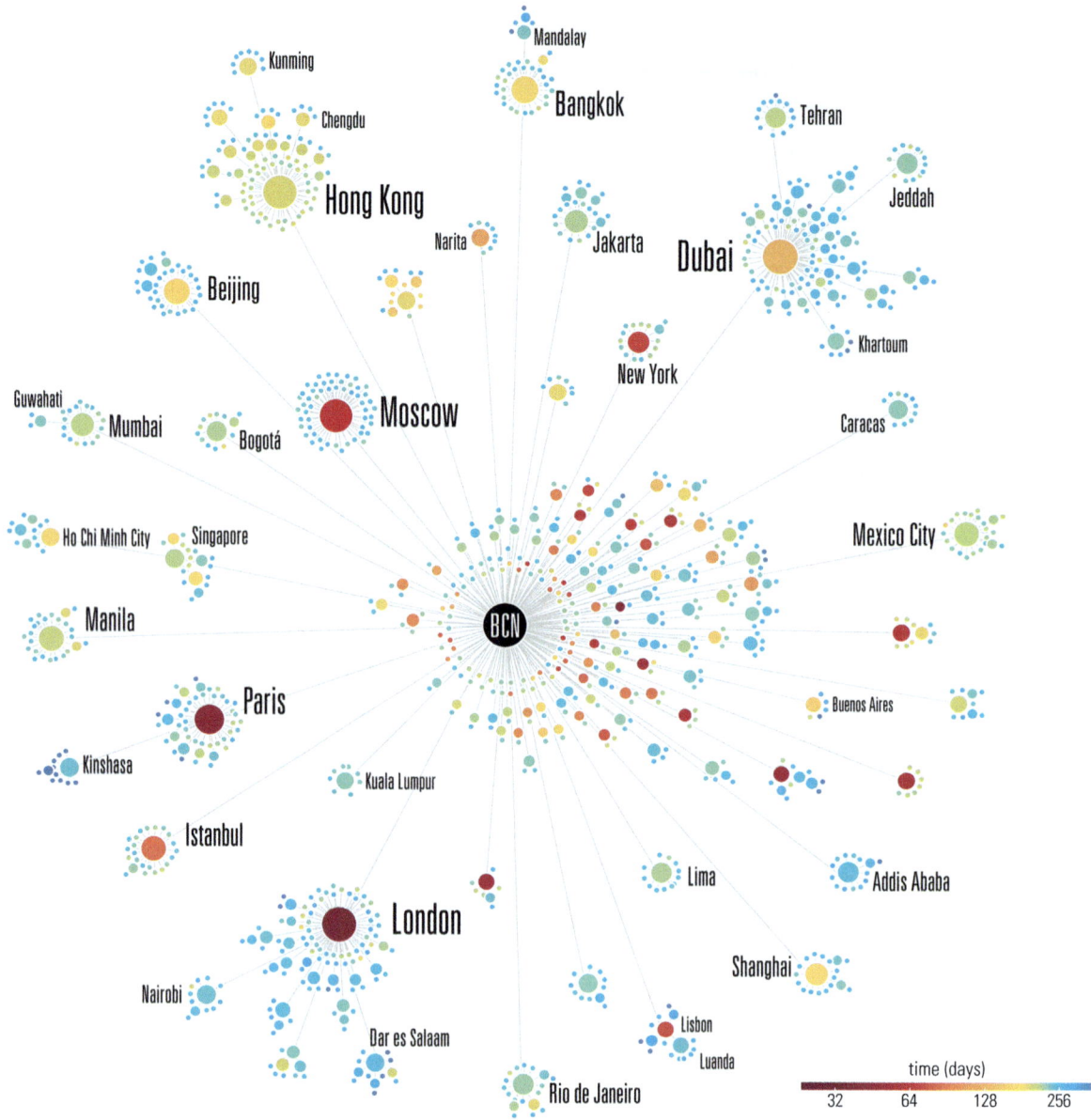

Figure 4.3 | Invasion tree graph

Each node represents an urban area, and each link describes an infection path along which the epidemic has moved from one population to the other. The color of the node refers to the time lag with respect to the start of the epidemic.

indicates the time ordering of the epidemic invasion tree. The invasion tree is a proxy of the epidemic path that is statistically more likely to occur in the epidemic's early stage, as well as of the timing of the disease's arrival in the different areas of the world.

FIGURE 4.3 provides an alternative visualization of the invasion tree that does not consider the geographical location in the final layout. Instead, it uses a force-directed algorithm[3] that positions the nodes in a two-dimensional space allowing exploration of the hierarchical structure of the invasion tree. From a network science perspective, we can observe that there are highly connected

---

3 Mathieu Jacomy, Tommaso Venturini, Sebastien Heymann, and Mathieu Bastian, "Forceatlas2, a continuous graph layout algorithm for handy network visualization designed for the gephi software," PloS One **9**, e98679 (2014).

nodes (hubs) that play a relevant role in the diffusion of the outbreak. Once again, the colors of the nodes here represent the arrival time of the epidemic to a given subpopulation and the size of the circles is proportional to the number of subpopulations that are infected by each one of them.

EPIDEMIC PROFILES AND ACTIVITY PEAK

In each subpopulation, once the epidemic starts, computational models keep track of the number of individuals who are in the different stages of the disease. Generally, the quantities routinely monitored are the number of new clinical cases per day or week (incidence), usually expressed as a rate per 1,000 individuals. This information provides the epidemic profiles that show the evolution of the disease as a function of time; they can be generated at different levels of aggregation, such as continents or countries or down to the scale of single urban areas or subpopulations.

For example, numerical simulations can be used to estimate the different timing of the epidemic in different places of the world. In **FIGURE 4.4**, we show the epidemic profiles of a baseline set of major cities located in different WHO regions and how the epidemic is affecting each place at different times and with different intensities. Similar plots can be obtained for different compartments, such as the number of exposed or hospitalized individuals. As said before, the numerical output of stochastic realizations defines an ensemble of statistically possible epidemic profiles; if not stated otherwise, we report the "median" epidemic profile. In some cases, shaded areas are added around the median to indicate the confidence interval within which we can expect single stochastic realizations to occur. This is a measure of the amount of variability that we can expect with respect to single occurrences of the epidemic profile. For the sake of simplicity, we do not report stochastic variability in the figures of this book.

Generally, the incidence curves show a maximum of activity—the peak of the epidemic—that is indicative of when to expect the most stress on the healthcare system. The peak time is identified by the period in which the incidence is larger than 80% of the incidence peak. The peak time of the epidemic

THE PEAK TIME OF THE EPIDEMIC PROVIDES VERY RELEVANT INFORMATION IN PLANNING FOR CONTAINMENT AND MITIGATION POLICIES.

Figure 4.4 | Urban level incidence
Daily incidence for six major urban areas.

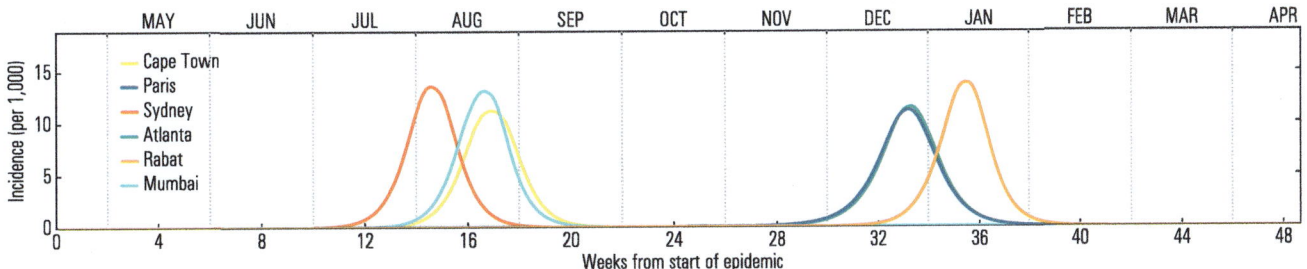

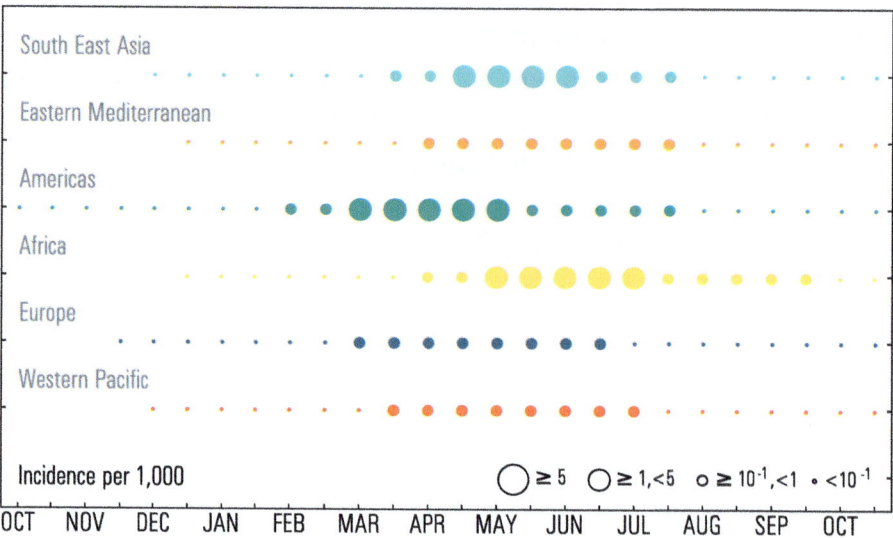

Figure 4.5 | Epidemic activity

Incidence at the WHO region level.

Figure 4.6 | Cumulative incidence map

Cumulative incidence of clinical cases.

in different places of the world, thus, provides very relevant information in planning for containment and mitigation policies. Peak times can be considered as the actual time scale of the epidemic impact on the population. The peak time also depends on the spatio temporal initial conditions of the epidemic. In **FIGURE 4.5**, we show the timing of the influenza pandemic in six regions of the world. Sporadic activity occurs when the average prevalence reports 10^{-2} cases per 1,000 people, while epidemic activity is typically defined by more than one case per 1,000. It is important to remark that seasonal forcing and other factors may induce multiple activity peaks in the same place, identified by the relative local maxima of the incidence profile.

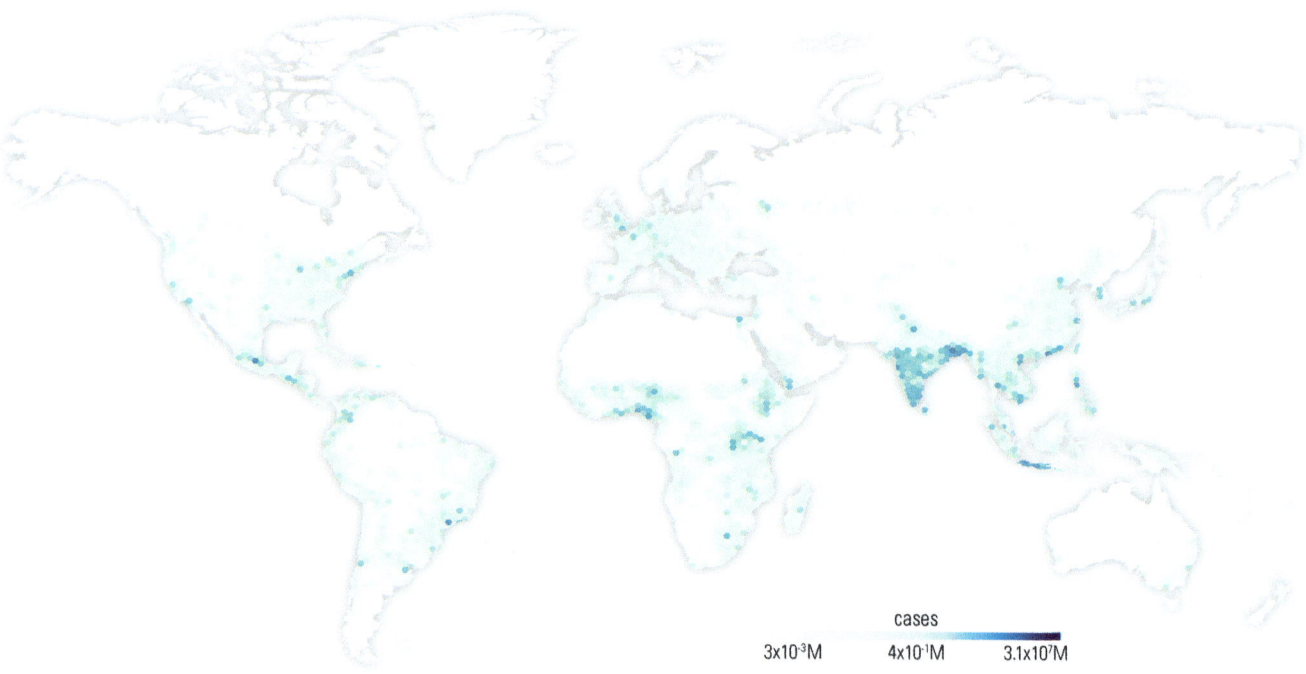

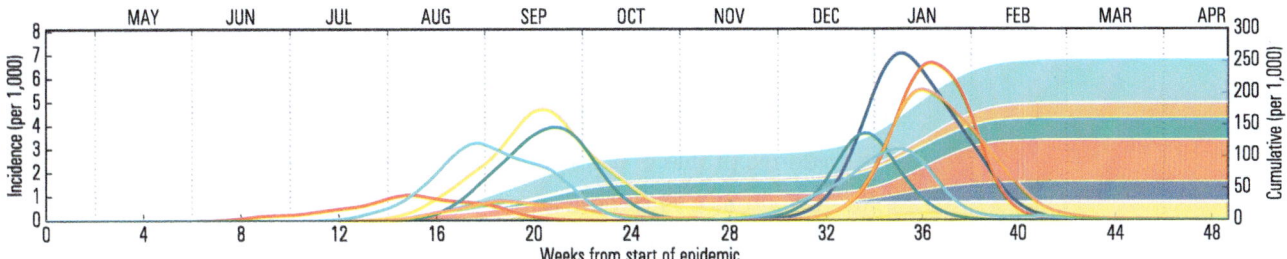

ATTACK RATE

Another quantity of interest is the cumulative incidence of clinical cases, which provides at each point in time the total number of individuals symptomatically affected by the disease. In **FIGURE 4.6**, we map the number of clinical cases of a given epidemic geographically; this kind of visualization readily provides the classic geolocalized perception of the impact of an epidemic at a given date. In order to disentangle the net impact of the disease in different regions of the world, data can be aggregated at different geographical levels. In this respect, in **FIGURE 4.7**, we show the evolution of the incidence composition per geographical region as a function of time. The treemap of **FIGURE 4.8** summarizes the composition of the attack rate after one year. The total area of the map represents the total number of clinical cases worldwide. The smaller areas represent the share of each region to the worldwide attack rate.

Figure 4.7 | Cumulative incidence and regional epidemic peaks

The colored regions describe the cumulative incidence of the six WHO regions. The solid lines indicate the respective epidemic incidence profiles.

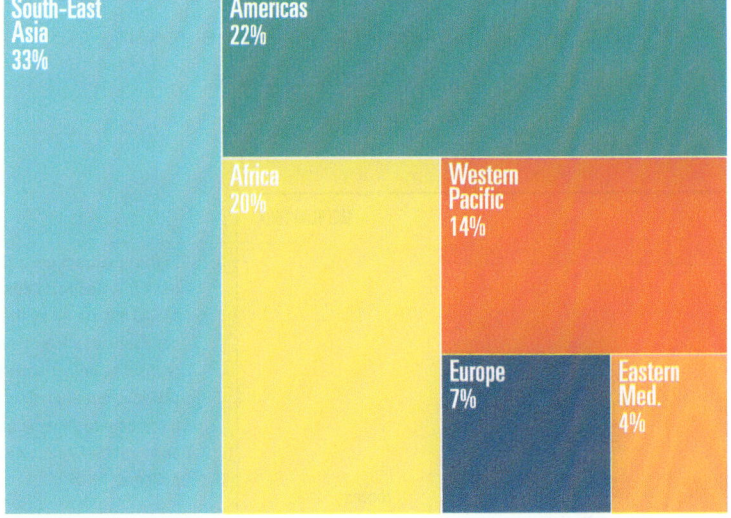

Figure 4.8 | Attack rate composition

The geographical impact of the disease considering the share of clinical cases in each of the six WHO regions.

RISK INDICATORS AND SEVERITY ASSESSMENT

Numerical simulations may contribute to the estimation of quantitative indicators for the assessment of severity of global epidemics. In practice, there is a basket of indicators, including transmissibility, virulence, impact on economic productivity, impact on the healthcare sector, etc. Obviously, some of those factors are summarized by the pathogen's intrinsic properties such as basic reproduction number, fatality rate, etc. Others, however, depend also on the spatio temporal initial conditions of the epidemic and the timing of the epidemic profile in each country. For instance, in the case of a pandemic influenza virus, it is generally possible to develop a vaccine within 6 to 8 months from the virus' isolation. Thus, countries or geographical locations for which the pandemic peak is forecasted prior to the vaccine deployment are at higher risk of suffering a more severe outbreak. Analogously, the larger the incidence at the peak relative to the number of available hospital beds in a country, the larger the epidemic impact on the healthcare system will be. While all indicators provide a continuum severity profile for each geographical place, it is possible to combine different indicators, as provided by the output of numerical models, to define coarse grained severity scales for the epidemic. In **FIGURE 4.9**, we report along the horizontal axis the incidence per 10,000 individuals divided by the number of hospital beds per 10,000 people; the vertical axis instead shows the epidemic-peak time. In order to measure the risk associated to each country, we divide the plane into four quadrants, using for the horizontal axis a mark given by the worldwide incidence divided by the average value of hospital beds per 10,000 individuals in the world, while for the vertical axis, the horizontal mark corresponds to 7 months after the virus identification—which can be considered a sensible estimated timing for virus isolation, vaccine development, production, and distribution. Of course, that estimated timing can vary according to expected development times, production shortages, etc. The quadrants show a decreasing order of risk according to the background color (from gray to white).

Figure 4.9 | Risk quad chart

We study the position of each country in a 2-D space corresponding to the number of clinical cases divided by the number of hospital beds, per 10,000 of the population (horizontal axis), and the peak time (vertical axis). By considering the average of the two quantities, we define four quadrants associated with different risk levels.

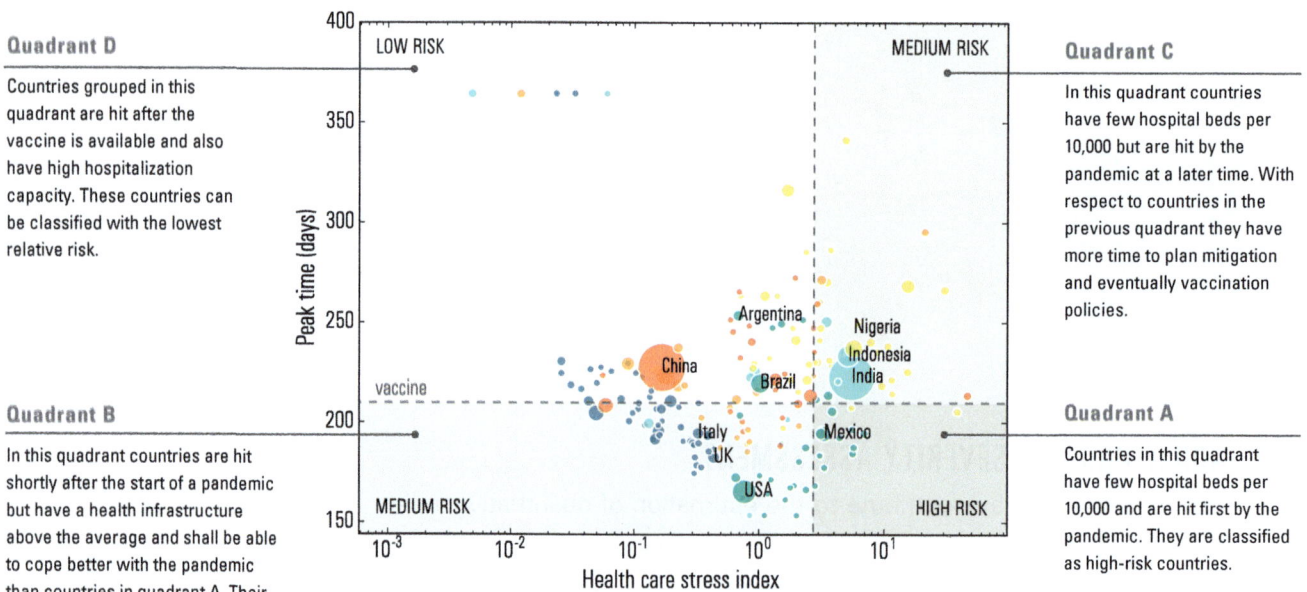

Quadrant D

Countries grouped in this quadrant are hit after the vaccine is available and also have high hospitalization capacity. These countries can be classified with the lowest relative risk.

Quadrant C

In this quadrant countries have few hospital beds per 10,000 but are hit by the pandemic at a later time. With respect to countries in the previous quadrant they have more time to plan mitigation and eventually vaccination policies.

Quadrant B

In this quadrant countries are hit shortly after the start of a pandemic but have a health infrastructure above the average and shall be able to cope better with the pandemic than countries in quadrant A. Their risk is still substantial.

Quadrant A

Countries in this quadrant have few hospital beds per 10,000 and are hit first by the pandemic. They are classified as high-risk countries.

It is important to stress that the severity scale used here is relative, not absolute, i.e., it just expresses a relative risk ranking among the countries, as mortality, hospitalization rates, etc., are not factored in. Analogous severity/risk scales can be developed by considering indicators such as the number of doctors, the country's Gross Domestic Product, etc.

CONTAINMENT AND MITIGATION SCENARIOS

Computational modeling also allows for the simulation of different intervention strategies, including pharmaceutical measures (such as vaccinations and the use of antiviral drugs) and non-pharmaceutical measures (such as travel restrictions and social distancing). In this case, the model has to make assumptions such as the degree of adherence to the vaccination campaign or the ability of a country to distribute the vaccine and inoculate a given fraction of the population in a given time window. Similarly, the effectiveness of a specific drug on a new pathogen may not be precisely assessed. For this reason, numerical models are able to provide "what-if" scenarios for the evolution of the epidemic according to containment measures and their implementation as assumed in the models.

For instance, mass vaccination aims at reducing: 1) the susceptibility to infection, 2) the infectiousness if infection occurs, and 3) the probability of developing clinical symptoms. Vaccination is considered in the model as a dynamic process simulating the inoculation of the vaccine, while taking into account the differences between countries in the availability and the deployment of vaccines and the starting dates for the vaccine stockpile availability. The efficacy of the vaccine is quantified by specific parameters that are included in the transmission model. Once the vaccination campaigns are included, we observe that vaccinations are able to produce a relative reduction of the incidence that depends on the effectiveness and timing of the vaccination campaign.

Analogously, it is possible to numerically simulate the treatment of clinical cases with drugs aimed at reducing the severity of the disease and the transmissibility while infectious. Lastly, control strategies implemented at the level of social groups and/or single individuals, such as social distancing measures or school closure, can be introduced into the model by an effective rescaling of the basic reproduction number for a given time period and in a specific geographic region. Also, in those cases, it is possible to visualize the effect of the implemented policies as the relative reduction of the incidence or clinical attack rate at a given point in time.

The possibility of including containment and mitigation scenarios transforms the computer into a laboratory for experimentation of policies which could hardly be feasible in reality for both practical and ethical reasons. Numerical simulations are then a formidable tool in support of the decision-making process during infectious disease threats.

COMPUTATIONAL MODELS ALLOW THE SIMULATIONS OF BOTH PHARMACEUTICALS AND NON-PHARMACEUTICALS INTERVENTION STRATEGIES.

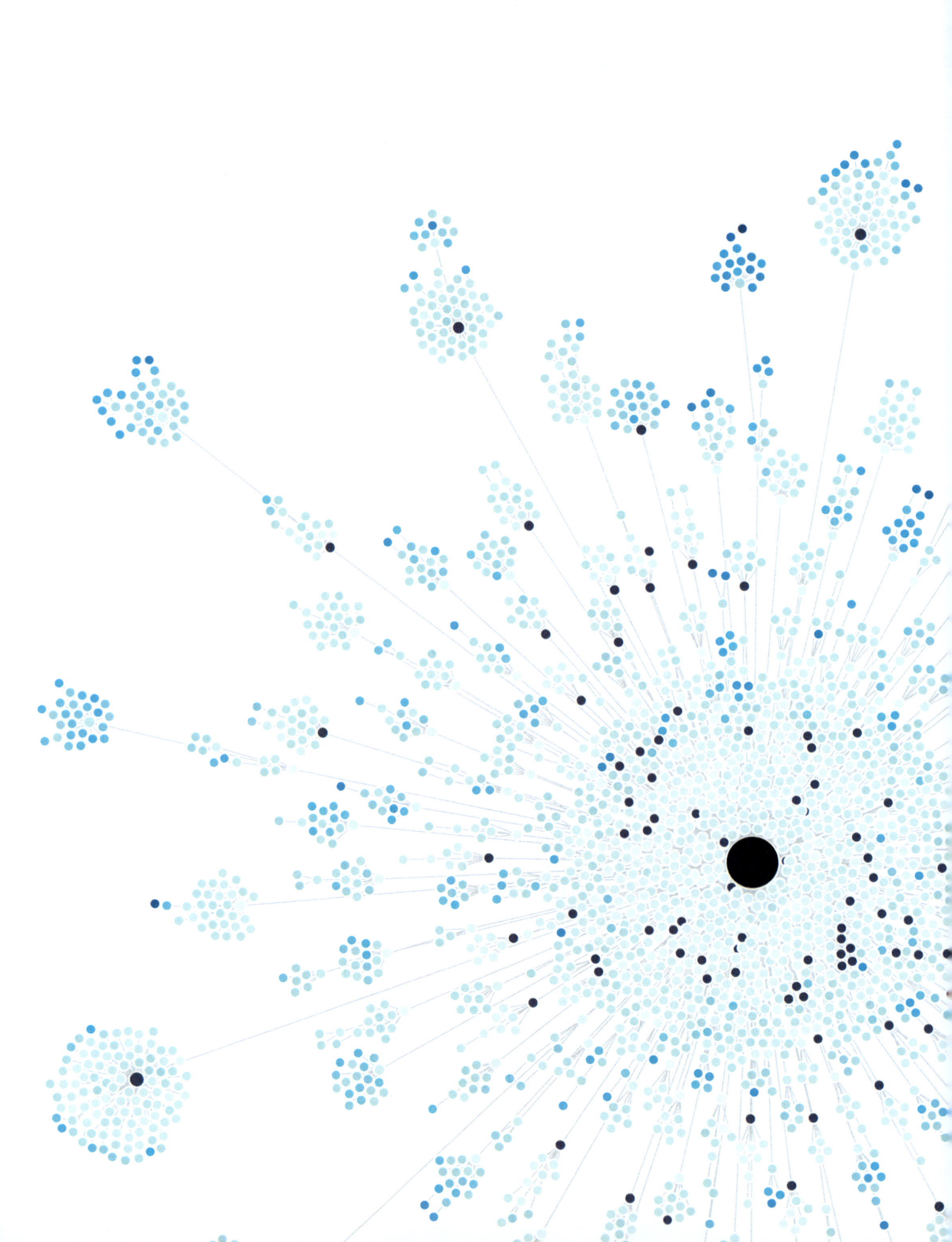

CHAPTER 5

THE NUMERICAL FORECAST OF PANDEMIC SPREADING: THE CASE STUDY OF THE 2009 A/H1N1 PDM

THE METHODOLOGICAL APPROACH TO EPIDEMIC MODELING presented in the previous chapters can be used to provide two different types of predictions. The first type concerns the classic scenario and "what-if" analysis. In this case, prototypical values for the basic disease parameters and other key parameters, such as time of implementation of specific policies, are assumed in the mathematical and computational model to produce plausible scenarios for the epidemics and to evaluate containment/mitigation procedures as a function of the explored parameter space. This is the kind of exercise we will perform in the next chapters: we will explore possible pandemic evolutions under different starting conditions and containment plans.

In the second type, we find a more challenging exercise, since the model is used for the real-time forecasting of unfolding epidemics. This is akin to employing atmospheric models to produce meteorological forecasts for future times at given locations. The model has to be calibrated using statistical estimates based on the analysis of epidemic outbreak data for as many key parameters as possible and likely by matching less crucial parameters with published historical data. One major technical problem for real-time forecasting, however, is that all modeling approaches contain a number of assumptions, such as those introduced by the model structure, scale, and implementation techniques.

For this reason, models have to be validated using datasets that are independent from those used for the model calibration. The validation process is generally performed by using data from historical epidemics that are simulated with the model. Even more stringent is the validation achieved by comparing real-time predictions with the actual unfolding of an epidemic outbreak. Only a few events in recent history have offered the possibility of a posteriori validation of the real-time forecasting of data-driven computational models using rich and high-quality datasets. One of those events is the 2009 A/H1N1 flu pandemic (H1N1pdm). During the course of the pandemic, which originated in Mexico in February 2009, models allowed estimation of unknown epidemiological parameters, description of the observed epidemic propagation, interpretation of surveillance data, exploration of possible scenarios, estimation of the efficacy of intervention measures, and predictions of future influenza activity. The data gathered during the course of the 2009 pandemic can now be used to make comparisons with the estimates calculated by models, and thus these represent an unprecedented opportunity to validate and assess the results obtained by mathematical and computational modeling approaches.

MODELING THE 2009 H1N1 PANDEMIC

With the emergence of the novel H1N1 influenza virus in 2009, the GLEAM framework has been used to study the spread of the pandemic in real time and thus evaluate specific public health actions and provide stochastic forecasts of its future unfolding. The H1N1 infection dynamics and transmissibility were calibrated by using the early data on H1N1 spreading. By July 2009, the H1N1 calibrated model predicted that the disease activity would peak in the Northern Hemisphere during the autumn of 2009, in contrast to most other years in which seasonal influenza activity peaks in January or February. This early timing of the peak was significant: it coincided with the start of planned large-scale vaccination campaigns, making those largely ineffective at the population level for mitigation purposes[1].

---

1   These findings were published in September 2009: Duygu Balcan, Hao Hu, Bruno Gonçalves, Paolo Bajardi, Chiara Poletto, José J Ramasco, Daniela Paolotti, Nicola Perra, Michele Tizzoni, Wouter Van den Broeck, Vittoria Colizza, and Alessandro Vespignani, "Seasonal transmission potential and activity peaks of the new influenza A (H1N1): a Monte Carlo likelihood analysis based on human mobility," BMC medicine **7**, 45 (2009).

DATA GATHERED DURING THE COURSE OF THE 2009 H1N1 PANDEMIC CAN NOW BE USED TO MAKE COMPARISONS WITH THE ESTIMATES CALCULATED BY MODELS.

Here, we report, using the diagrams and charts defined in Chapter 4, the results obtained over the real-time simulation campaign carried out at the time of the 2009 A/H1N1 pandemic. This represents an important exercise to understand the kind of data and "disease intelligence" that can be obtained by data-driven numerical models, as well as the problems and difficulties inherent to the real-time calibration of the computational models. Most importantly, the data gathered during the course of the pandemic can be used to provide a validation test of the model and an assessment of the reliability of the results obtained.

CALIBRATION AND PARAMETER ESTIMATES FOR SPECIFIC OUTBREAKS

For forecasting purposes, the parameters of the model are determined from early data about the epidemic outbreak. Given the number of parameters that a realistic epidemic model contains, a full calibration at the early stage of an outbreak is made difficult by the lack of timely data about the number of cases, the temporal description of the disease in hosts (e.g., the incubation period and infectious period), the presence of asymptomatic infections, and their transmissibility rate. When considering an influenza pandemic, however, it is possible to distinguish between two classes of parameters: 1) key parameters, such as the specific transmissibility of the disease, which can be determined from the data, and 2) parameters that, from a sensitivity analysis, have been found to be less crucial for defining the spatiotemporal pattern of the disease propagation and for which plausible and relatively stable values can generally be found in the literature.

In the analysis of the H1N1pdm virus, we consider the standard compartmental model depicted in **INFOBOX 3.3** and its variation to accommodate vaccination programs. The basic model parameters (transmissibility and seasonality) were obtained with a Monte Carlo Maximum Likelihood (MCML)-based approach using the chronological data on the pandemic invasion up to June 18, 2009; a schematic explanation of this approach is provided in **INFOBOX 5.1**. All the other parameters are set to the values reported in the literature for influenza infections (e.g., asymptomatic rate and relative infectiousness and vaccine and antiviral efficacy) or to those available from the analysis of the early outbreak of the 2009 A/H1N1 pandemic. In order to ensure the stability of the forecasts provided by the model, we performed a sensitivity analysis of the way the results change by varying the parameters over plausible ranges.

Each calibration of the model is computationally expensive, as millions of stochastic simulations are needed to generate a suitable appropriate ensemble describing the spatiotemporal statistical properties that define the corresponding likelihood function. Thus, real-time calibrations often require supercomputing resources. Furthermore, the calibration procedure must be repeated if any new information on the disease is introduced or if specific data not initially available can affect the initial conditions or the early phase of the outbreak.

FORECAST OUTPUT

Once the model has been calibrated, it can be used to provide epidemic forecasts by generating a large number of nominally identically initialized numerical stochastic simulations of an epidemic's global progression. In the case of the H1N1 pandemic, the baseline calibration integrates data accumulated before June 18, 2009. This calibration is then used to provide forecasts of the unfolding of the pandemic during the following months. In particular, the GLEAM framework was considered for developing a model able to output forecasts of the activity peak of the fall/winter wave of the H1N1 pandemic in the Northern Hemisphere, along with other quantities of interest.

5.1 | INFOBOX
MODEL CALIBRATION

Estimates of the reference value of the basic parameters can be performed using a Monte Carlo Maximum Likelihood (MCML) approach based on the early chronology of the epidemic. For the H1N1 epidemic, the model is initialized with the data from an outbreak near La Gloria (in the state of Veracruz, Mexico) on February 18, 2009, according to the official reporting and literature on the origin of the outbreak. The arrival time of infected individuals in the countries seeded by Mexico is clearly a combination of the number of cases present in the originating country (Mexico) and the mobility network, both within Mexico and connecting Mexico with other countries. By relying on the explicit modeling of the travel behavior of individuals based on real data, it is possible to shift the estimation of R_0 from the incidence data in the seed country to the timing of the early invasion pattern, with the aim of reducing the errors induced by possible underestimation of cases by surveillance sources.

The MCML analysis is schematically depicted in **FIGURE 5.1.1.** Being fully stochastic, GLEAM allows for the

simulation of a statistical ensemble of epidemic evolutions. The set of arrival times t_i for all the countries can be considered the statistical observable and the infection parameters $\{p\}$ as the (unknown) input parameters. In the case of the H1N1 pandemic, those parameters were the reproduction number, the generation time, and the

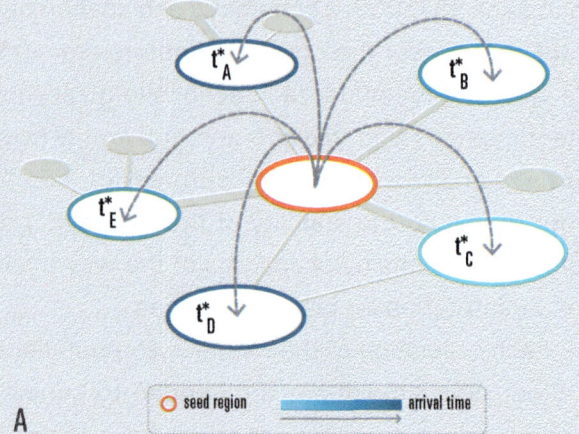

A

INVASION TREE

In the early stage of the 2009 A/H1N1 pandemic, the worldwide air transportation network was the main dissemination mechanism from Mexico to the rest of the world. By using the GLEAM computational approach, it is possible to provide the invasion tree and the timeline of the spreading of the infection in 220 countries and dependent territories. In **FIGURE 5.1** and **FIGURE 5.2**, we report the invasion tree (as defined in Chapter 4) and a diagram showing the number of countries affected by the epidemic in different world regions as a function of time. It is possible to see that initially the pandemic was mostly radiating from Mexico, while the US and the big European hubs played a major role in the pandemic spreading as time passed. The data on the arrival of the disease were meticulously collected during the first months of the pandemic, but after June/July 2009 border monitoring was relaxed, making it impossible to provide a full validation of the corresponding invasion tree.

THE WORLDWIDE AIR TRANSPORTATION NETWORK WAS THE MAIN DISSEMINATION MECHANISM FROM MEXICO TO THE REST OF THE WORLD.

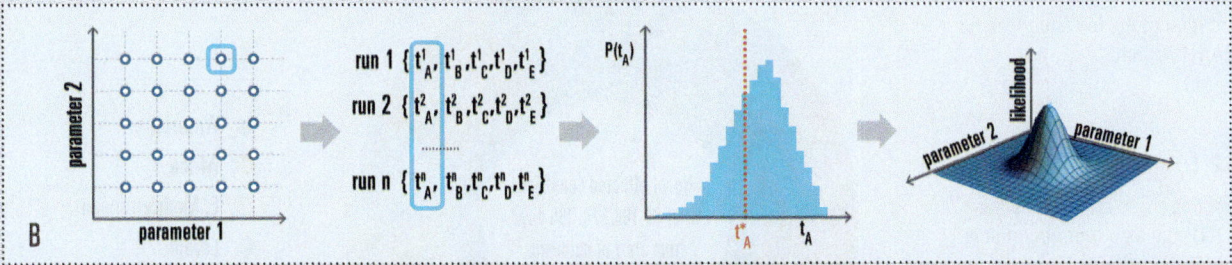

seasonal rescaling. The numerical simulations are used to generate a statistical ensemble describing the epidemic evolution for each point in the parameter space {p}. For each of these points, it is possible to compute numerically the probability distribution of the arrival time $P(t_i)$, which represents the probability that the first infectious individual enters the country i at time t_i. At the point defined by the empirical dataset $\{t_i^*\}$, this probability defines the likelihood function $L(\{p\}) = P(\{t_i^*\} \mid \{p\})$, which by using a set of arrival times in different countries, assume a factorized expression. The best estimate for the disease parameters is obtained by the maximum likelihood value in the parameter space. The analysis of the phase space of both transmissibility and generation interval needs particularly intensive computational power, as thousands of realizations of the synthetic epidemic have to be simulated for each point of the parameter space.

Figure 5.1.1 | Monte Carlo Maximum Likelihood (MCML) method used to estimate the transmission potential of the A/H1N1 pandemic

(A) Schematic visualization of the invasion dynamics of an emerging infectious disease from the subpopulation origin of the outbreak (orange patch) to the neighboring subpopulations. The color code refers to the time of arrival of the first infectious individual in the corresponding subpopulation. Links of different width represent mobility connections characterized by large or small traffic volumes.

(B) Diagram illustrating the Monte Carlo Maximum Likelihood (MCML) method. First, 2,000 stochastic realizations, all with the same initial conditions, are generated for each point of the parameter space. Second, the probability distribution built using the simulated arrival times is compared with the actual arrival times for each country. Finally, the likelihood function in the parameter space is obtained, whose maximum value corresponds to the set of parameters that best fit the data.

Adapted from M. Tizzoni et al, BMC Medicine 10, 165 (2012)

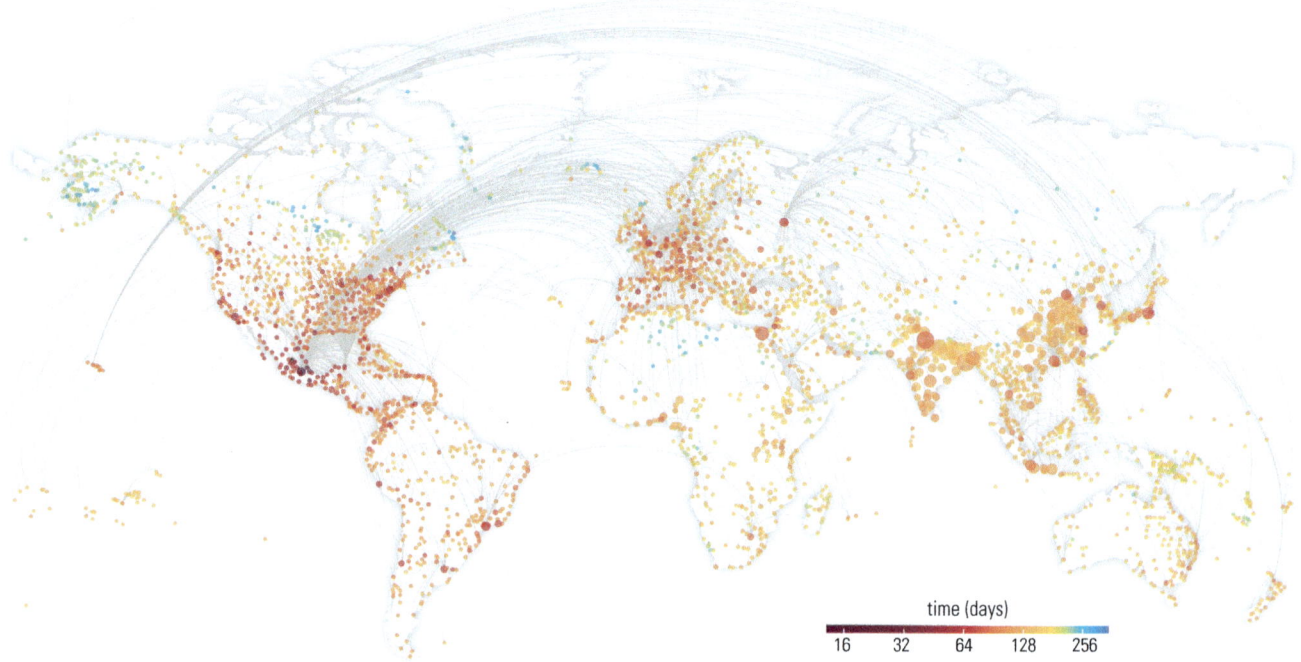

time (days)
16 32 64 128 256

Figure 5.1 | Invasion tree

Representation of the invasion tree for the 2009 A/H1N1pdm.

Figure 5.2 | Affected countries

Number of countries affected by the 2009 A/H1N1pdm as a function of time for the six WHO regions.

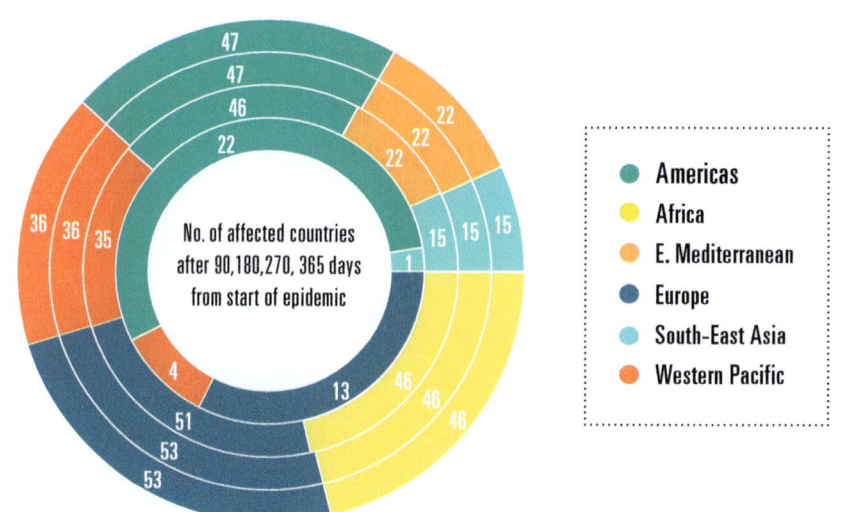

No. of affected countries after 90,180,270, 365 days from start of epidemic

- Americas
- Africa
- E. Mediterranean
- Europe
- South-East Asia
- Western Pacific

PANDEMIC ACTIVITY PEAKS

The numerical results from the GLEAM model were used during the summer of 2009 to forecast the pandemic activity in the autumn/winter of 2009 and 2010. In **FIGURE 5.3** and **FIGURE 5.4**, we show the activity profiles for a select number of cities around the world and the pandemic incidence in the WHO surveillance regions, respectively. It is interesting to note that, according to the numerical forecast, the pandemic activity peaks in the Northern Hemisphere would occur in October/November; this early timing of the peaks was remarkable as it stood in contrast to most other years in which seasonal influenza activity peaks in January or February. These numerical results are highly non trivial

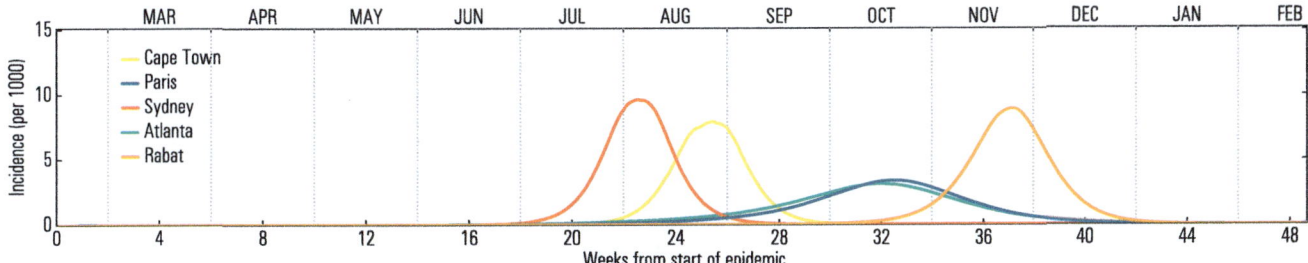

because the GLEAM model does not alter the timing of the seasonal forcing that could intuitively generate an activity peak in mid-January. The anticipated peaks are thus a genuine result originating from the initial conditions of the pandemic, the transmissibility estimate, and the spreading pattern generated by the human mobility integrated into the model. The numerical forecast can be compared with the empirical data of the influenza activity collected from 48 countries in the Northern Hemisphere (some of which lie across the Northern Hemisphere and the Tropical region)[2]. In **FIGURE 5.5**, we compare the numerical forecast, including statistical uncertainties, with empirical data from the field from more than 40 countries.

The close match between simulated results and actual monitoring data provides support for the use of numerical epidemic forecast for predicting epidemic spread due to the modern human mobility infrastructure.

Figure 5.3 | Urban level profiles

Daily incidence for major urban areas during the 2009 A/H1N1pdm.

2 A full technical analysis of the results of GLEAM and their comparison with empirical data can be found in: Michele Tizzoni, Paolo Bajardi, Chiara Poletto, José J Ramasco, Duygu Balcan, Bruno Gonçalves, Nicola Perra, Vittoria Colizza, and Alessandro Vespignani, "Real-time numerical forecast of global epidemic spreading: case study of 2009 A/H1N1pdm," BMC Medicine **10**, 165 (2012).

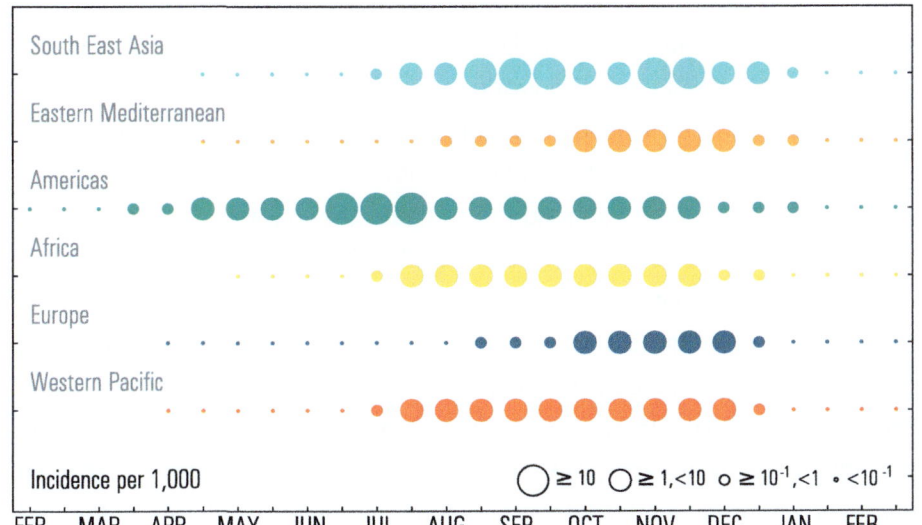

Figure 5.4 | Epidemic activity

Incidence at the WHO region level during the 2009 A/H1N1pdm.

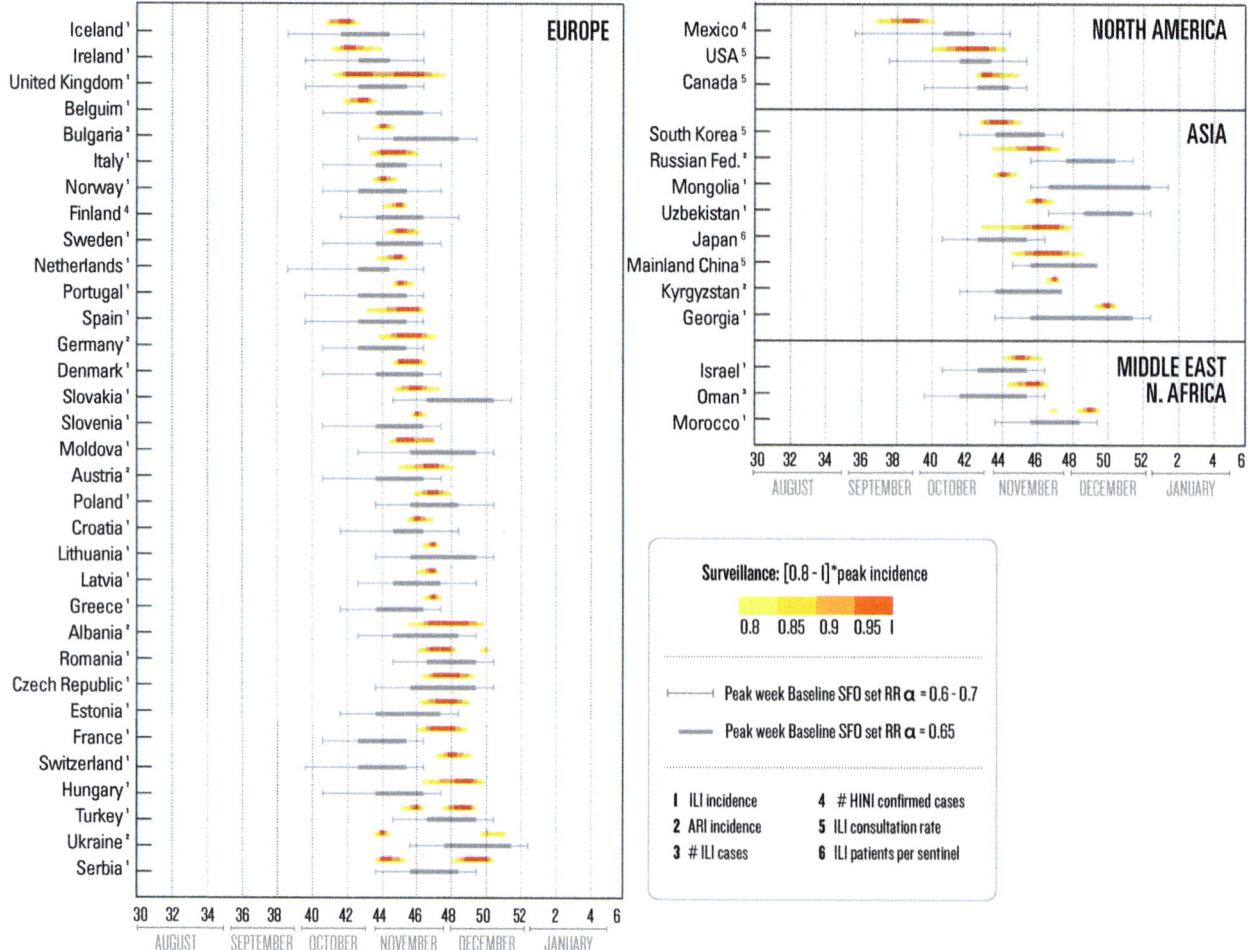

Figure 5.5 | Peak timing in the Northern Hemisphere: simulations and real data

The reference ranges of the simulated peak week were obtained by analysis of 2,000 stochastic realizations considering parameters sensitivity analysis. The peak weeks reported by the surveillance for the fall/winter wave are shown as a color gradient, whose limits correspond to the time interval at which an incidence greater than 80% of the maximum incidence was observed. The numbered weeks of the year correspond to the calendar used by the US Center for Disease Control and Prevention.

Adapted from M. Tizzoni et al., BMC Medicine 10, 165 (2012)

RISK AND SEVERITY ASSESSMENT

The numerical forecast allows a risk analysis that considers the peak timing versus the capacity of the health infrastructure (hospital beds per 10,000 individuals) in each country. In **FIGURE 5.6**, we show a risk map that considers four quadrants identified by the line corresponding to 7 months after the start of the pandemic and the world average of hospital beds per 10,000 individuals. Countries whose pandemic peak is well above the 7-month line are at a relatively lower risk than other countries because they can potentially use a mass vaccination campaign to mitigate the impact of the pandemic. Countries with a better than average health infrastructure absorb the blow of the pandemic more efficiently, thus also being at a lower risk than other countries. The four quadrants in the figure immediately identify Mexico and a few other Latin American countries as those with the most substantial risk. The pandemic started in the Americas, and the Southern Hemisphere winter was favoring a very early peak. It is possible to readily observe that most European countries are located in a region of relatively moderate or low risk because of the peak timing and their health infrastructure. It is important to

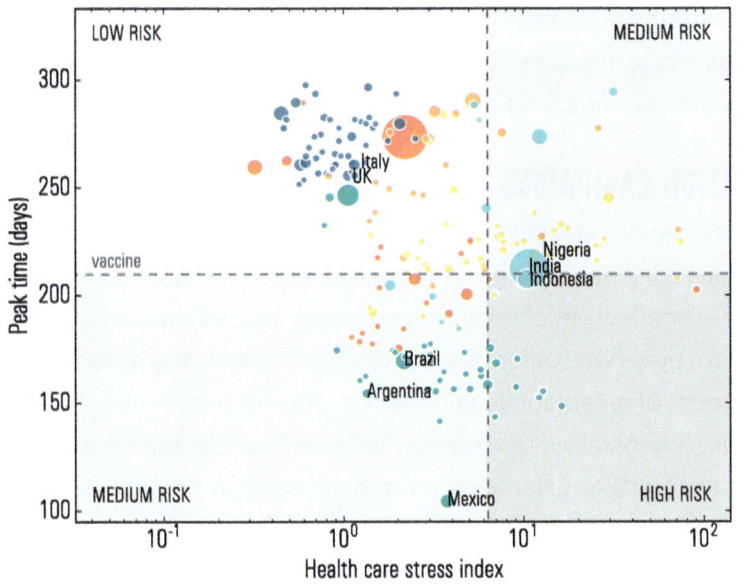

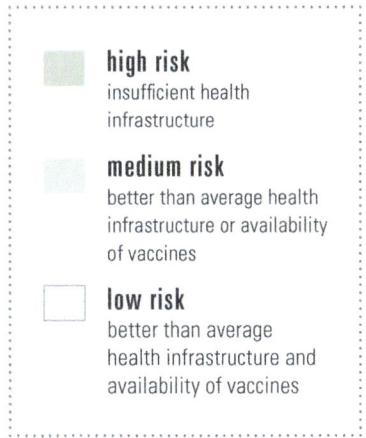

stress that this risk assessment is not an absolute risk map but just a measure of each country's relative risk with respect to the others.

Another measure of risk assessment is the number of infected people in the population at the end of the pandemic in different countries and world regions (see **FIGURE 5.7**). Despite the fact that these numbers can be easily accessed from the numerical forecast, they cannot be readily validated, since measuring the people actually exposed to the virus is extremely difficult. Individuals with mild symptoms do not seek medical care, as well as many symptomatic individuals who do not have complications; moreover, this behavior varies a lot between countries and cultures. Serological analysis can provide high-quality estimates of the people who have been exposed to the virus but are very expensive and generally made on a very limited sample of the population. From a numerical stand point, the estimated attack rate also depends on the level of detail at which the transmission dynamic is modeled within the population: for instance, it is well known that homogeneous models generally overestimate the attack rate because of the lack of clustering and structure in the population. In order to have more accurate estimates of attack rates, the

Figure 5.6 | Risk quad chart

We map the risk for each country by looking at the timing of the epidemic peak and the incidence at the peak per hospital beds. Each quadrant is colored according to the relative risk.

Figure 5.7 | Cumulative incidence and regional epidemic peaks

The colored regions describe the cumulative incidence of the six WHO regions during the 2009 A/H1N1pdm. The solid lines indicate the respective epidemic profiles.

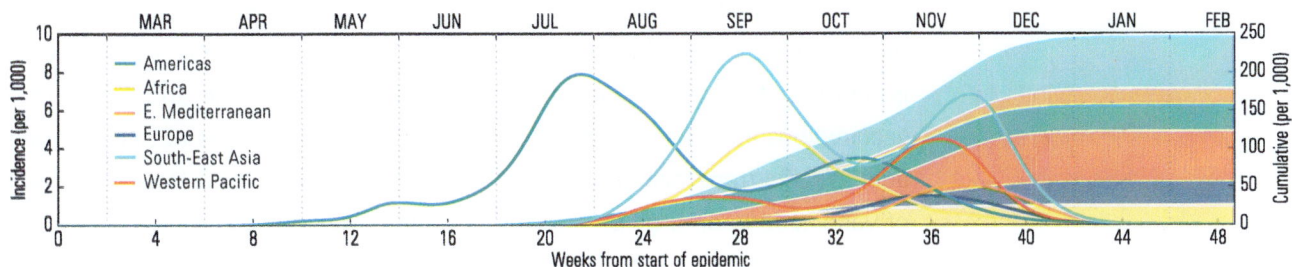

different patterns of contacts between individuals with different ages, and the country-dependent features of households, schools, and workplaces must be considered in the modeling approach.

VACCINATION CAMPAIGNS

The early timing of the pandemic peaks in the Northern Hemisphere coincided with the start of planned large-scale vaccination campaigns, making many of those largely ineffective for mitigation purposes. Indeed, vaccination campaigns need to take place well before the activity peak to have any meaningful impact on the control of an epidemic and the reduction of the number of cases.

Although generally implemented too late to affect the timing of the pandemic in the Northern Hemisphere, the vaccination campaigns adopted by several countries might have helped to accelerate the decline of the pandemic and reduce its final attack rate. With the computational approach it is possible to provide counter factual analysis in which the attack rate with no vaccination is compared with the attack rate of simulations incorporating the data of the actual vaccination campaigns. In this way, it is possible to provide a numerical estimate of the effective reduction of cases yielded by the introduction of the vaccine. The final vaccine uptake differed widely between countries in the Northern Hemisphere, ranging from 0.6% of the population in the Czech Republic to about half of the population or more in some northern European countries (Sweden, Finland, Iceland, Norway) and in Canada. Notwithstanding the large uptakes reached in some countries, the effect of the mass vaccination campaigns on the timing of the epidemic was negligible because most of the vaccine doses were not deployed before November 2009.

In **FIGURE 5.8**, we visualize the relative reduction of the attack rate by country. The model predictions show that the effect of vaccination on the final attack rates was non-negligible, particularly in those countries where an early start was possible. As expected, the largest relative reductions in the attack rate are found in those countries that adopted prompt and rapid administration of vaccines (the United States, Hungary, and Sweden). In these cases, vaccination was able to achieve a relative reduction ranging from about 8% to 16%. A late start or a slow pace of vaccine distribution predicts a much smaller reduction, as shown by the cases of Italy, the Czech Republic, and Norway, which were characterized by a low final uptake (Italy and the Czech Republic) or late start of the vaccination campaign (the Czech Republic and Norway). The average relative reduction obtained by considering all countries was 3.6%. However, it is important to note that estimates about reduction of the final attack rate from vaccinations are limited due to the fact that the model does not incorporate age-specific transmission and vaccination strategies. Therefore, the effect of vaccinations on a given age group could differ greatly from the average reduction in the final attack rate observed at the country level.

REAL-TIME FORECASTS ARE A CHALLENGE BECAUSE THEY DEPEND ON THE TIMELY ESTIMATE OF THE PARAMETERS OBTAINED BY USING DATA FROM SURVEILLANCE SYSTEMS.

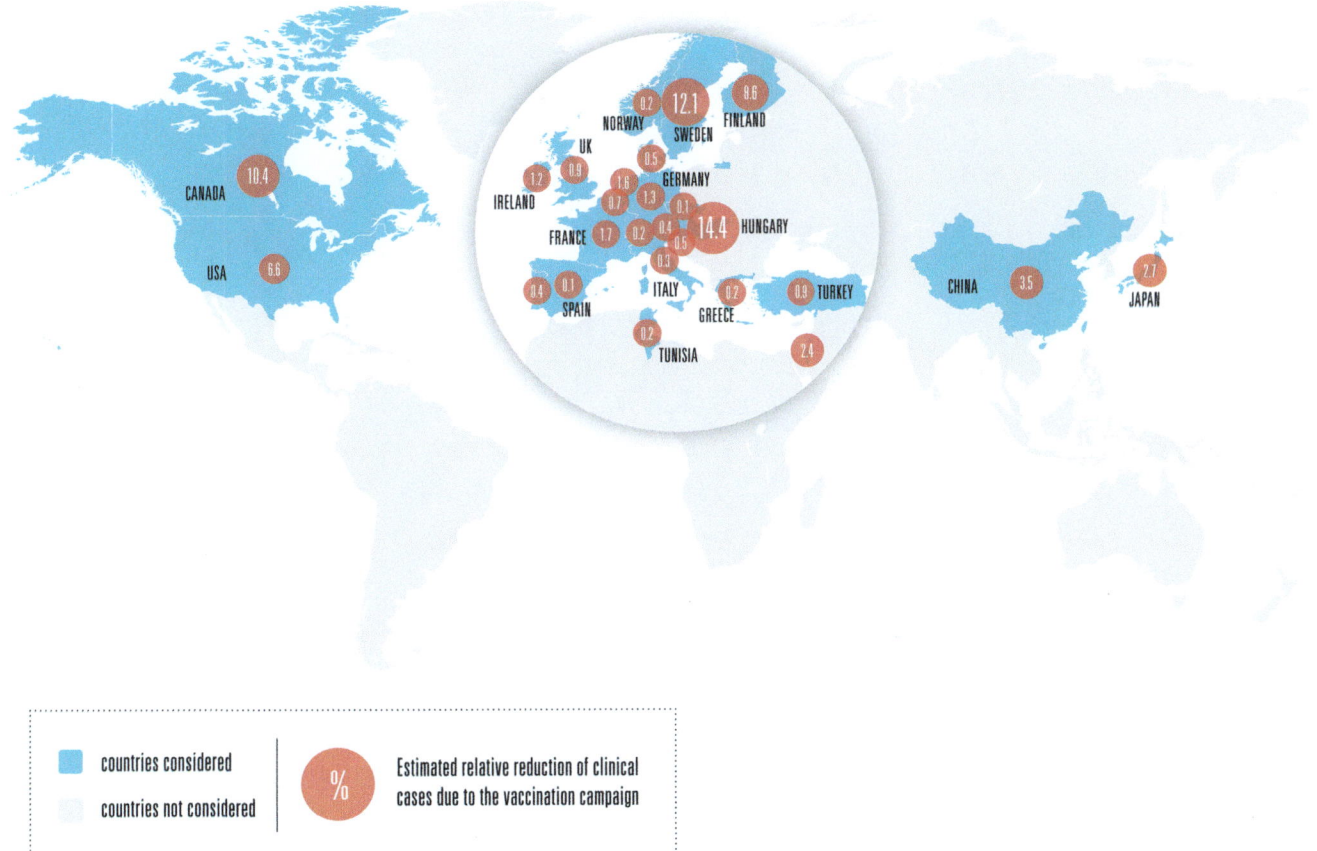

countries considered

countries not considered

% — Estimated relative reduction of clinical cases due to the vaccination campaign

REAL-TIME FORECAST WITH COMPUTATIONAL EPIDEMIC MODELS

The results of the real-time analysis of the 2009 A/H1N1 pandemic with the computational model discussed in this chapter indicate a very good agreement in the prediction obtained for a large number of countries. The results are encouraging and support advocating for the use of large-scale computational approaches in providing real-time forecast and scenario analysis of epidemic outbreaks. It is important to stress that real-time forecasts are a challenge because they depend on the timely estimate of the parameters obtained by using data from surveillance systems and National and International agencies. This constraint is not present in the case of scenario analysis, where a wide range of parameters and policies can be explored numerically in order to generate a portfolio of possible pandemic events and map their worldwide impact and dynamic. These scenarios can be especially useful in the preparation of pandemic contingency plans and the definition of a catalog of possible outcomes, to be used at the onset of real pandemic emergencies, when data are not yet sufficient for a careful estimate of the model parameters.

Figure 5.8 | Impact of vaccination campaigns

Attack rate reduction by country when considering vaccination campaigns.

Source: M. Tizzoni et al., BMC Medicine 10, 165 (2012)

PART II

PANDEMIC CHARTS

PANDEMIC CHARTS

. . . .

In this part, we want to use the numerical simulations generated by GLEAM to create an atlas of charts depicting possible pandemic evolutions. In principle, one could consider each day of the year as a possible starting date and every place in the world as the geographic origin of a pandemic. For each infectious disease, the full atlas of possible epidemic trajectories would therefore consist of millions of maps generated under the different initial conditions. A huge database such as this could be stored and used to aid in possible pattern identification in the case of actual epidemic outbreaks, but its graphical representation in the pages of a book would simply be impossible, as well as not readily useful, to say the least.

However, it is possible to group initial conditions by specific regions and periods of the year, in order to provide general patterns that allow the understanding of major trends, seasonal differences, and the timing of the pandemic's unfolding.

As we have discussed previously, no model fits all diseases; therefore, we have to generate specific simulations for different infectious illnesses. Each disease is also sensitive to different determinants and/or containment measures. For this reason, we define specific sections and use different charts and infographics for each disease we consider: influenza, coronavirus, and the Ebola virus. A map of the six WHO regions along with the pandemic's origin cities and starting conditions we considered can be found on the following pages.

The ensuing atlas of pandemics does not consider the most detailed level of modeling in each country. Analyzing the effects of specific local socioeconomic features on the attack rate of the disease and other epidemic indicators would require a major computational effort that is beyond the scope of this atlas. Furthermore, we consider only a few of the possible interventions that can be used to contain or mitigate epidemic outbreaks. Additionally, we do not include the possible behavioral changes induced by the awareness of the disease in the population. Such an effect would be increasingly relevant for diseases with a high case fatality rate. In other words, the patterns provided in this book are only a small sample of how an epidemic could unfold in a population. Nevertheless we hope that the present atlas conveys the richness and variability of such patterns and the importance of their understanding in the fight against infectious diseases.

SYMBOL DEFINITIONS

INITIAL CONDITIONS

 **R₀ value**

R_0 value

For every starting condition, we consider one or more possible values of the pandemic reproduction number R_0.

 Starting date

For every geographical intitial condition, we consider one or more different starting dates.

SEASONALITY

 Mild seasonality

The basic reproduction number varies up to a factor 2 across seasons.

 High seasonality

The basic reproduction number varies up to a factor 10 across seasons.

 No seasonality

The basic reproduction number remains constant across seasons.

ADDITIONAL SCENARIOS

 Air Traffic Reduction

We consider the implementation of travel restrictions.

 Vaccinations

We study the effects of implementing a vaccination campaign.

New York City

AMERICAS

ORIGIN CITIES

Influenza

Barcelona, Spain
Buenos Aires, Argentina
Hanoi, Vietnam
Johannesburg, South Africa
Melbourne, Australia
New York City, USA

Coronavirus

Barcelona, Spain
Guangzhou, China
Jeddah, Saudi Arabia

Ebola virus

Arua, Uganda
Kisangani, Democratic Republic of the Congo
Lagos, Nigeria

Buenos Aires

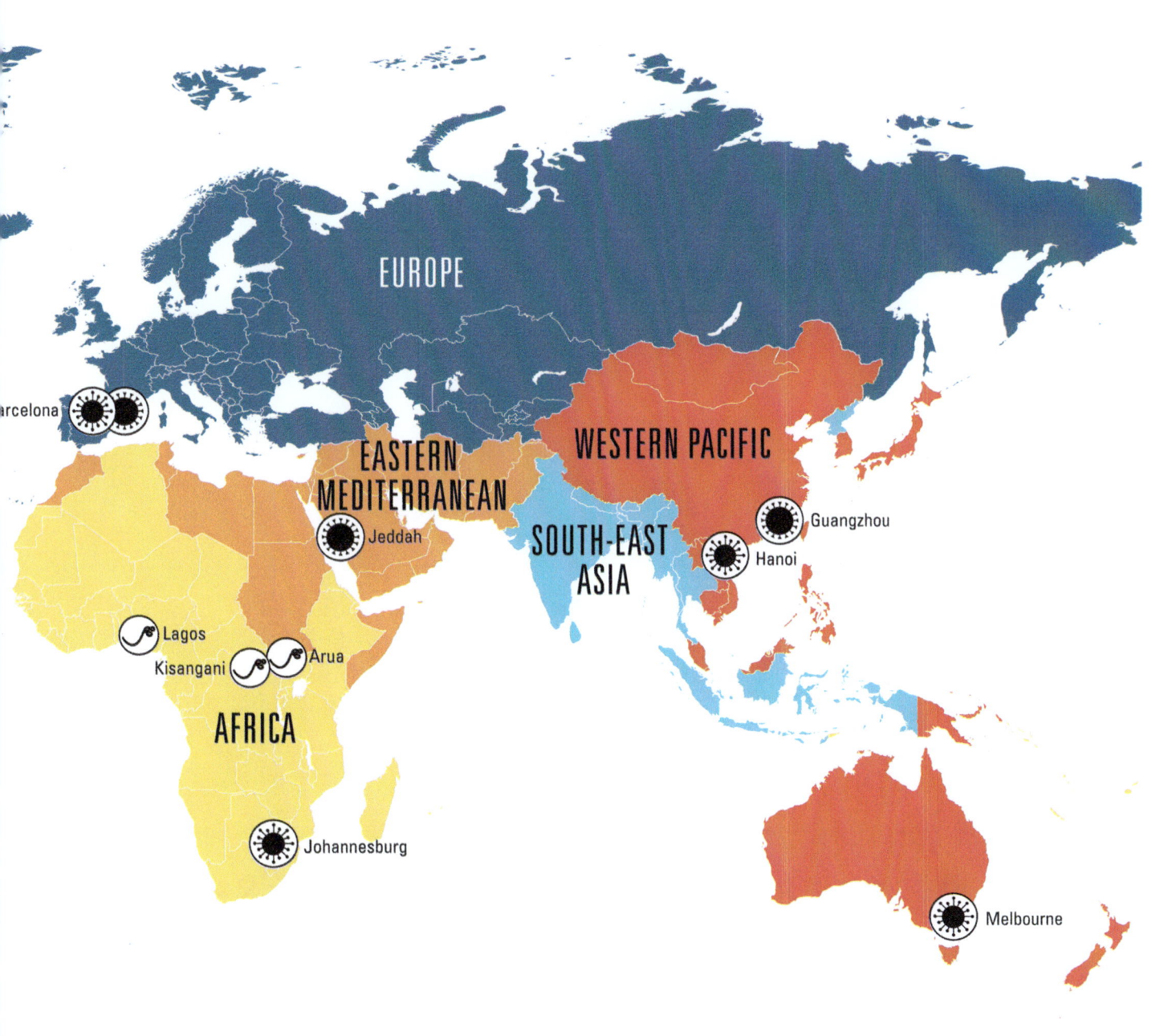

Barcelona

EUROPE

WESTERN PACIFIC

EASTERN
MEDITERRANEAN

Jeddah

SOUTH-EAST
ASIA

Guangzhou

Hanoi

Lagos

Kisangani

Arua

AFRICA

Johannesburg

Melbourne

▲ WHO REGIONS + PANDEMIC ORIGIN CITIES

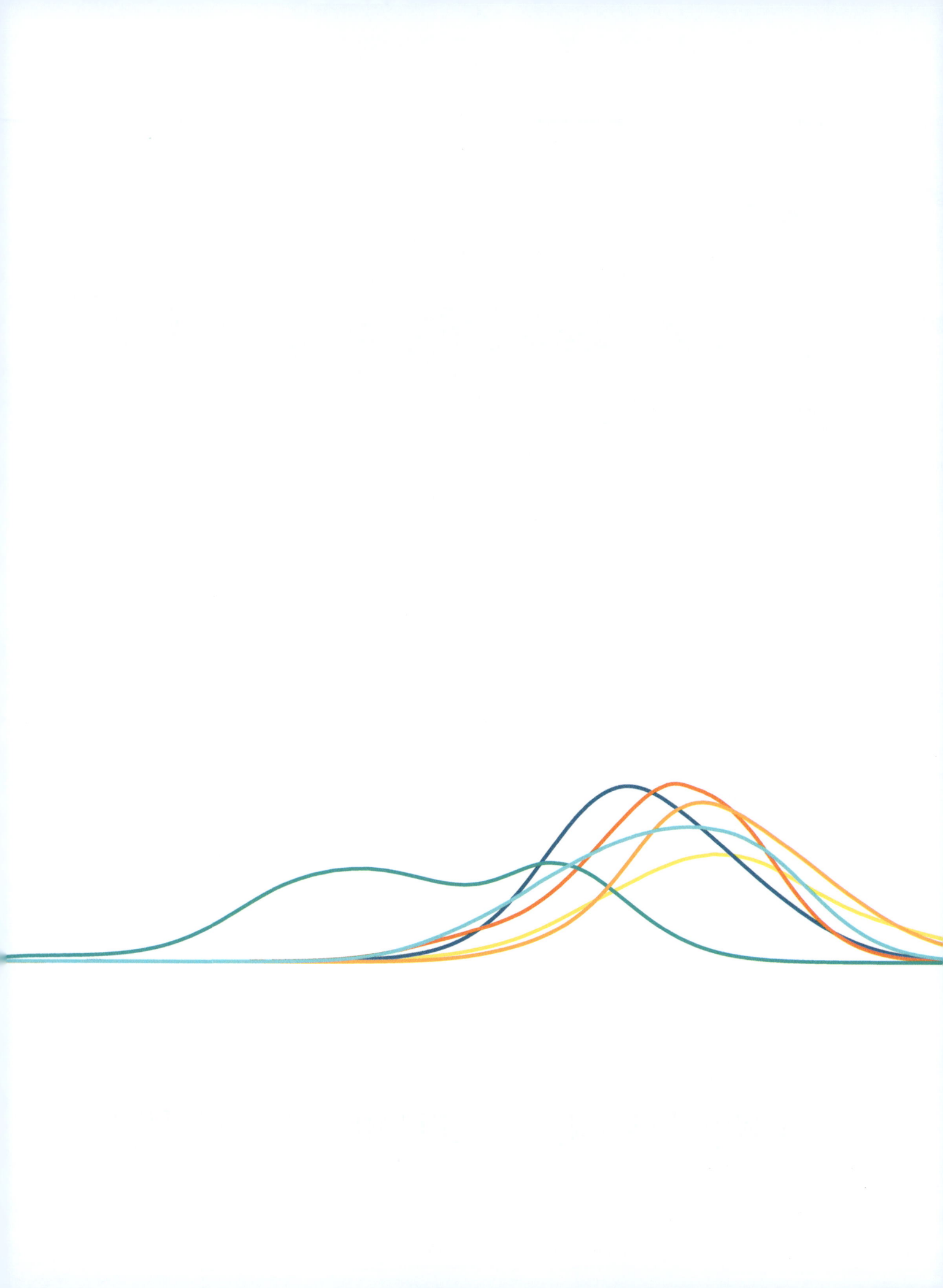

PANDEMIC INFLUENZA

Flu virus transmission is subject to seasonal forcing, and therefore it is particularly important to consider different starting periods for a disease of this type. The interplay between the timing and location of the first cases of the pandemic will translate into different unfolding patterns at the global level.

In particular, for the simulations we use the standard sinusoidal forcing that provides a transmission peak in the winter of each hemisphere (January for the Northern hemisphere, July for the Southern hemisphere). The virus is unaffected by seasonality in the tropical regions. We consider starting conditions for representative countries in each WHO surveillance region. For every geographical initial condition, we consider two different starting times of the year, April 15 and October 15.

In addition, for every starting condition, we consider two possible values of the pandemic reproduction number R_0. Namely, we simulate pandemics with $R_0 = 1.5$ and 2.0 values, corresponding to a mild transmissibility as observed in the H1N1 pandemic of 2009 and a high transmissibility similar to the 1918 pandemic, respectively.

We also report the effects of a worldwide vaccination campaign starting 7 months after the beginning of the outbreak and the implementation of travel restrictions to and from infected areas.

This yields a total of 22 different scenarios summarized in the specific infographic charts. We provide the typical influenza chart layouts and a quick "how to read it" guide on the following pages.

STARTING CONDITIONS

- $R_0 = 1.5$ and 2.0
- **Starting date:** April 15, October 15

SCENARIOS

- Baseline
- Effects of vaccination campaigns
- Implementation of travel restrictions
- Mild seasonality (R_0 varies up to a factor **2** across seasons)
- Strong seasonality (R_0 varies up to a factor **10** across seasons)

ORIGIN CITIES

Barcelona, Spain
15 million passengers/year

Buenos Aires, Argentina
8.7 million passengers/year

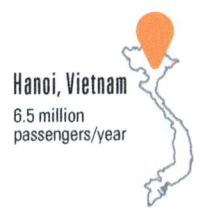

Hanoi, Vietnam
6.5 million passengers/year

Melbourne, Australia
13.3 million passengers/year

Johannesburg, South Africa
7.3 million passengers/year

New York City, USA
46 million passengers/year

CHART LAYOUT: PANDEMIC INFLUENZA

Location of pandemic origin

Infection tree

Each link describes an infection path along which the epidemic moved from one population to the other. The color of the nodes refers to the time lag with respect to the initial spreading.

Cumulative incidence and epidemic peaks

The colored regions describe the cumulative incidence at the regional level. The solid lines represent the regional epidemic profiles.

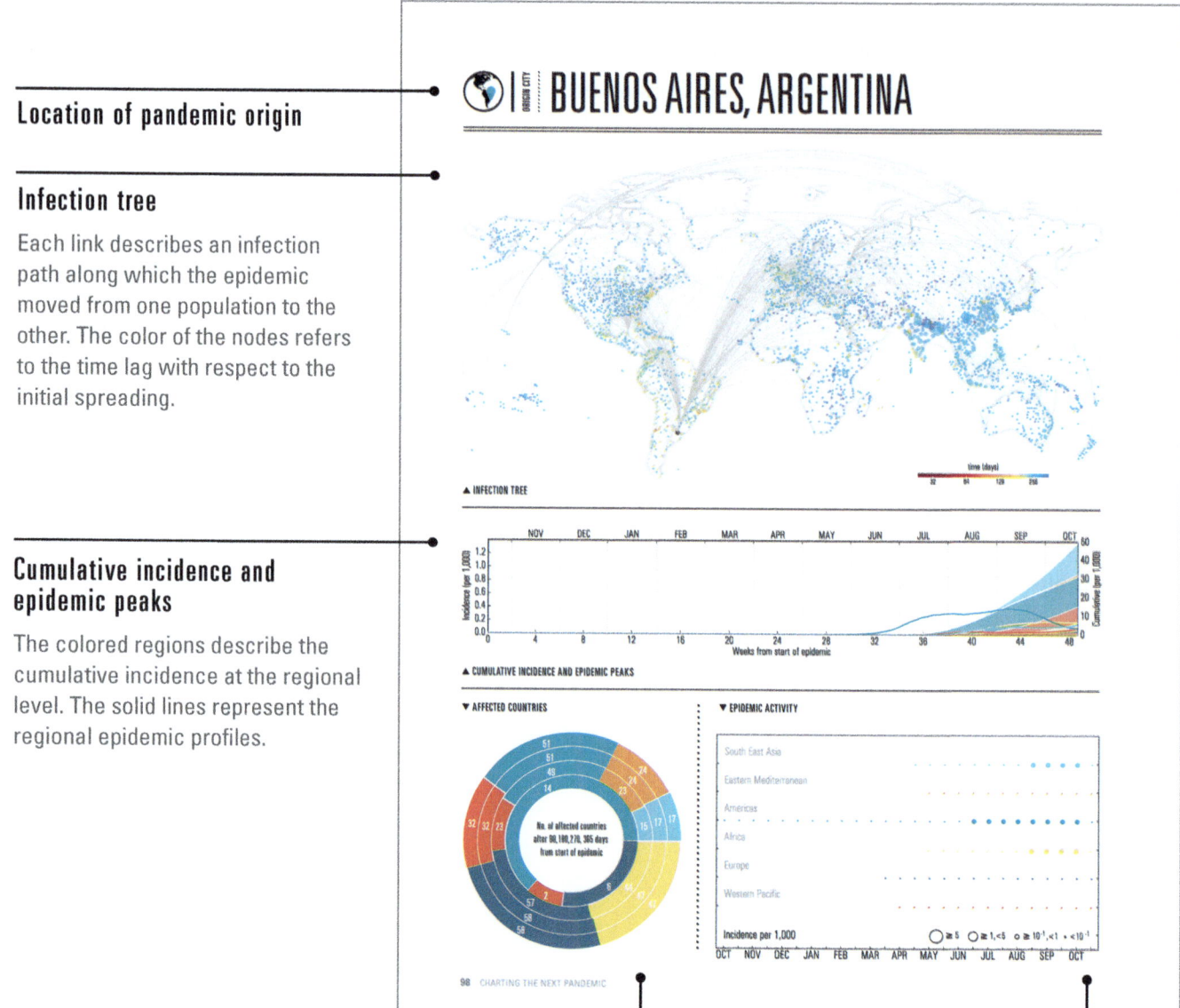

BUENOS AIRES, ARGENTINA

▲ INFECTION TREE

▲ CUMULATIVE INCIDENCE AND EPIDEMIC PEAKS

▼ AFFECTED COUNTRIES

No. of affected countries after 90, 180, 270, 365 days from start of epidemic

▼ EPIDEMIC ACTIVITY

South East Asia
Eastern Mediterranean
Americas
Africa
Europe
Western Pacific

Incidence per 1,000

98   CHARTING THE NEXT PANDEMIC

Affected countries as a function of time

The concentric circles refer to 90, 180, 270, and 365 days from the start of the spreading, respectively. In each circle we report the number of countries affected by the outbreak in each region.

Epidemic activity

Epidemic activity at the regional level. The size of the circles indicates the level of activity, that is proportional to the number of cases.

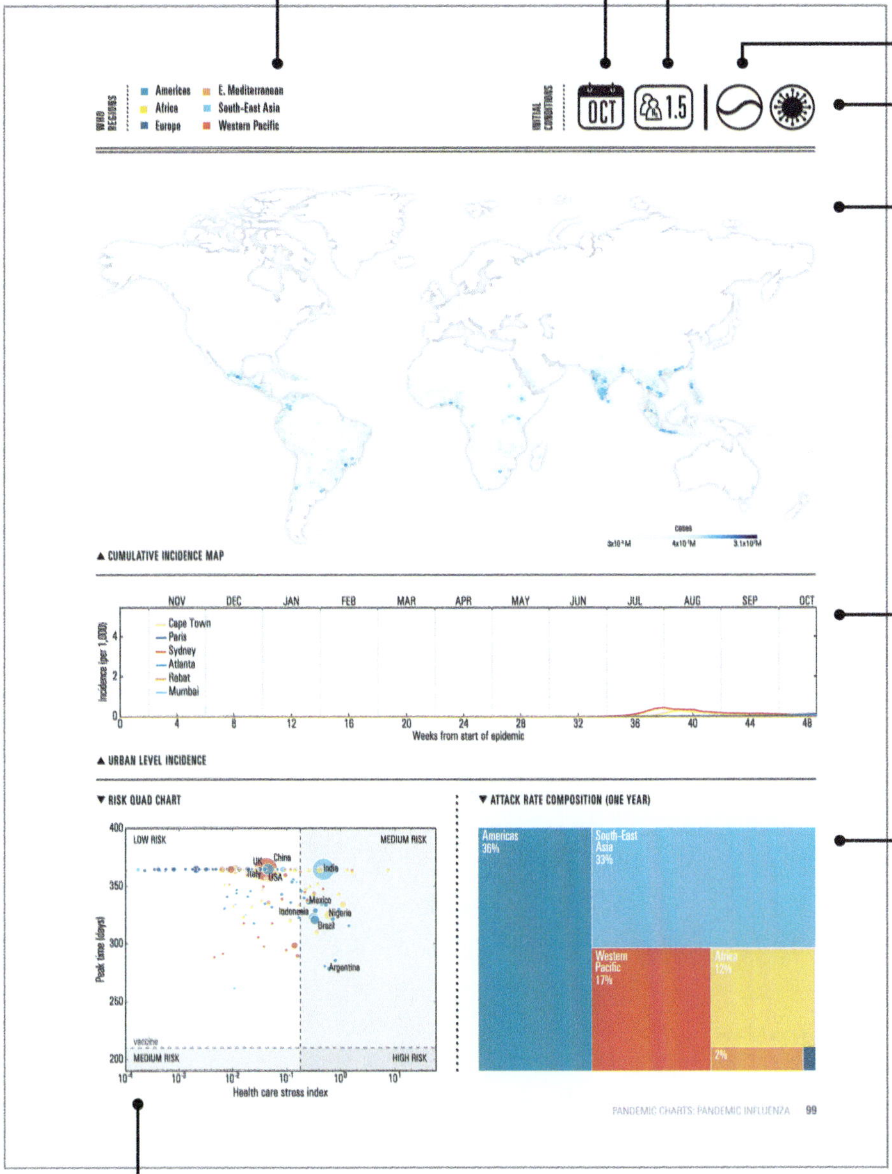

WHO REGIONS
- Americas
- Africa
- Europe
- E. Mediterranean
- South-East Asia
- Western Pacific

INITIAL CONDITIONS
OCT | 1.5

Cumulative incidence map
Cumulative incidence of clinical cases one year after the epidemic started.

▲ CUMULATIVE INCIDENCE MAP

cases
0x10^4 M   4x10^5 M   3.1x10^6 M

NOV DEC JAN FEB MAR APR MAY JUN JUL AUG SEP OCT

Incidence (per 1,000)
- Cape Town
- Paris
- Sydney
- Atlanta
- Rabat
- Mumbai

0 4 8 12 16 20 24 28 32 36 40 44 48
Weeks from start of epidemic

▲ URBAN LEVEL INCIDENCE

Urban level incidence
Daily incidence for six different urban areas.

▼ RISK QUAD CHART

LOW RISK   MEDIUM RISK
400

350   UK China
       India
       USA Mexico
       Indonesia Nigeria
       Brazil
300
                  Argentina

250

vaccine
200
MEDIUM RISK   HIGH RISK
10^-4 10^-3 10^-2 10^-1 10^0 10^1
Health care stress index

Peak time (days)

▼ ATTACK RATE COMPOSITION (ONE YEAR)

Americas 36%
South-East Asia 33%
Western Pacific 17%
Africa 12%
2%

Attack rate composition
We illustrate the geographical impact of the disease after one year considering the share of clinical cases in each WHO region.

Risk quad chart
We study the position of each country in a 2-D space corresponding to the clinical cases divided by the number of hospital beds per 10,000 of population (x-axis) and the peak time (y-axis). By considering the average of the two quantities, we define four quadrants associated with different risk levels.

CHART LAYOUT: PANDEMIC INFLUENZA

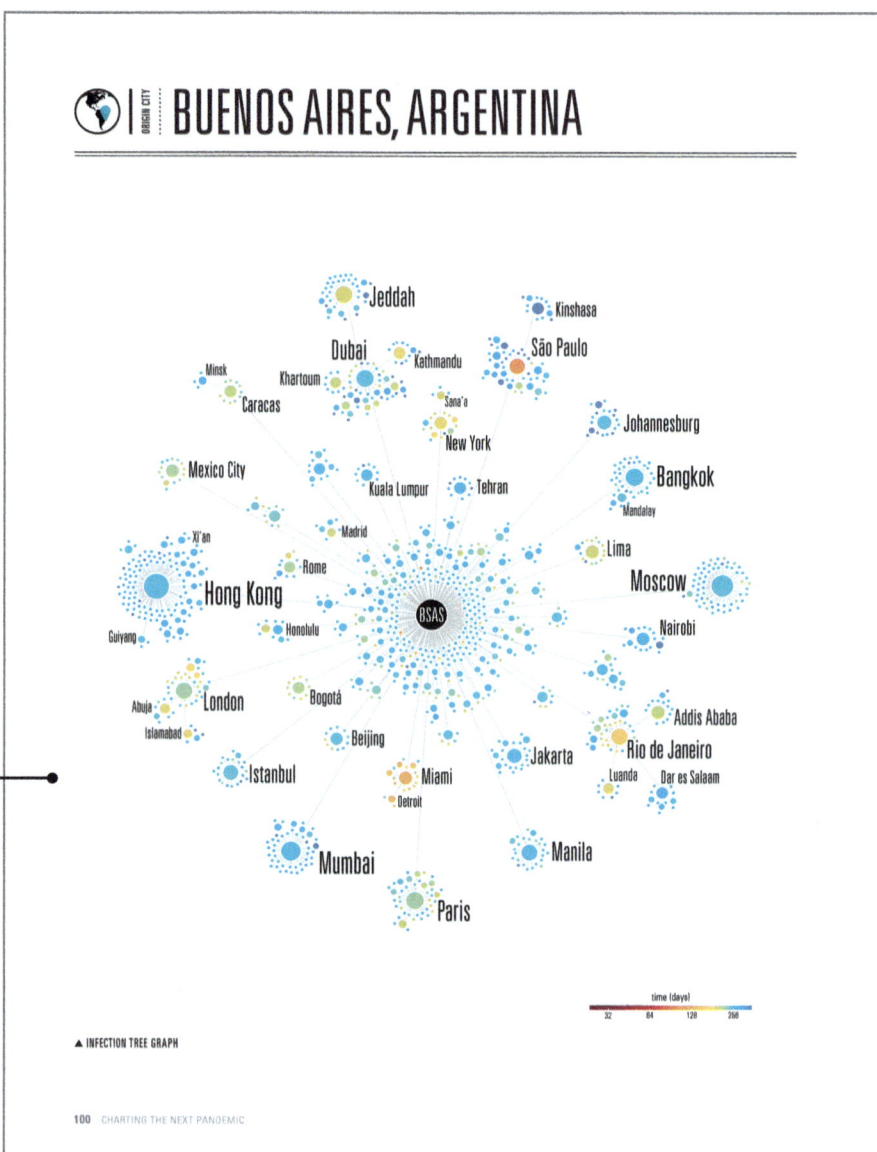

ORIGIN CITY | **BUENOS AIRES, ARGENTINA**

▲ INFECTION TREE GRAPH

Infection tree graph

Each link describes an infection-path along which the epidemic has moved from one population to the other. The color of the nodes refers to the time lag with respect to the start of the epidemic.

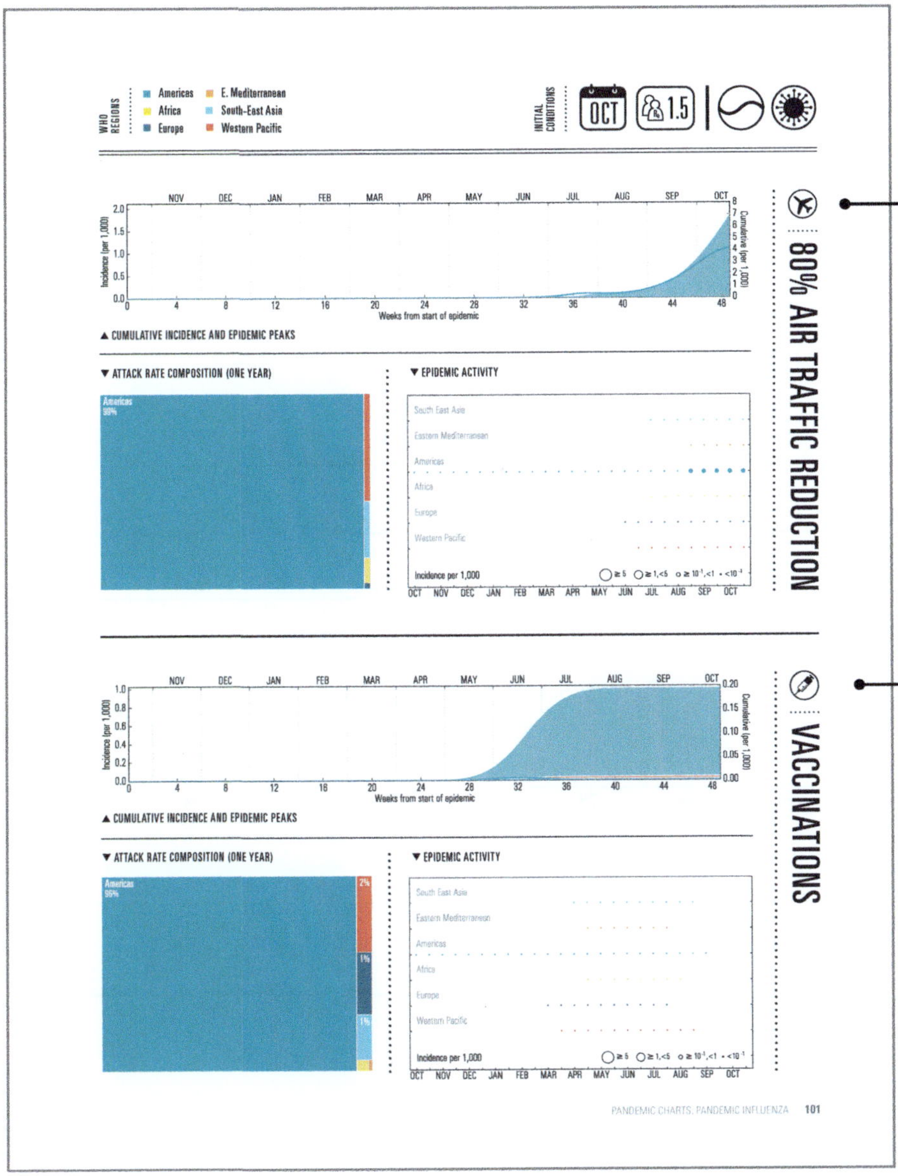

WHO REGIONS
- ■ Americas
- ■ Africa
- ■ Europe
- ■ E. Mediterranean
- ■ South-East Asia
- ■ Western Pacific

INITIAL CONDITIONS: OCT | 1.5

80% AIR TRAFFIC REDUCTION

▲ CUMULATIVE INCIDENCE AND EPIDEMIC PEAKS

▼ ATTACK RATE COMPOSITION (ONE YEAR)

Americas
98%

▼ EPIDEMIC ACTIVITY

South East Asia
Eastern Mediterranean
Americas
Africa
Europe
Western Pacific

Incidence per 1,000   ○ ≥ 5 ○ ≥ 1,<5 ○ ≥ 10⁻¹,<1 · <10⁻¹

Air Traffic Reduction

This scenario considers 80% air traffic reduction as a non-pharmaceutical intervention to mitigate the epidemic outbreak.

VACCINATIONS

▲ CUMULATIVE INCIDENCE AND EPIDEMIC PEAKS

▼ ATTACK RATE COMPOSITION (ONE YEAR)

Americas
96%

▼ EPIDEMIC ACTIVITY

South East Asia
Eastern Mediterranean
Americas
Africa
Europe
Western Pacific

Incidence per 1,000   ○ ≥ 5 ○ ≥ 1,<5 ○ ≥ 10⁻¹,<1 · <10⁻¹

Vaccinations

This scenario considers the introduction of vaccine campaigns to mitigate the epidemic outbreak.

time (days)

32   64   128   256

▲ INFECTION TREE

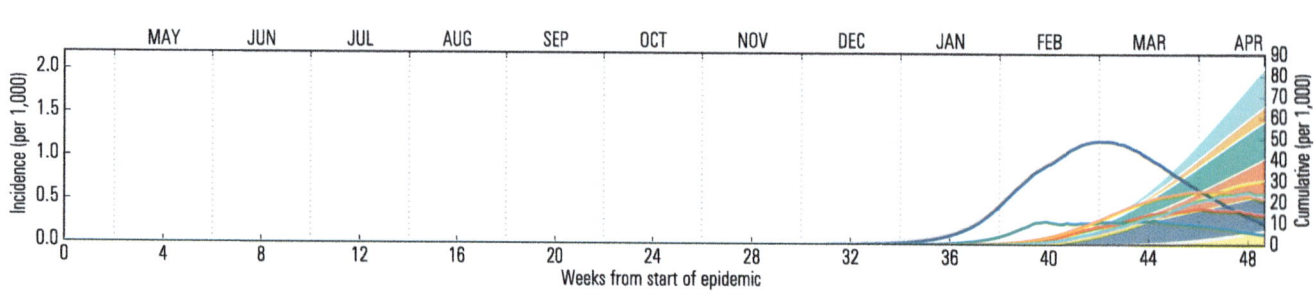

▲ CUMULATIVE INCIDENCE AND EPIDEMIC PEAKS

▼ AFFECTED COUNTRIES

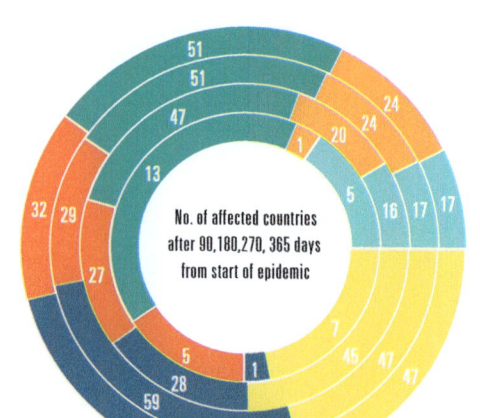

No. of affected countries
after 90, 180, 270, 365 days
from start of epidemic

▼ EPIDEMIC ACTIVITY

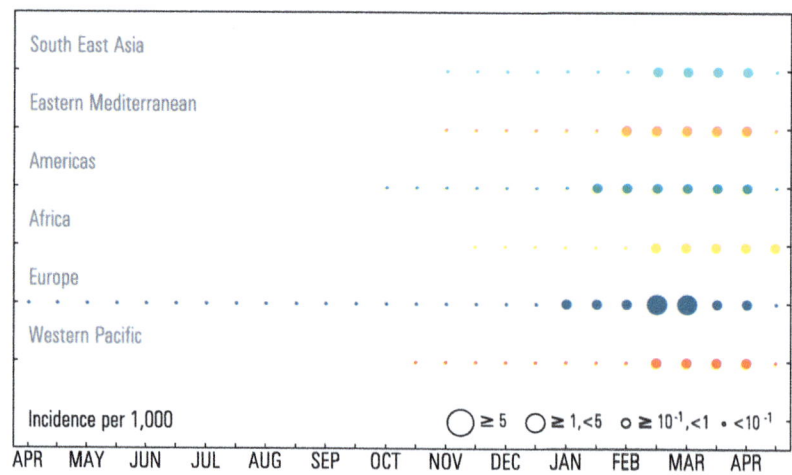

Incidence per 1,000   ◯ ≥ 5   ◯ ≥ 1,<5   ○ ≥ 10⁻¹,<1   · <10⁻¹

WHO REGIONS
- Americas
- Africa
- Europe
- E. Mediterranean
- South-East Asia
- Western Pacific

INITIAL CONDITIONS

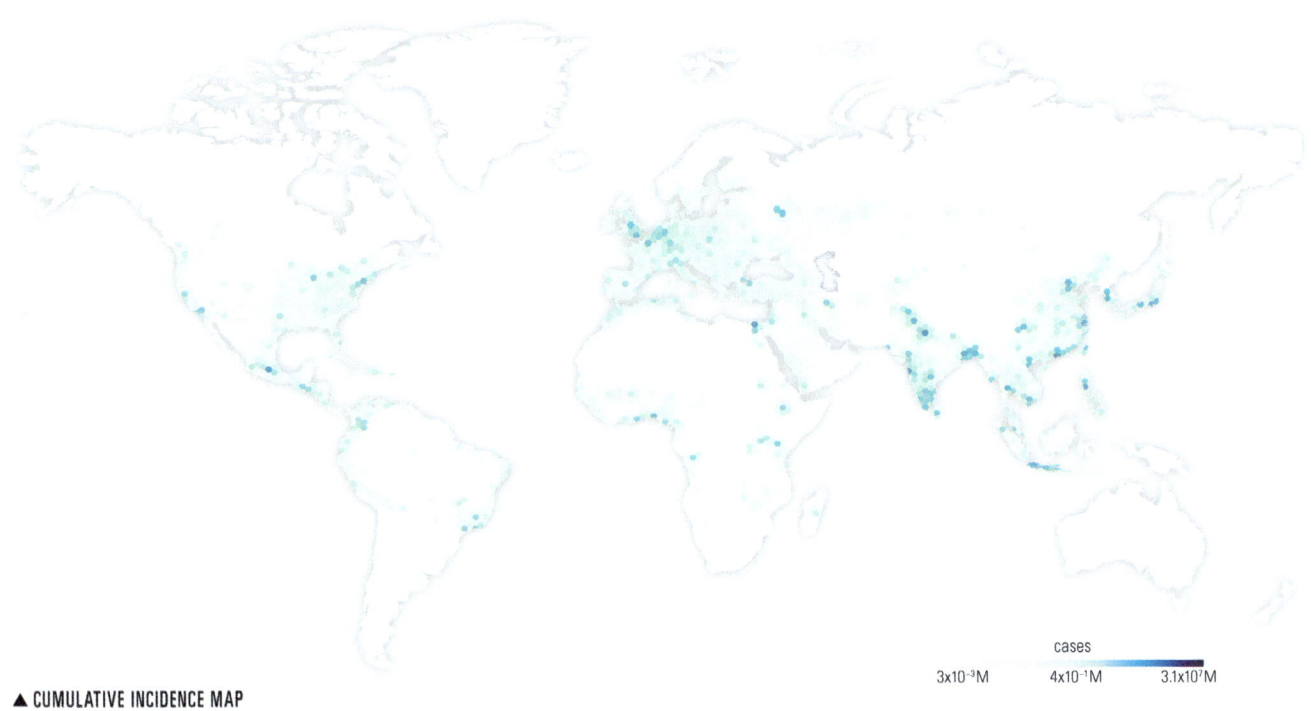

cases
3x10⁻³M   4x10⁻¹M   3.1x10⁷M

▲ CUMULATIVE INCIDENCE MAP

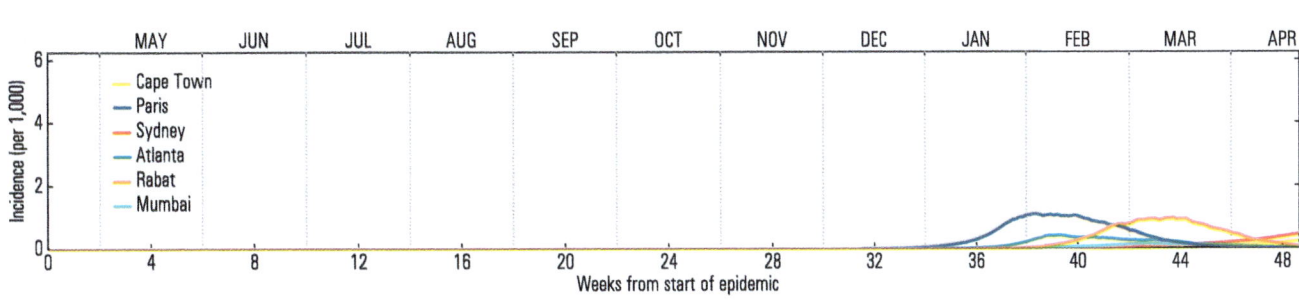

Cape Town
Paris
Sydney
Atlanta
Rabat
Mumbai

Incidence (per 1,000)

Weeks from start of epidemic

▲ URBAN LEVEL INCIDENCE

▼ RISK QUAD CHART

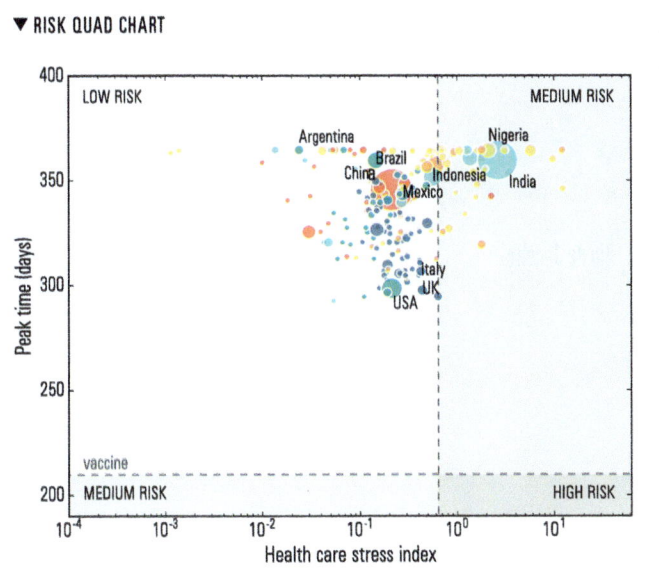

LOW RISK   MEDIUM RISK

Argentina
Nigeria
Brazil
China
Indonesia
Mexico
India
Italy
UK
USA

vaccine
MEDIUM RISK   HIGH RISK

Peak time (days)

Health care stress index

▼ ATTACK RATE COMPOSITION (ONE YEAR)

Americas 23%
South-East Asia 21%
Europe 20%
Western Pacific 20%
Eastern Mediterranean 8%
Africa 8%

BARCELONA, SPAIN

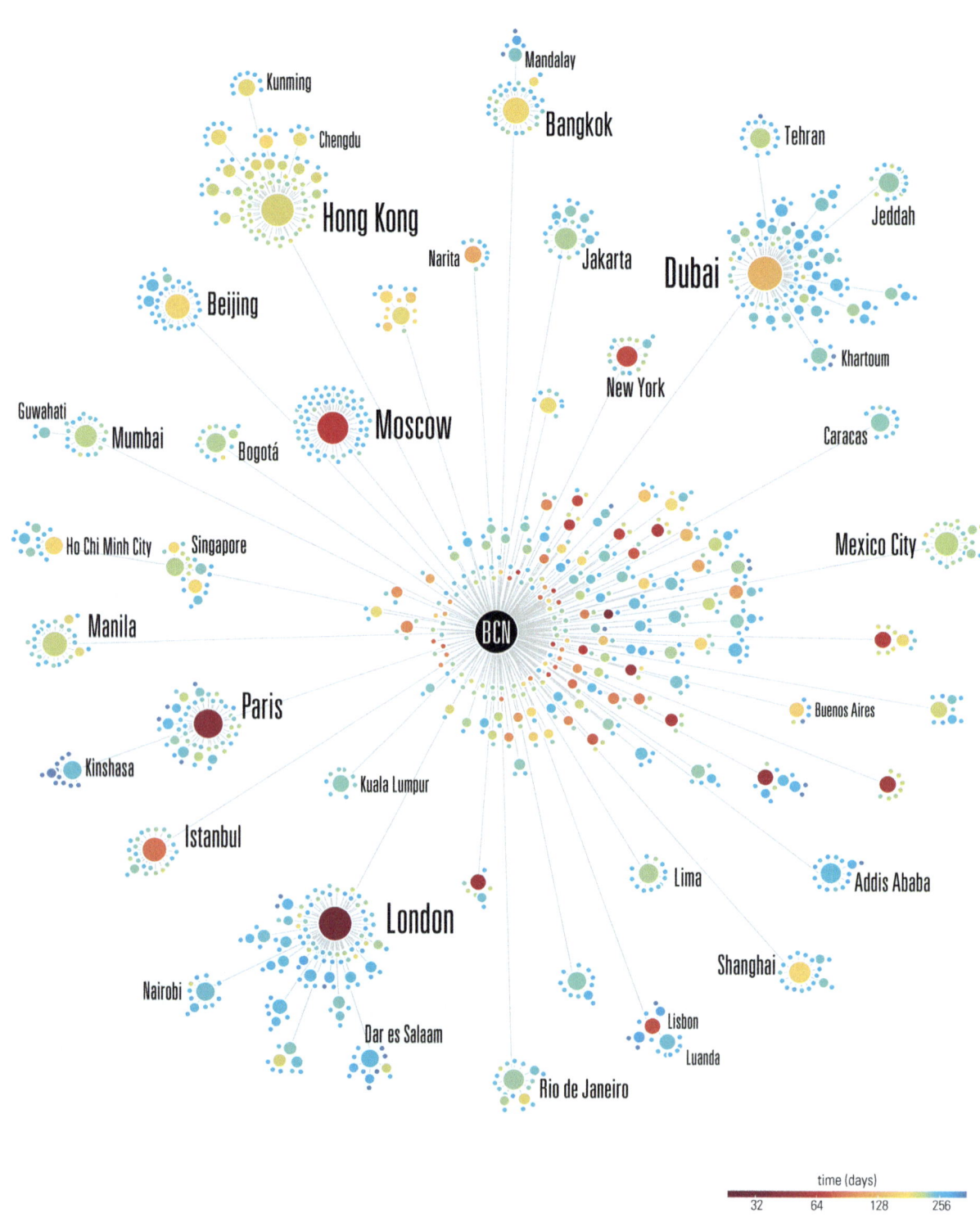

Kunming
Chengdu
Hong Kong
Beijing
Mandalay
Bangkok
Narita
Jakarta
Dubai
Tehran
Jeddah
Khartoum
New York
Caracas
Guwahati
Mumbai
Bogotá
Moscow
Ho Chi Minh City
Singapore
Mexico City
Manila
Buenos Aires
Paris
Kinshasa
Kuala Lumpur
Istanbul
Lima
Addis Ababa
London
Shanghai
Nairobi
Lisbon
Dar es Salaam
Luanda
Rio de Janeiro

BCN

time (days)
32   64   128   256

▲ INFECTION TREE GRAPH

- **Americas**
- **Africa**
- **Europe**
- **E. Mediterranean**
- **South-East Asia**
- **Western Pacific**

INITIAL CONDITIONS

 APR  1.5

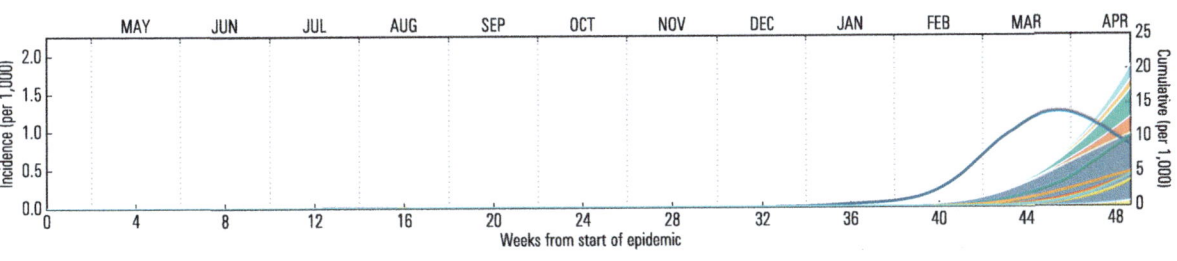

▲ CUMULATIVE INCIDENCE AND EPIDEMIC PEAKS

▼ ATTACK RATE COMPOSITION (ONE YEAR)

Europe 52%
Americas 19%
Western Pacific 11%
South-East Asia 9%
Eastern Mediterranean 6%
Africa 3%

▼ EPIDEMIC ACTIVITY

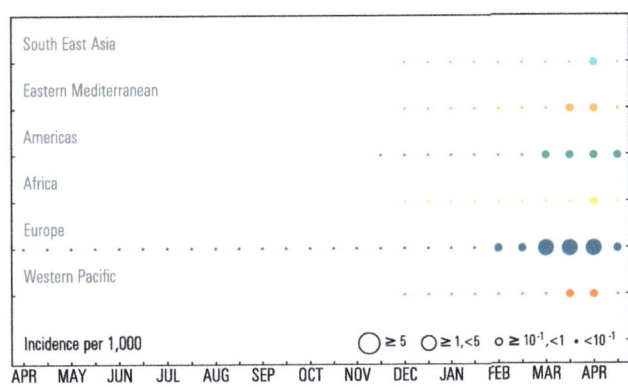

Incidence per 1,000   ◯ ≥ 5   ◯ ≥ 1,<5   ○ ≥ 10⁻¹,<1   · <10⁻¹

80% AIR TRAFFIC REDUCTION

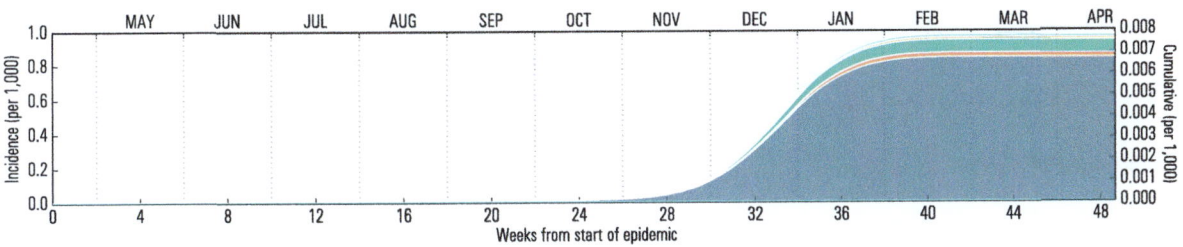

▲ CUMULATIVE INCIDENCE AND EPIDEMIC PEAKS

▼ ATTACK RATE COMPOSITION (ONE YEAR)

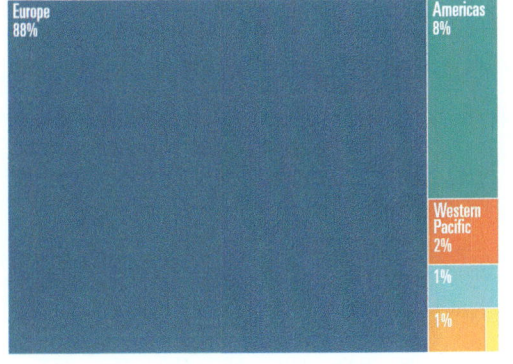

Europe 88%
Americas 8%
Western Pacific 2%
1%
1%

▼ EPIDEMIC ACTIVITY

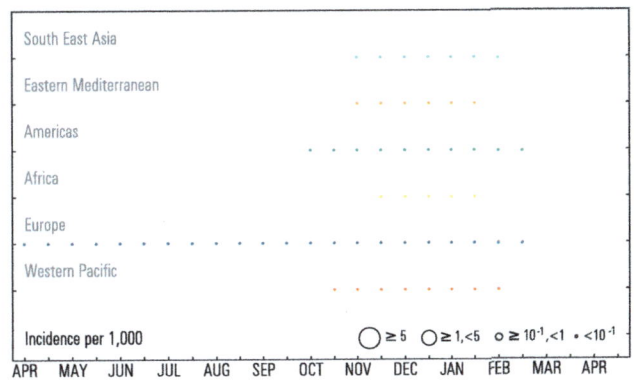

Incidence per 1,000   ◯ ≥ 5   ◯ ≥ 1,<5   ○ ≥ 10⁻¹,<1   · <10⁻¹

 VACCINATIONS

BARCELONA, SPAIN

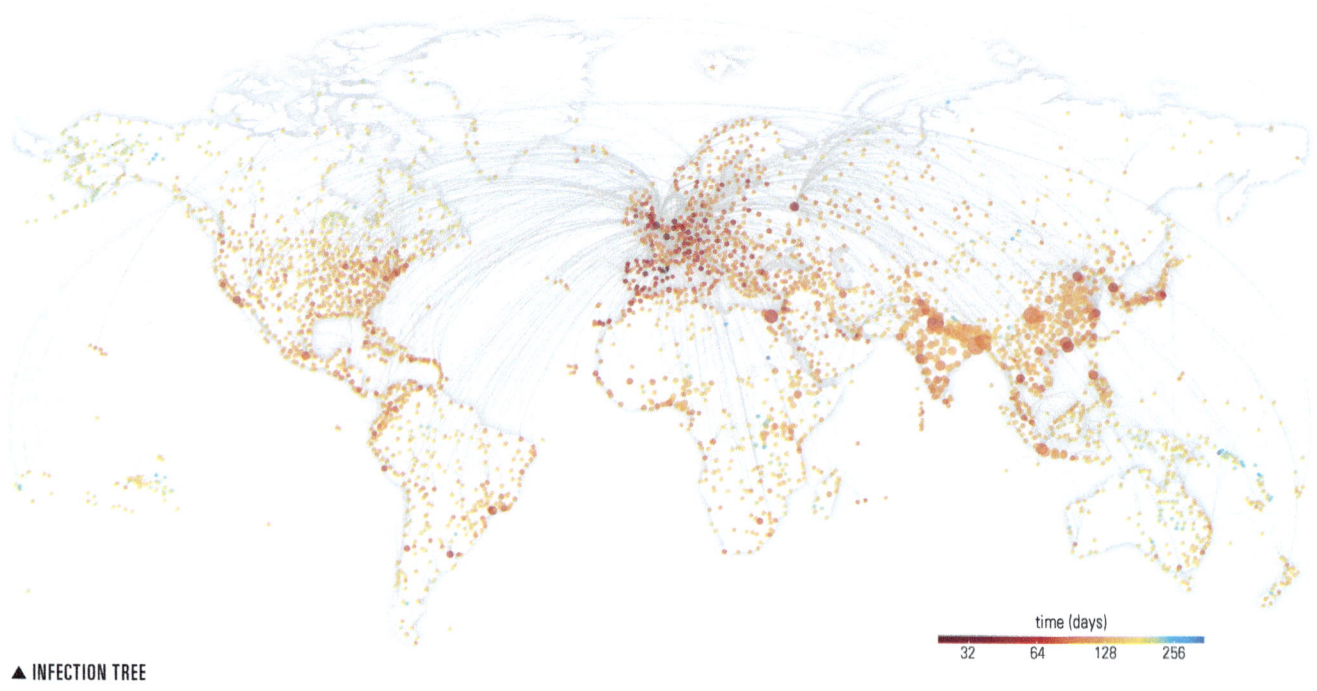

time (days)

32   64   128   256

▲ INFECTION TREE

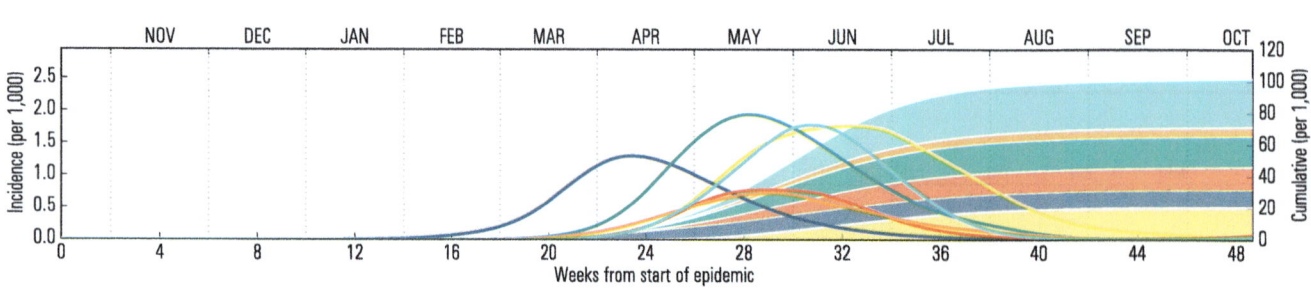

▲ CUMULATIVE INCIDENCE AND EPIDEMIC PEAKS

▼ AFFECTED COUNTRIES

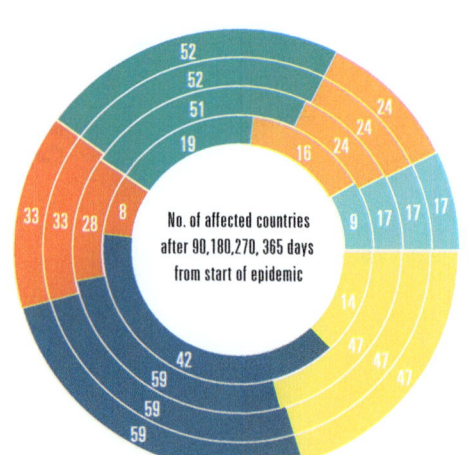

No. of affected countries
after 90,180,270, 365 days
from start of epidemic

▼ EPIDEMIC ACTIVITY

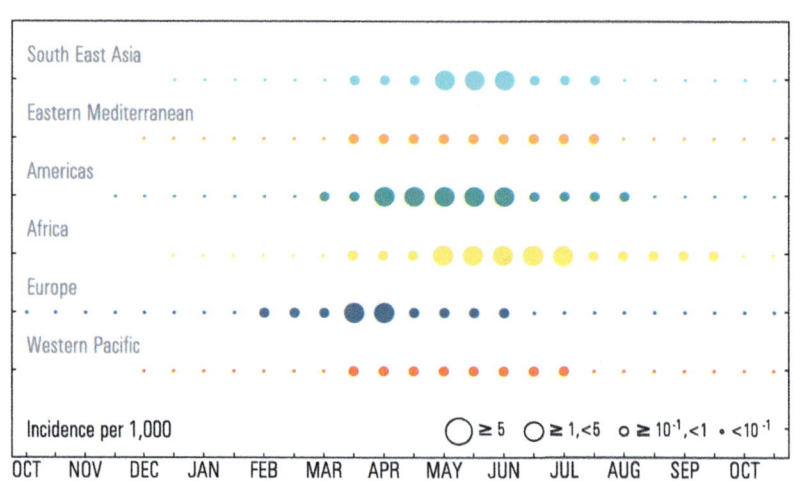

WHO REGIONS
- Americas
- Africa
- Europe
- E. Mediterranean
- South-East Asia
- Western Pacific

INITIAL CONDITIONS

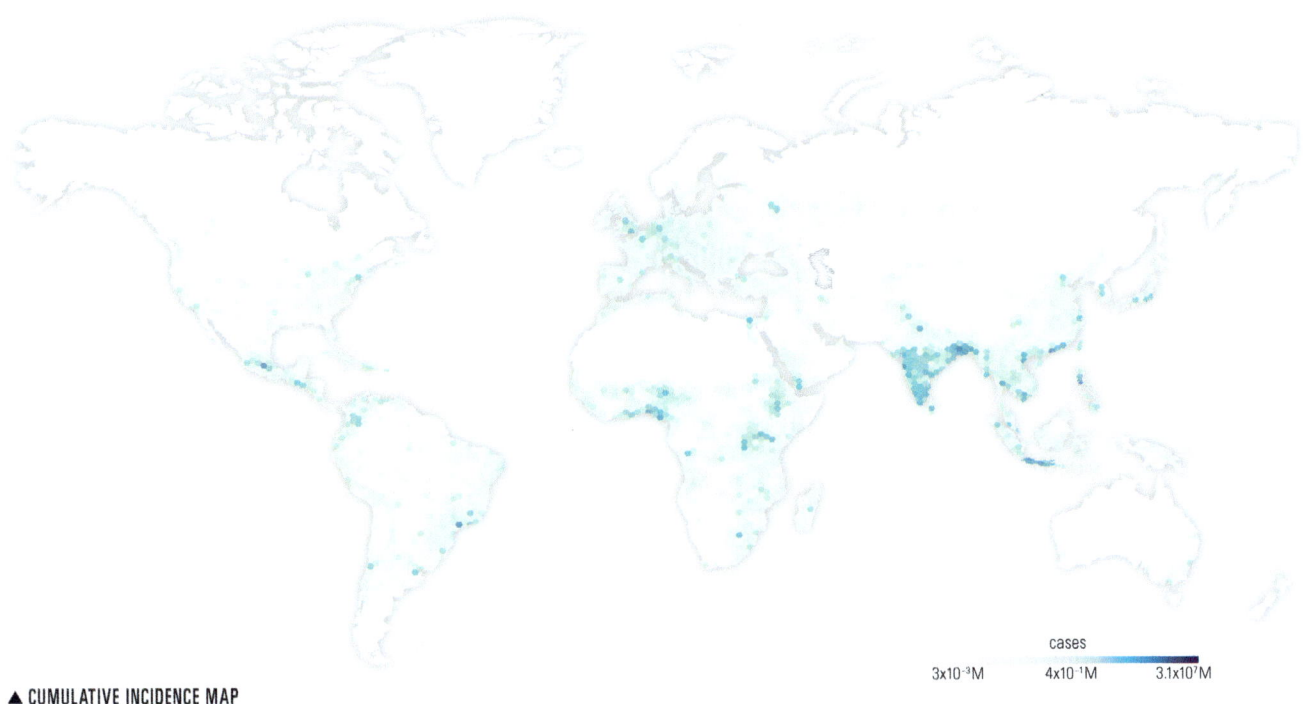

cases

3x10⁻³M   4x10⁻¹M   3.1x10⁷M

▲ CUMULATIVE INCIDENCE MAP

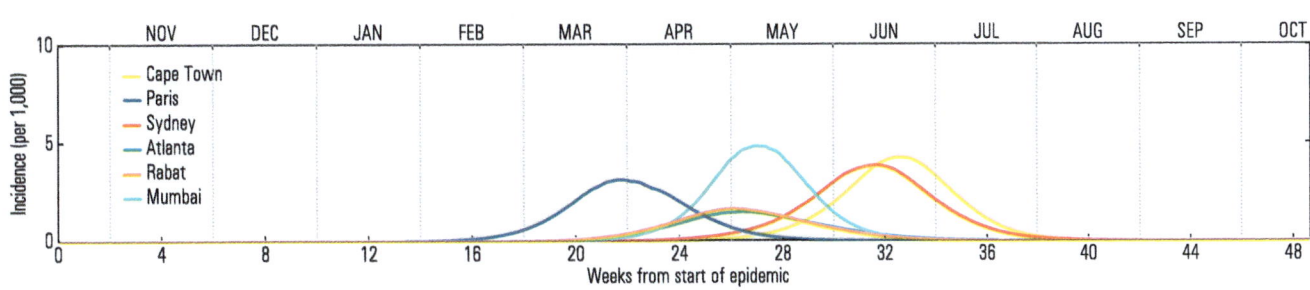

- Cape Town
- Paris
- Sydney
- Atlanta
- Rabat
- Mumbai

Incidence (per 1,000)

Weeks from start of epidemic

▲ URBAN LEVEL INCIDENCE

▼ RISK QUAD CHART

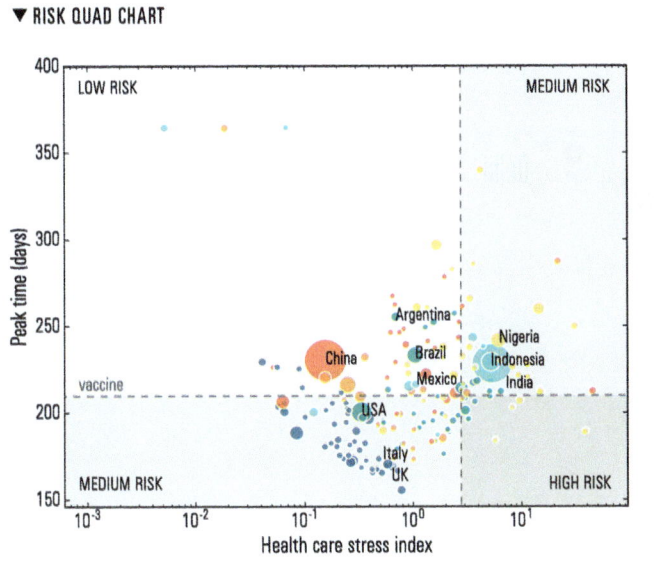

LOW RISK   MEDIUM RISK

Peak time (days)

vaccine

Argentina
Nigeria
China
Brazil
Mexico
Indonesia
India
USA
Italy
UK

MEDIUM RISK   HIGH RISK

Health care stress index

▼ ATTACK RATE COMPOSITION (ONE YEAR)

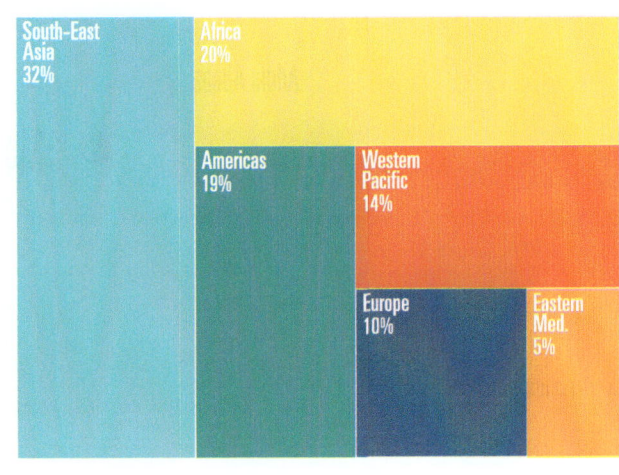

South-East Asia 32%
Africa 20%
Americas 19%
Western Pacific 14%
Europe 10%
Eastern Med. 5%

BARCELONA, SPAIN

Port Harcourt

Abuja

London

Mandalay

Bangkok

Narita

New York

Kuala Lumpur

Madrid

Tehran

Mexico City

Caracas

Riyadh

Buenos Aires

Luanda

Lisbon

Beijing

Paris

Bogotá

Jakarta

BCN

Istanbul

Manila

Shanghai

Kunming

Dubai

Nantong

Khartoum

Lucknow

Lima

Addis Ababa

Kinshasa

Delhi

Moscow

time (days)

32   64   128   256

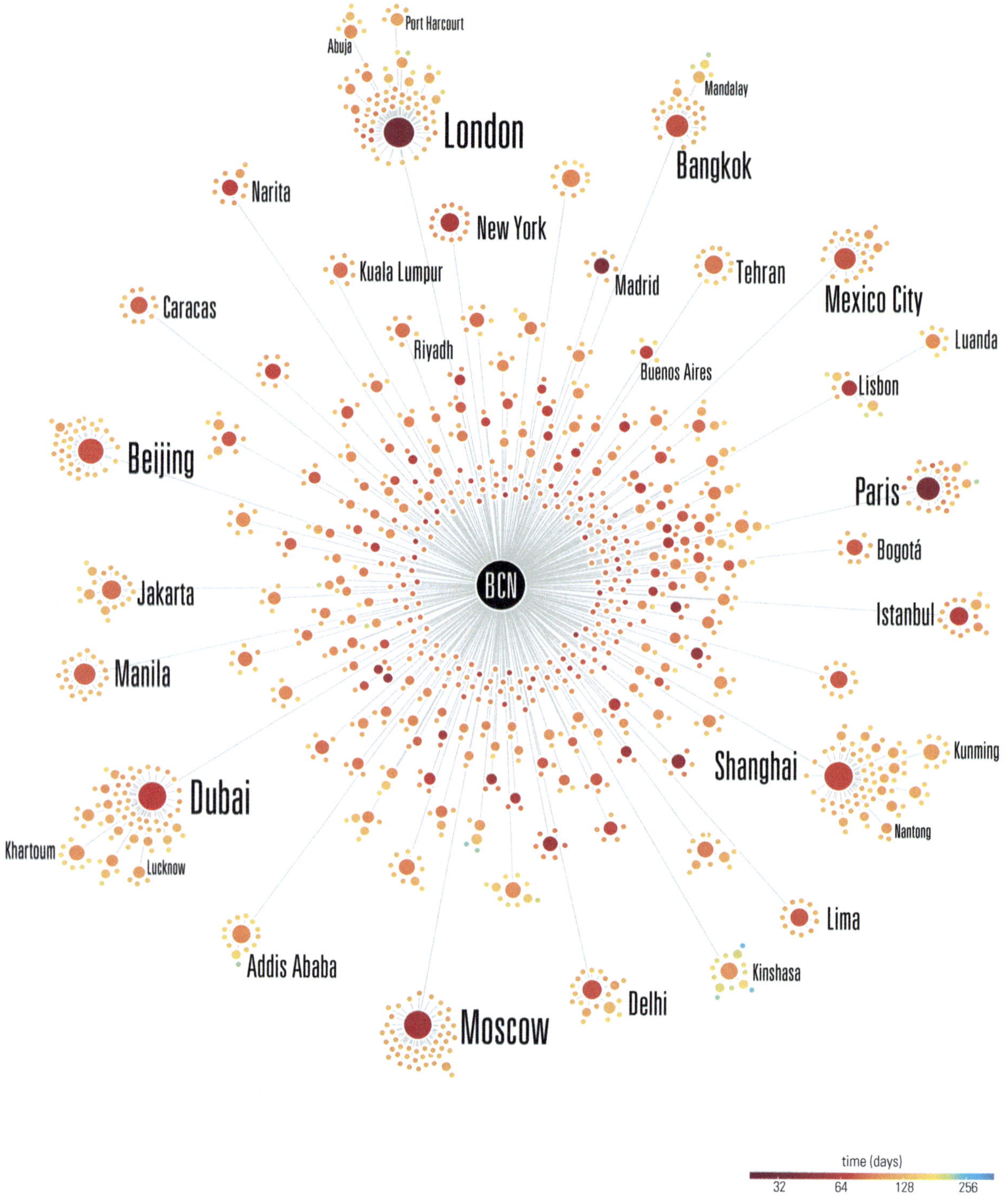

▲ INFECTION TREE GRAPH

ORIGIN CITY

WHO REGIONS

- Americas
- Africa
- Europe
- E. Mediterranean
- South-East Asia
- Western Pacific

INITIAL CONDITIONS

80% AIR TRAFFIC REDUCTION

▲ CUMULATIVE INCIDENCE AND EPIDEMIC PEAKS

▼ ATTACK RATE COMPOSITION (ONE YEAR)

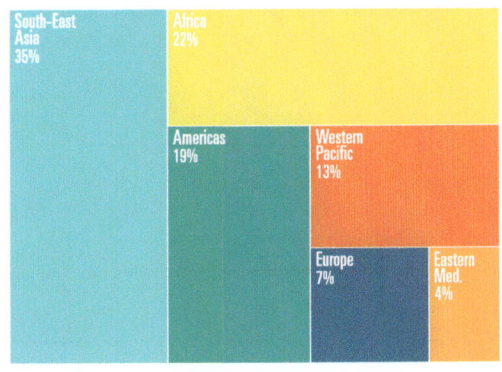

▼ EPIDEMIC ACTIVITY

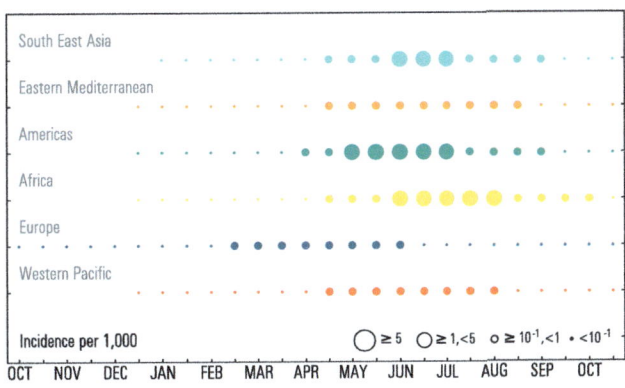

Incidence per 1,000   ● ≥ 5   ● ≥ 1,<5   ○ ≥ 10⁻¹,<1   · <10⁻¹

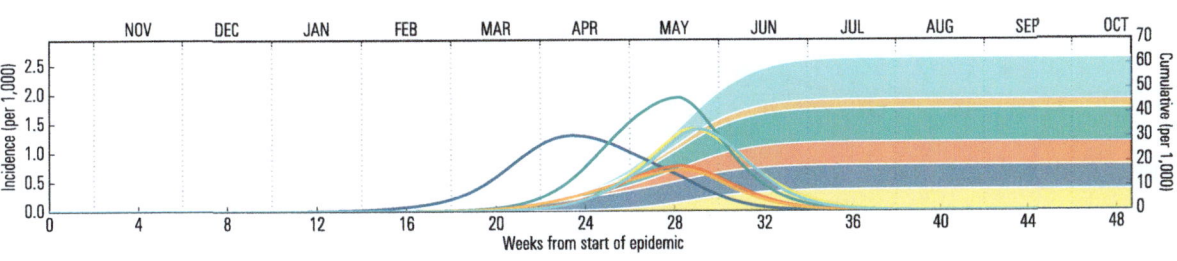

VACCINATIONS

▲ CUMULATIVE INCIDENCE AND EPIDEMIC PEAKS

▼ ATTACK RATE COMPOSITION (ONE YEAR)

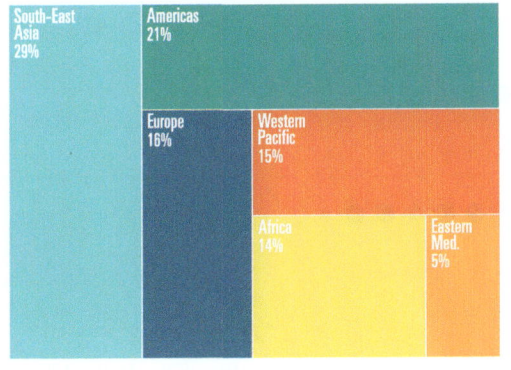

▼ EPIDEMIC ACTIVITY

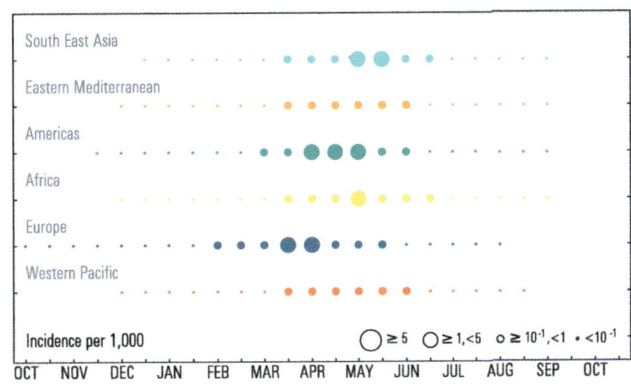

Incidence per 1,000   ● ≥ 5   ● ≥ 1,<5   ○ ≥ 10⁻¹,<1   · <10⁻¹

BARCELONA, SPAIN

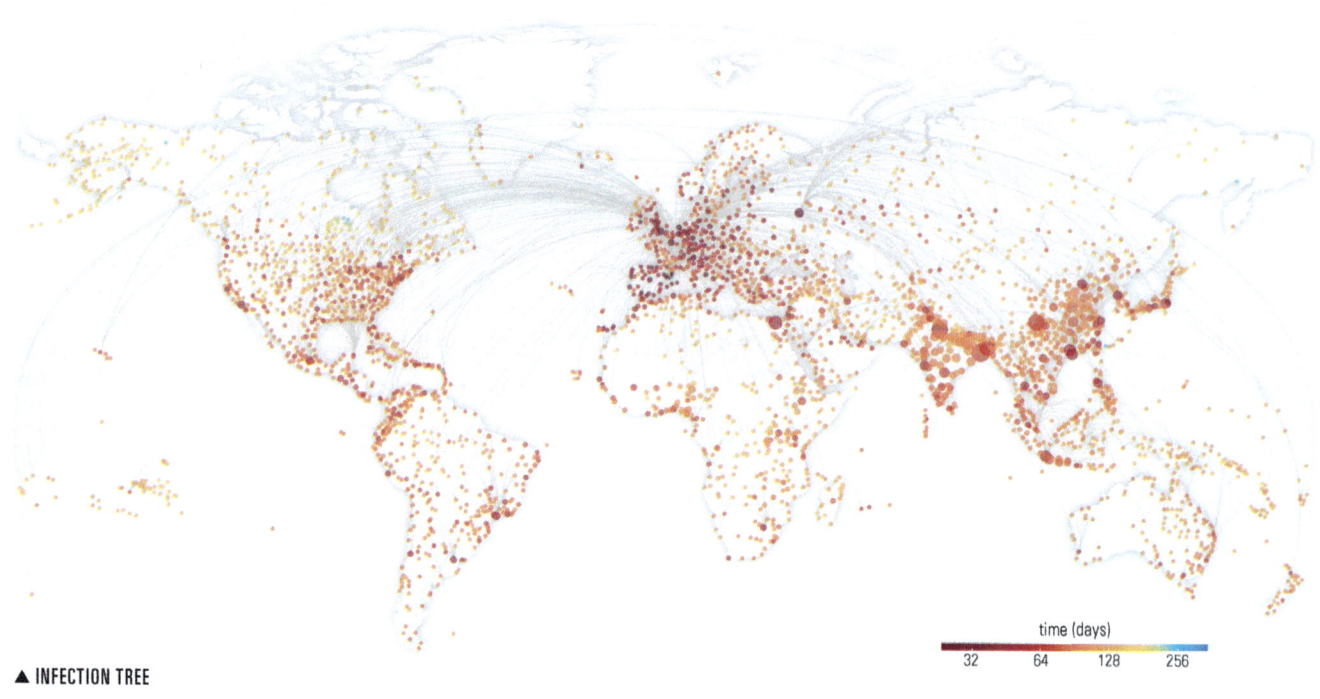

time (days)

32   64   128   256

▲ INFECTION TREE

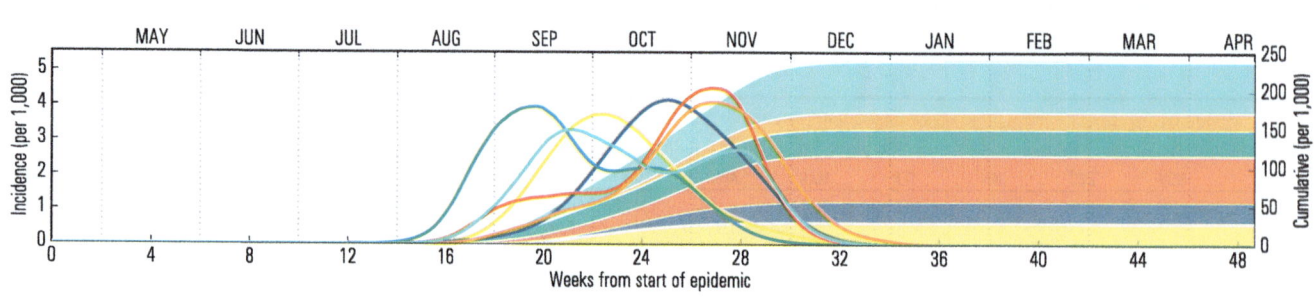

MAY   JUN   JUL   AUG   SEP   OCT   NOV   DEC   JAN   FEB   MAR   APR

Incidence (per 1,000)

Cumulative (per 1,000)

Weeks from start of epidemic

▲ CUMULATIVE INCIDENCE AND EPIDEMIC PEAKS

▼ AFFECTED COUNTRIES

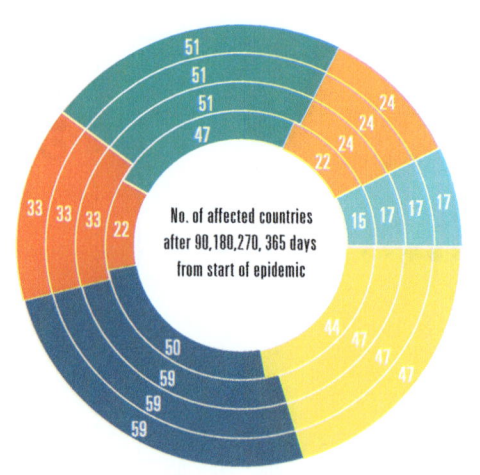

No. of affected countries after 90, 180, 270, 365 days from start of epidemic

▼ EPIDEMIC ACTIVITY

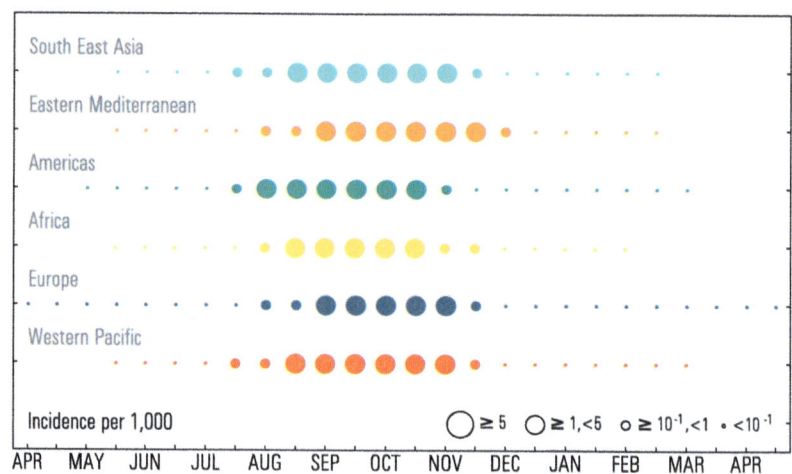

South East Asia

Eastern Mediterranean

Americas

Africa

Europe

Western Pacific

Incidence per 1,000   ◯ ≥ 5   ◯ ≥ 1,<5   ○ ≥ 10⁻¹,<1   · <10⁻¹

APR   MAY   JUN   JUL   AUG   SEP   OCT   NOV   DEC   JAN   FEB   MAR   APR

■ Americas   ■ E. Mediterranean
■ Africa   ■ South-East Asia
■ Europe   ■ Western Pacific

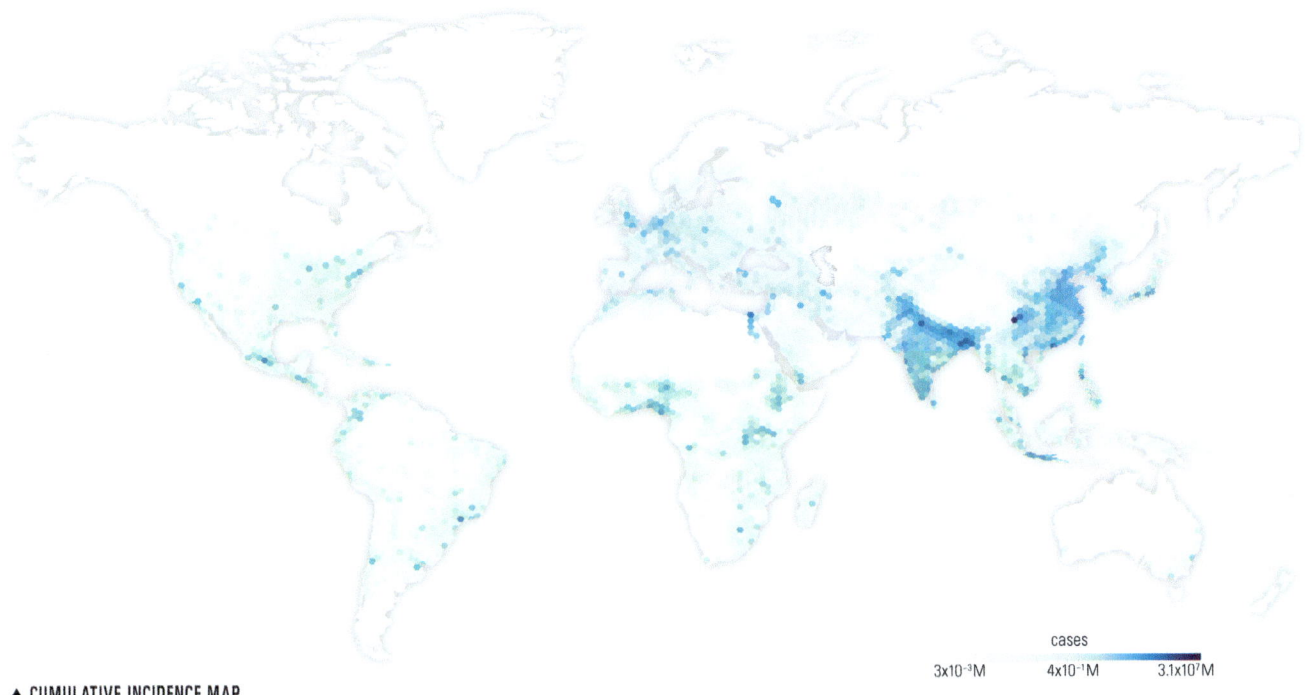

cases

3x10⁻³M   4x10⁻¹M   3.1x10⁷M

▲ CUMULATIVE INCIDENCE MAP

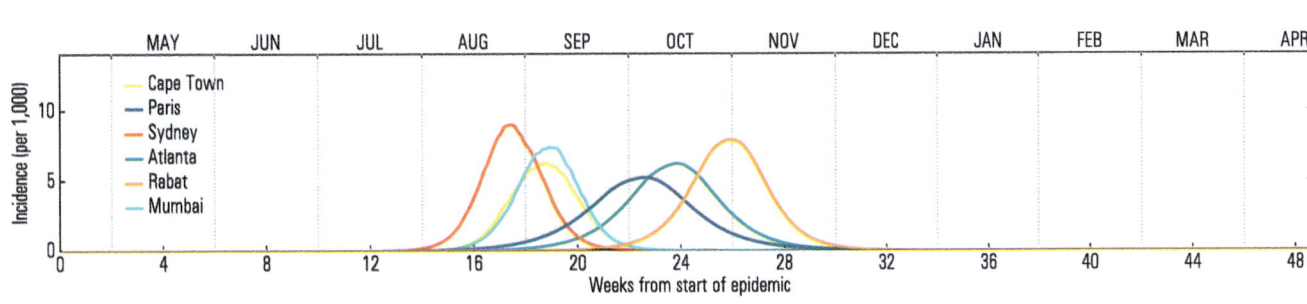

▲ URBAN LEVEL INCIDENCE

▼ RISK QUAD CHART

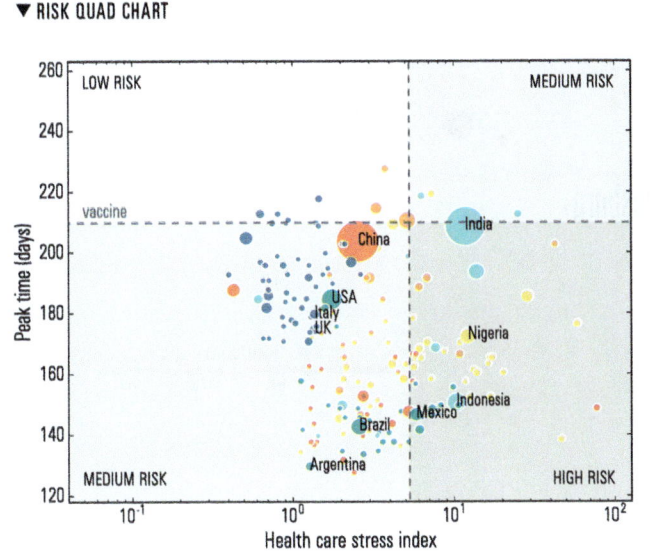

▼ ATTACK RATE COMPOSITION (ONE YEAR)

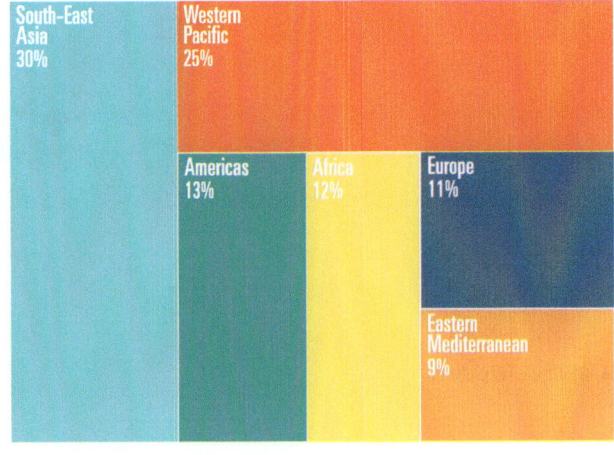

BARCELONA, SPAIN

Moscow

Mumbai

Mandalay

Abuja

Mexico City

Caracas

Bangkok

London

Bogotá

Kuala Lumpur

Luanda

Dubai

Addis Ababa

Paris

Kinshasa

BCN

Istanbul

Kunming

Nairobi

Hong Kong

Algiers

Jeddah

Singapore

Beijing

Khartoum

Lima

Buenos Aires

Manila

Jakarta

São Paulo

Johannesburg

time (days)

32   64   128   256

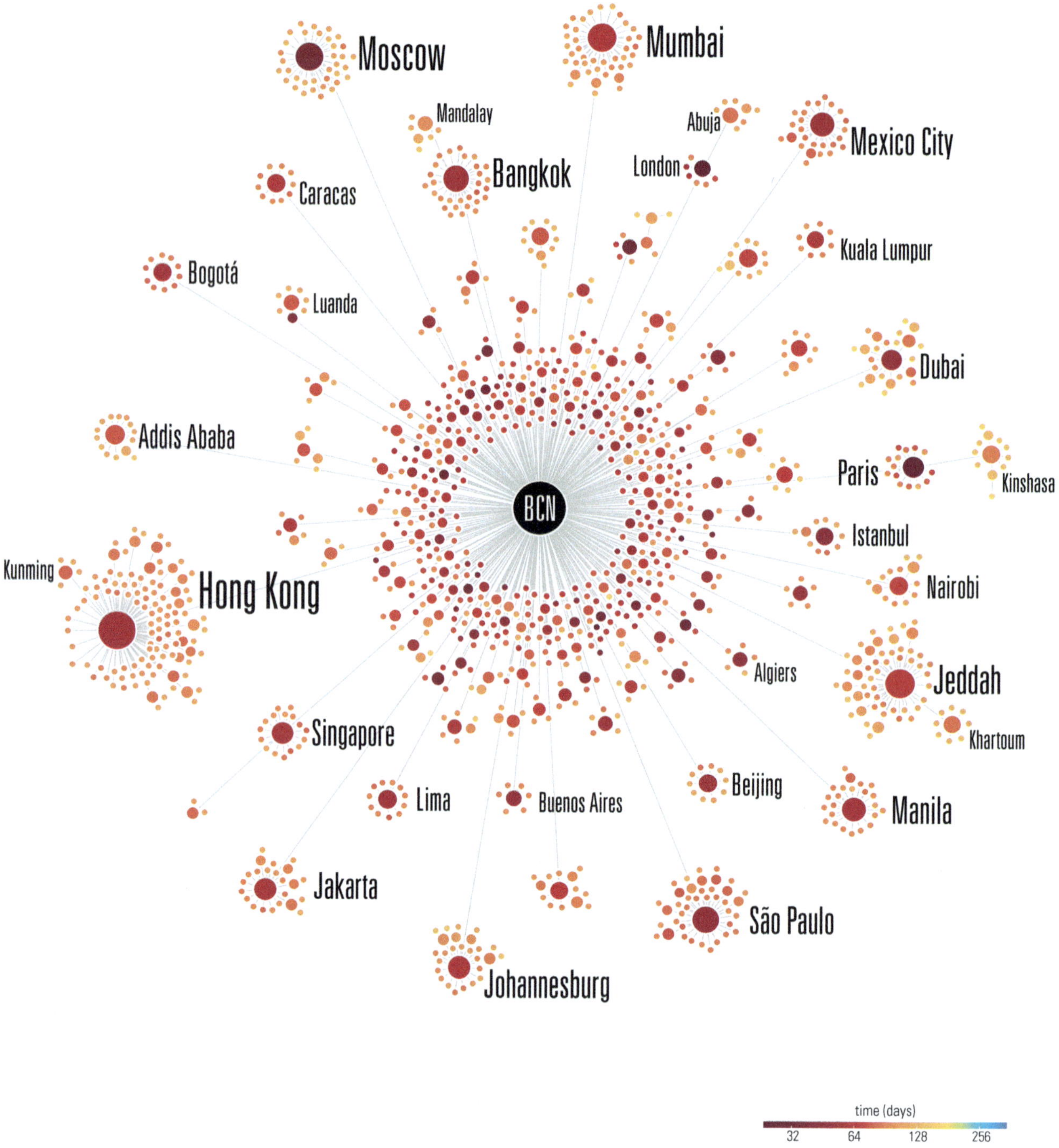

▲ INFECTION TREE GRAPH

■ Americas ■ E. Mediterranean
■ Africa ■ South-East Asia
■ Europe ■ Western Pacific

INITIAL CONDITIONS

 APR  2.0

 80% AIR TRAFFIC REDUCTION

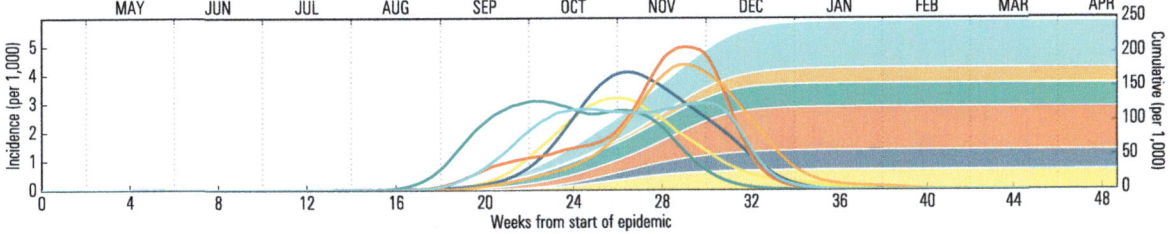

▲ CUMULATIVE INCIDENCE AND EPIDEMIC PEAKS

▼ ATTACK RATE COMPOSITION (ONE YEAR)

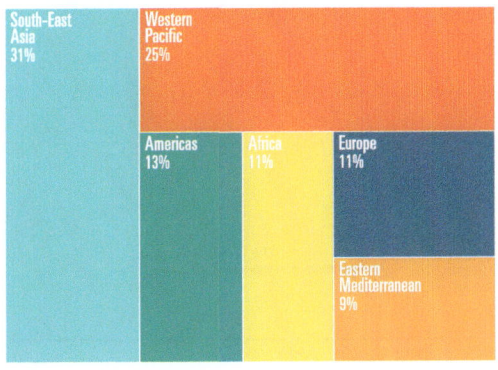

South-East Asia 31%
Western Pacific 25%
Americas 13%
Africa 11%
Europe 11%
Eastern Mediterranean 9%

▼ EPIDEMIC ACTIVITY

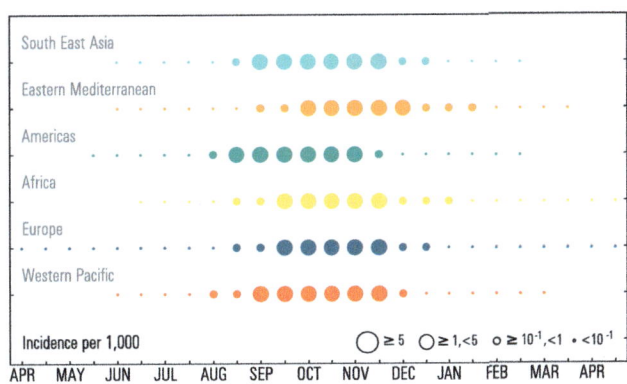

Incidence per 1,000   ○ ≥ 5 ○ ≥ 1,<5 ○ ≥ 10⁻¹,<1 • <10⁻¹

 VACCINATIONS

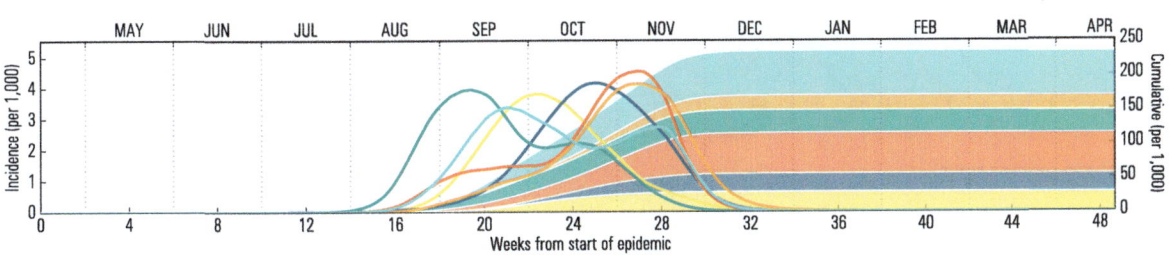

▲ CUMULATIVE INCIDENCE AND EPIDEMIC PEAKS

▼ ATTACK RATE COMPOSITION (ONE YEAR)

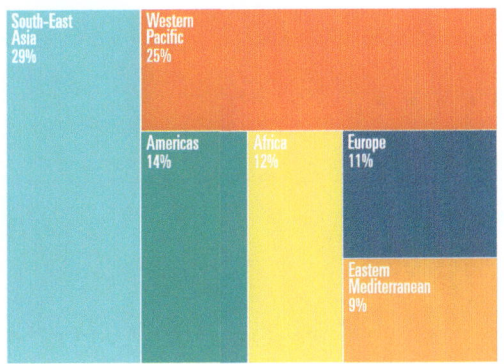

South-East Asia 29%
Western Pacific 25%
Americas 14%
Africa 12%
Europe 11%
Eastern Mediterranean 9%

▼ EPIDEMIC ACTIVITY

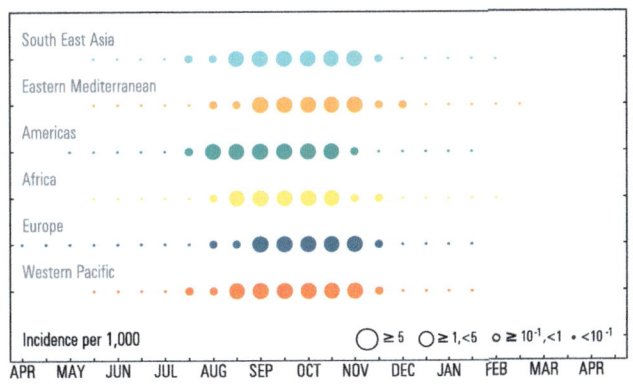

Incidence per 1,000   ○ ≥ 5 ○ ≥ 1,<5 ○ ≥ 10⁻¹,<1 • <10⁻¹

BARCELONA, SPAIN

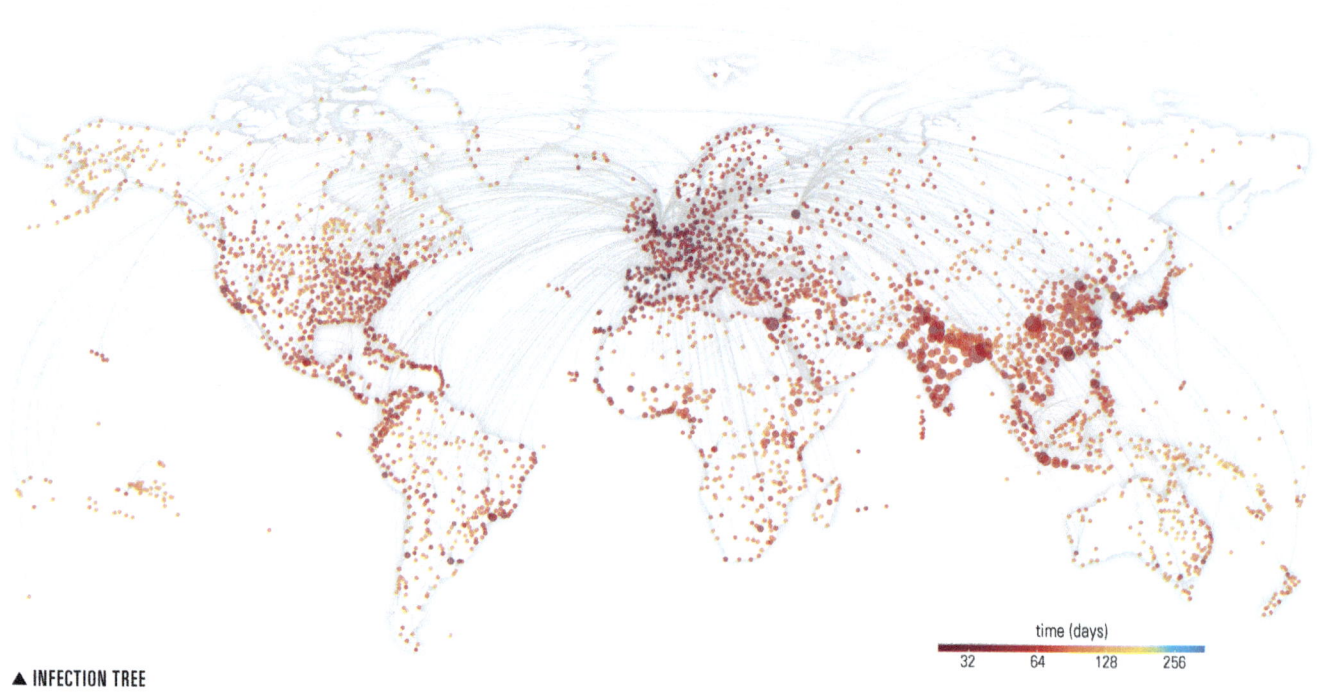

time (days)

32   64   128   256

▲ INFECTION TREE

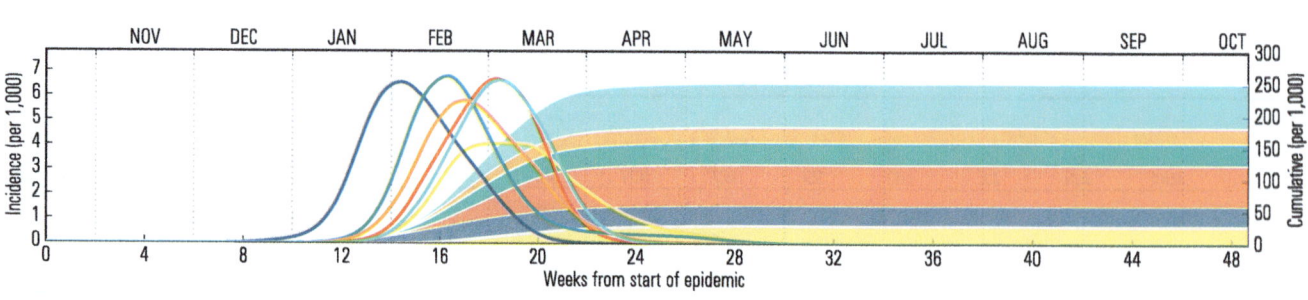

▲ CUMULATIVE INCIDENCE AND EPIDEMIC PEAKS

▼ AFFECTED COUNTRIES

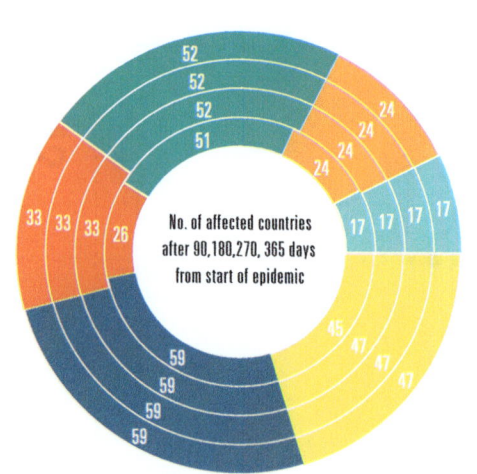

No. of affected countries after 90, 180, 270, 365 days from start of epidemic

▼ EPIDEMIC ACTIVITY

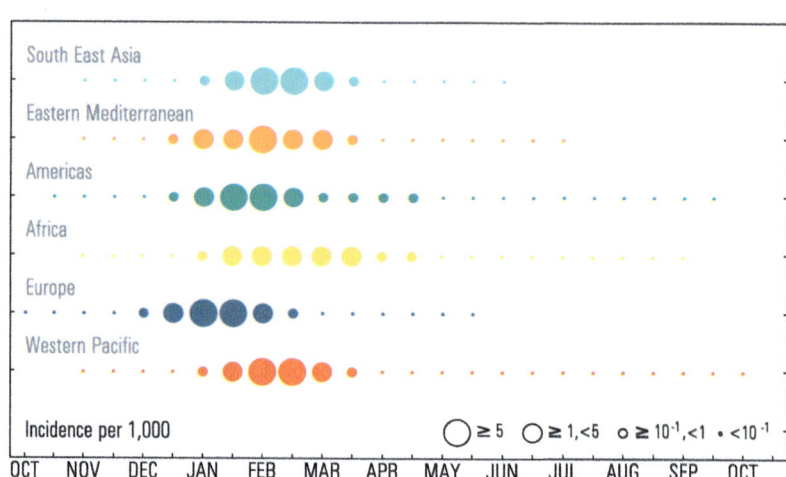

South East Asia

Eastern Mediterranean

Americas

Africa

Europe

Western Pacific

Incidence per 1,000

≥ 5 $\geq 1, < 5$ $\geq 10^{-1}, < 1$ $< 10^{-1}$

OCT NOV DEC JAN FEB MAR APR MAY JUN JUL AUG SEP OCT

WHO
REGIONS

■ Americas ■ E. Mediterranean
■ Africa   ■ South-East Asia
■ Europe   ■ Western Pacific

INITIAL
CONDITIONS

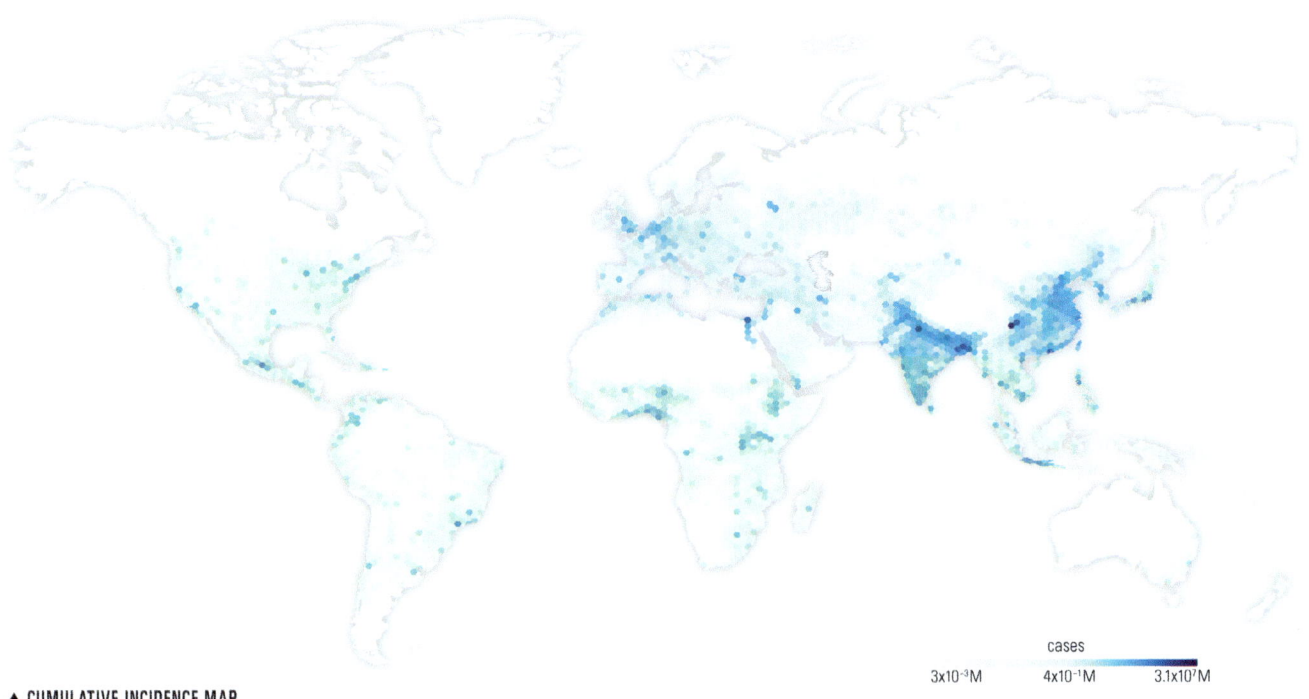

cases

3×10^{-3} M 4×10^{-1} M 3.1×10^{7} M

▲ CUMULATIVE INCIDENCE MAP

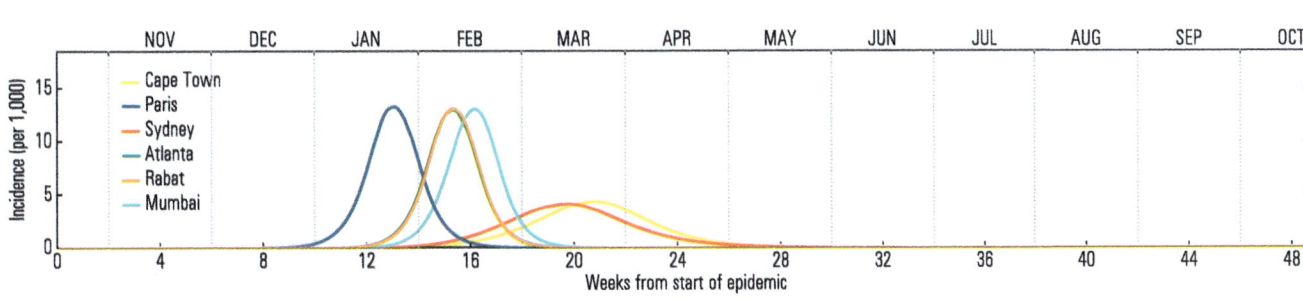

▲ URBAN LEVEL INCIDENCE

▼ RISK QUAD CHART

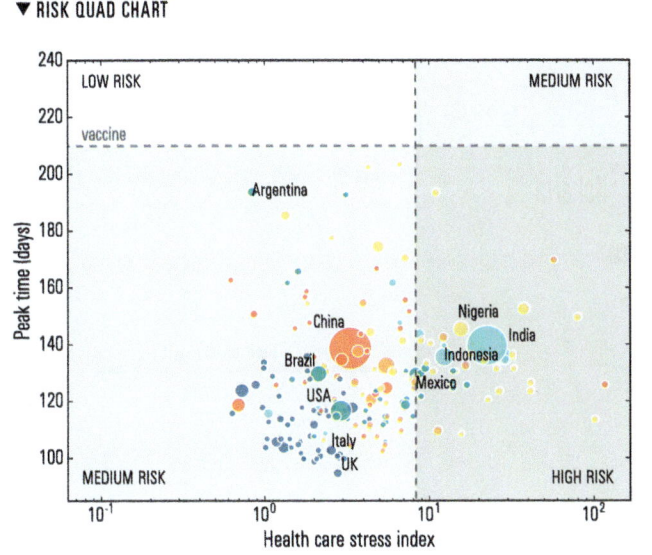

▼ ATTACK RATE COMPOSITION (ONE YEAR)

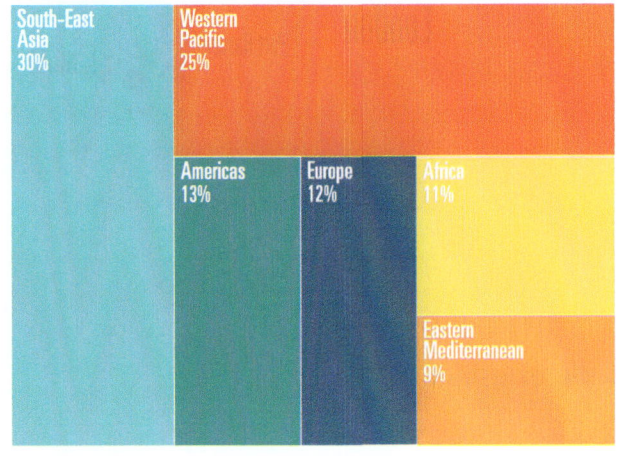

ORIGIN CITY | BARCELONA, SPAIN

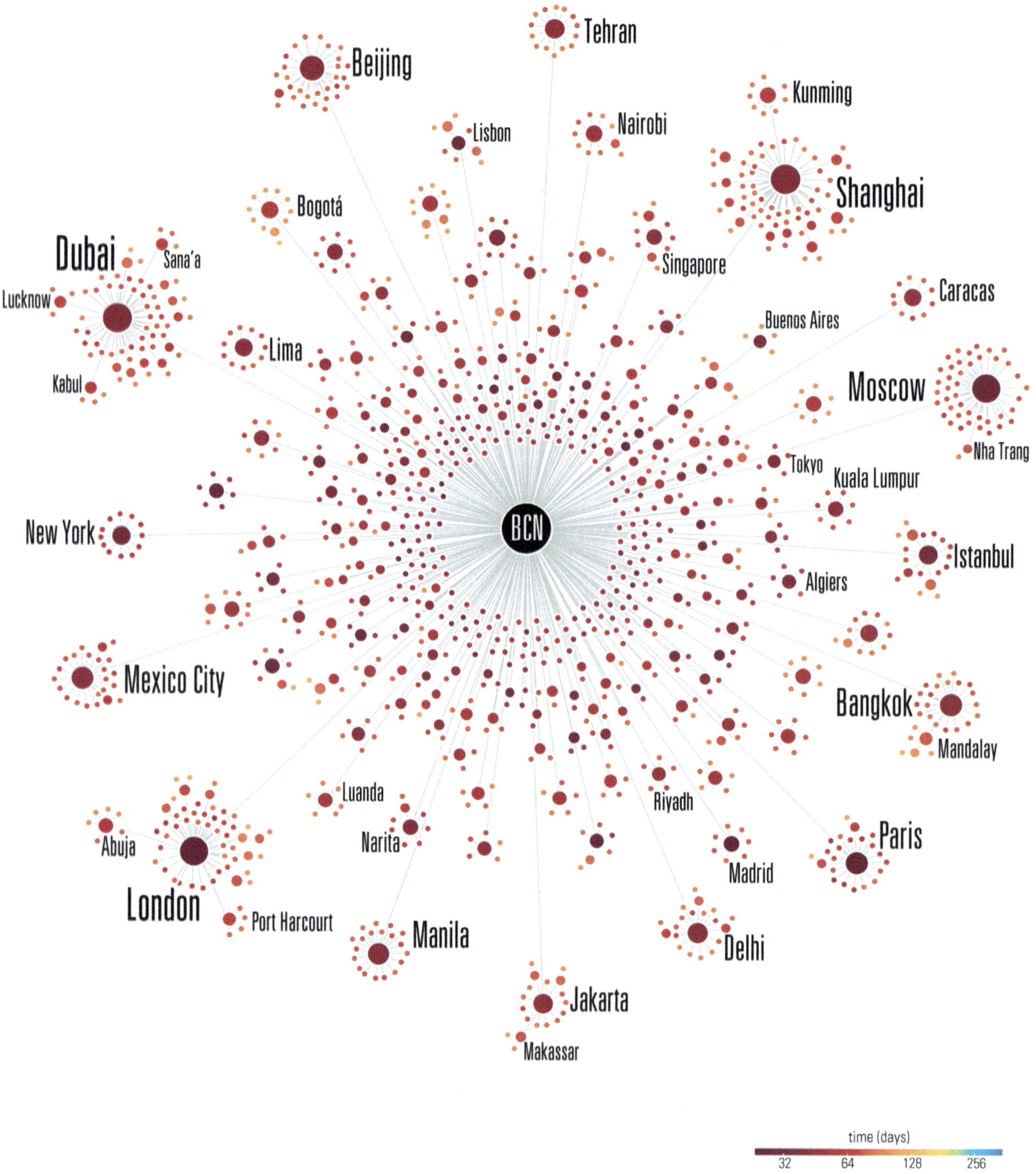

BCN
Tehran
Beijing
Kunming
Lisbon
Nairobi
Shanghai
Bogotá
Dubai
Sana'a
Singapore
Lucknow
Caracas
Buenos Aires
Lima
Moscow
Kabul
Nha Trang
Tokyo
Kuala Lumpur
New York
Istanbul
Algiers
Mexico City
Bangkok
Mandalay
Luanda
Riyadh
Paris
Abuja
Narita
Madrid
London
Port Harcourt
Manila
Delhi
Jakarta
Makassar
time (days)
32
64
128
256

▲ INFECTION TREE GRAPH

WHO REGIONS
- Americas
- Africa
- Europe
- E. Mediterranean
- South-East Asia
- Western Pacific

INITIAL CONDITIONS

80% AIR TRAFFIC REDUCTION

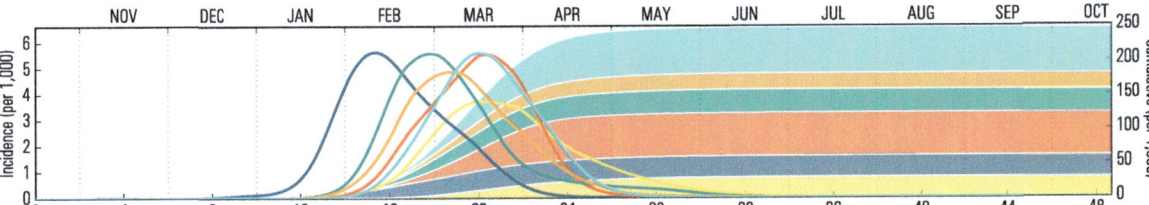

▲ CUMULATIVE INCIDENCE AND EPIDEMIC PEAKS

▼ ATTACK RATE COMPOSITION (ONE YEAR)

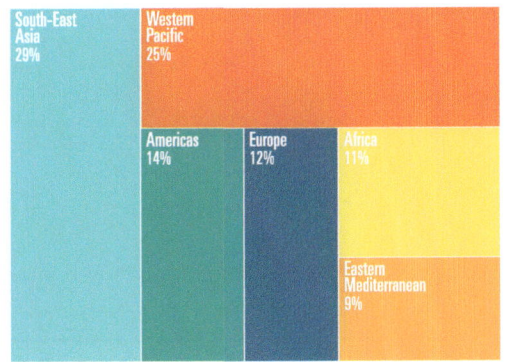

South-East Asia 29%
Western Pacific 25%
Americas 14%
Europe 12%
Africa 11%
Eastern Mediterranean 9%

▼ EPIDEMIC ACTIVITY

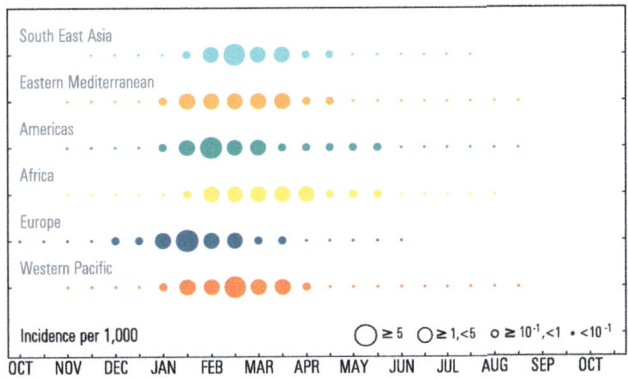

South East Asia
Eastern Mediterranean
Americas
Africa
Europe
Western Pacific

Incidence per 1,000   ○ ≥ 5   ○ ≥ 1,<5   ○ ≥ 10⁻¹,<1   • <10⁻¹

VACCINATIONS

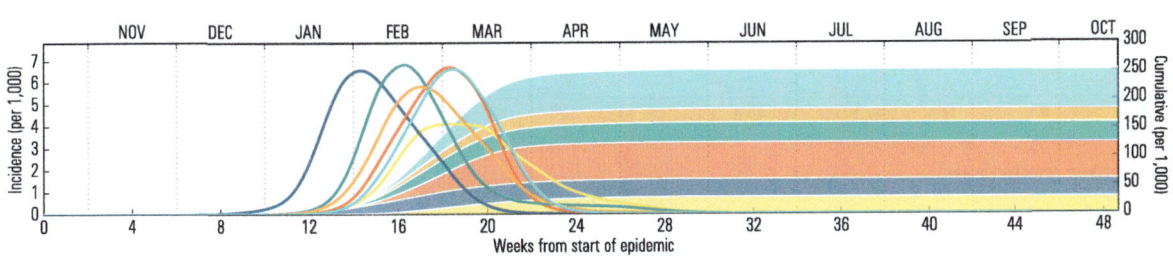

▲ CUMULATIVE INCIDENCE AND EPIDEMIC PEAKS

▼ ATTACK RATE COMPOSITION (ONE YEAR)

South-East Asia 30%
Western Pacific 25%
Americas 13%
Europe 12%
Africa 11%
Eastern Mediterranean 9%

▼ EPIDEMIC ACTIVITY

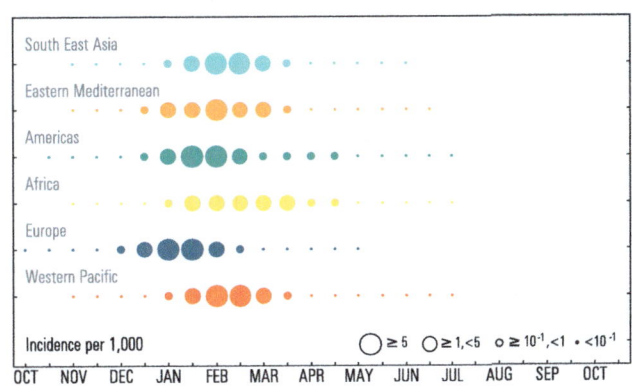

South East Asia
Eastern Mediterranean
Americas
Africa
Europe
Western Pacific

Incidence per 1,000   ○ ≥ 5   ○ ≥ 1,<5   ○ ≥ 10⁻¹,<1   • <10⁻¹

ORIGIN CITY

BUENOS AIRES, ARGENTINA

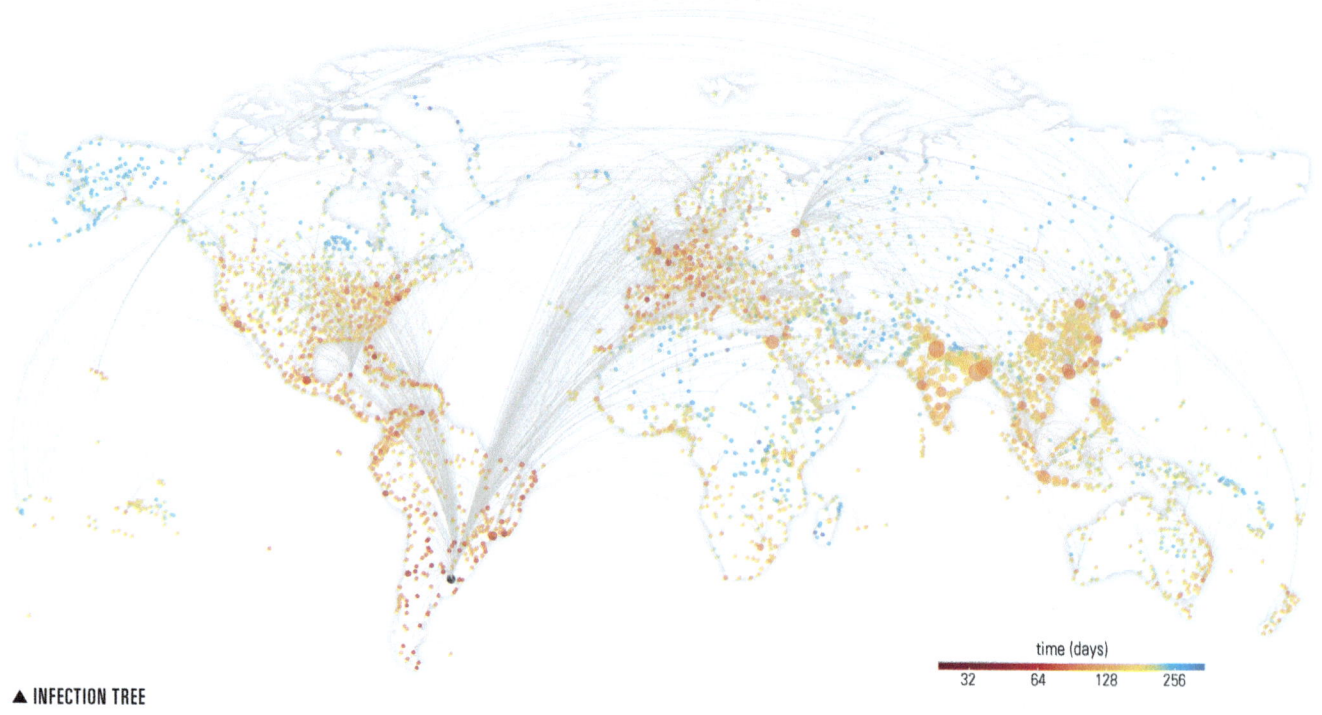

time (days)

| 32 | 64 | 128 | 256 |

▲ INFECTION TREE

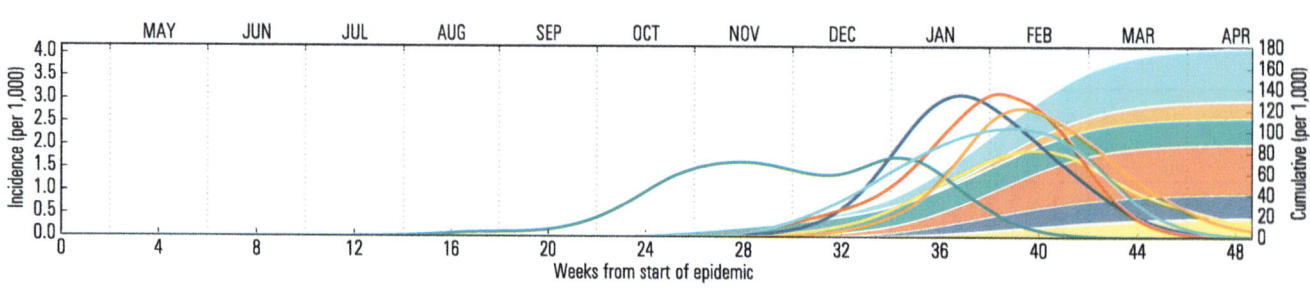

▲ CUMULATIVE INCIDENCE AND EPIDEMIC PEAKS

▼ AFFECTED COUNTRIES

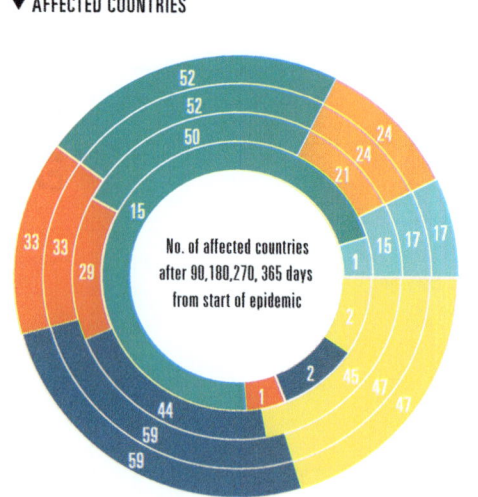

No. of affected countries after 90, 180, 270, 365 days from start of epidemic

▼ EPIDEMIC ACTIVITY

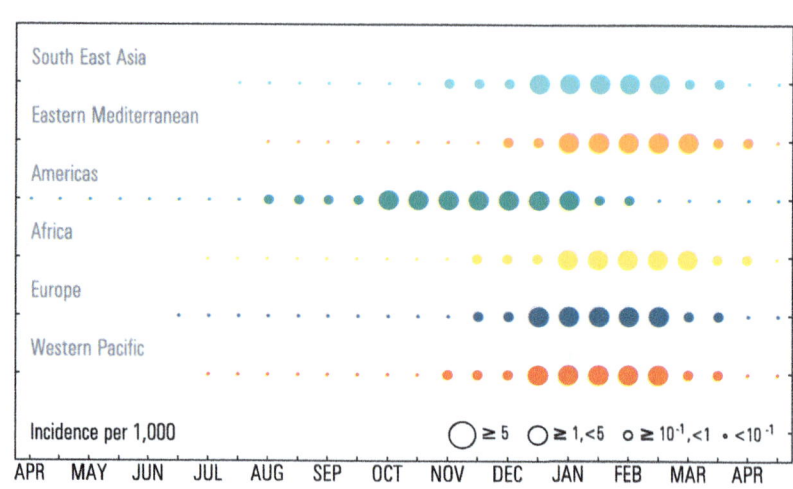

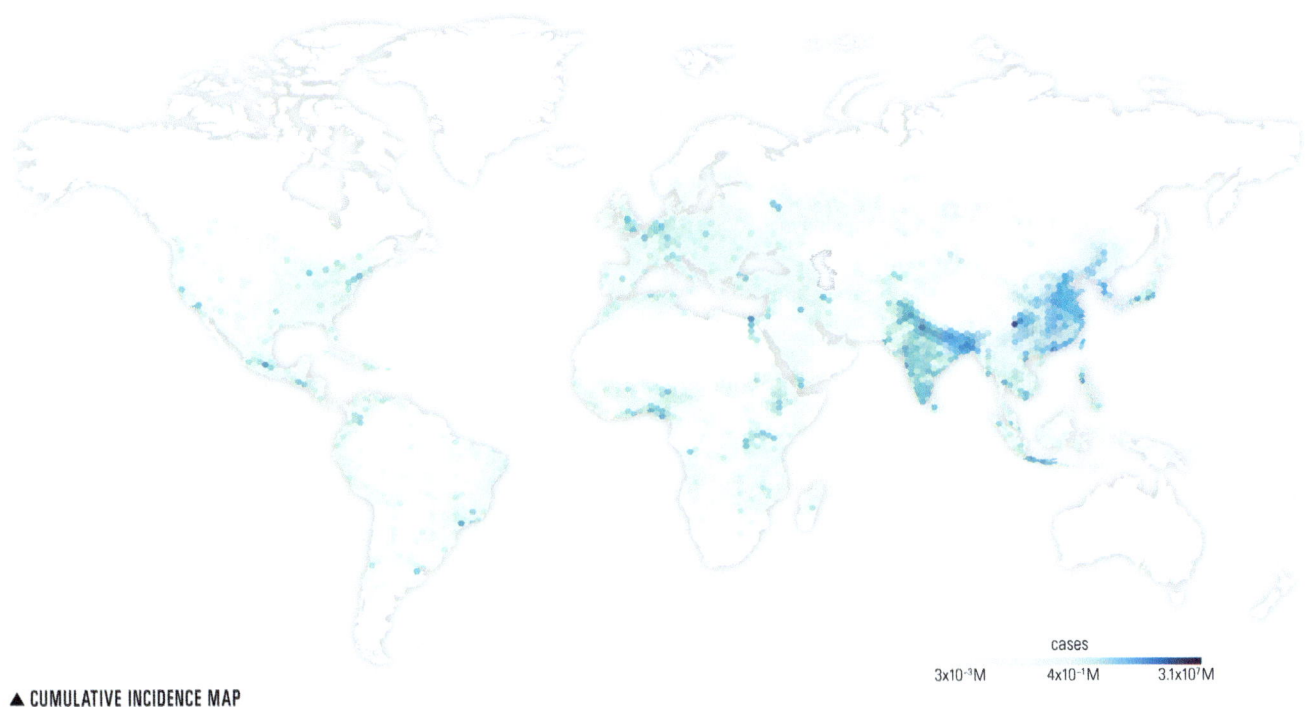

cases
3x10⁻³M   4x10⁻¹M   3.1x10⁷M

▲ CUMULATIVE INCIDENCE MAP

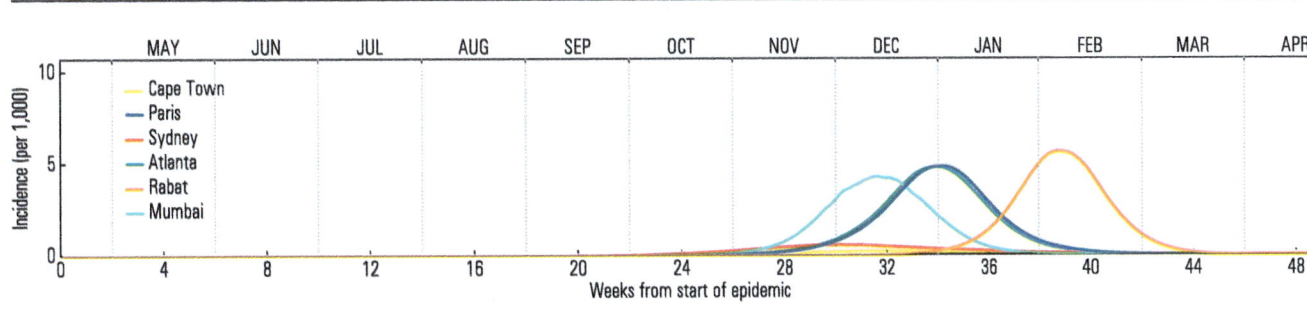

▲ URBAN LEVEL INCIDENCE

▼ RISK QUAD CHART

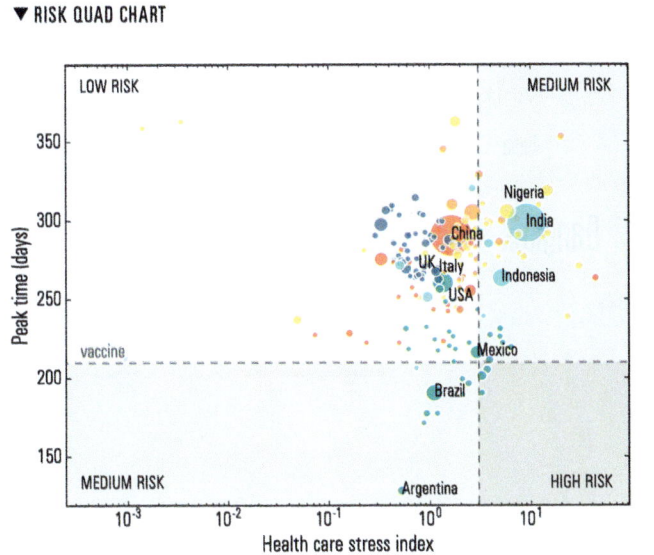

▼ ATTACK RATE COMPOSITION (ONE YEAR)

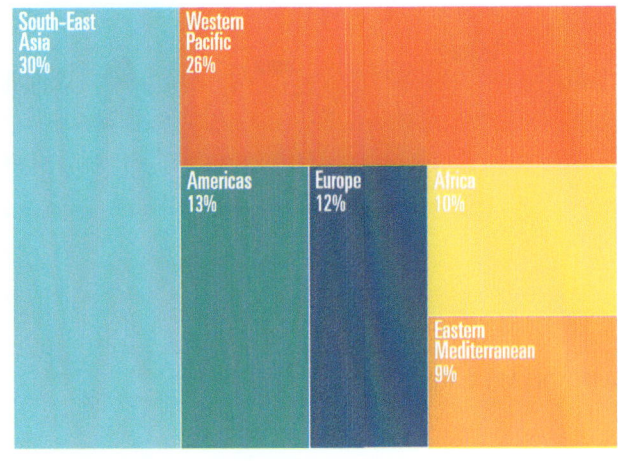

BUENOS AIRES, ARGENTINA

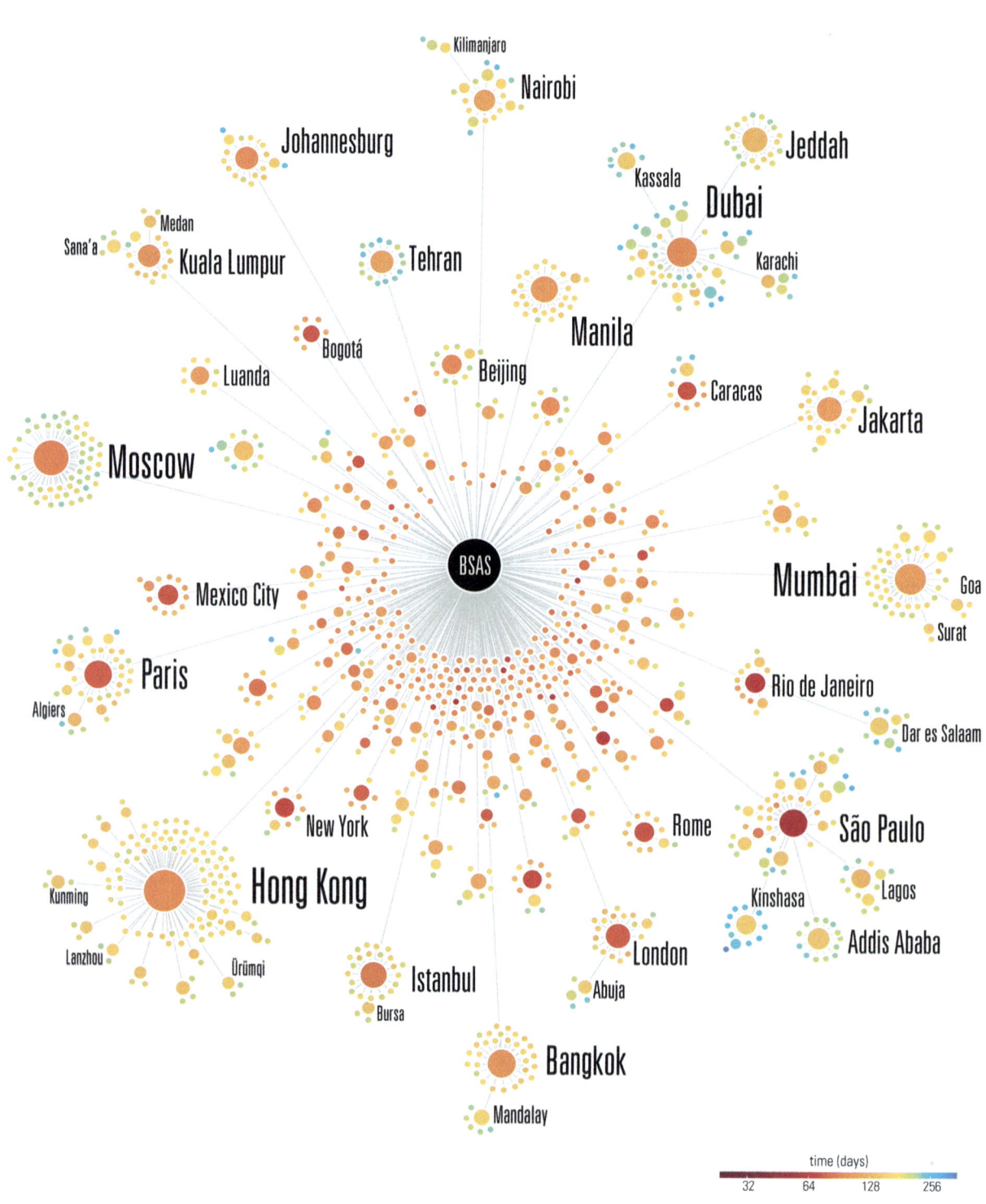

Kilimanjaro
Nairobi
Johannesburg
Jeddah
Kassala
Dubai
Medan
Sana'a
Kuala Lumpur
Tehran
Karachi
Bogotá
Manila
Luanda
Beijing
Caracas
Jakarta
Moscow
Mumbai
Goa
Mexico City
Surat
Paris
Rio de Janeiro
Algiers
Dar es Salaam
BSAS
New York
Rome
São Paulo
Lagos
Hong Kong
Kunming
Kinshasa
Addis Ababa
Lanzhou
London
Ürümqi
Istanbul
Abuja
Bursa
Bangkok
Mandalay

time (days)
32   64   128   256

▲ INFECTION TREE GRAPH

- Americas
- Africa
- Europe
- E. Mediterranean
- South-East Asia
- Western Pacific

INITIAL CONDITIONS

 APR  R₀ 1.5

80% AIR TRAFFIC REDUCTION

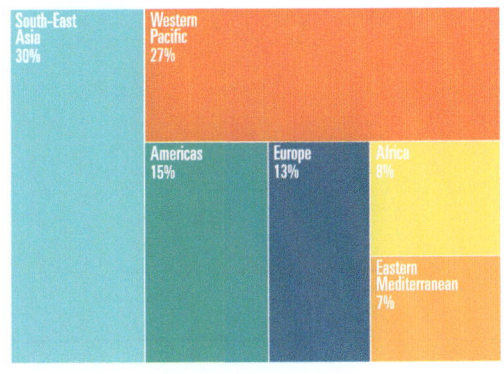

▲ CUMULATIVE INCIDENCE AND EPIDEMIC PEAKS

▼ ATTACK RATE COMPOSITION (ONE YEAR)

South-East Asia 30%
Western Pacific 27%
Americas 15%
Europe 13%
Africa 8%
Eastern Mediterranean 7%

▼ EPIDEMIC ACTIVITY

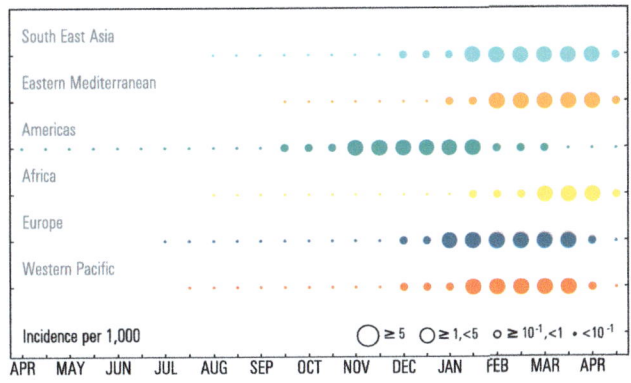

VACCINATIONS

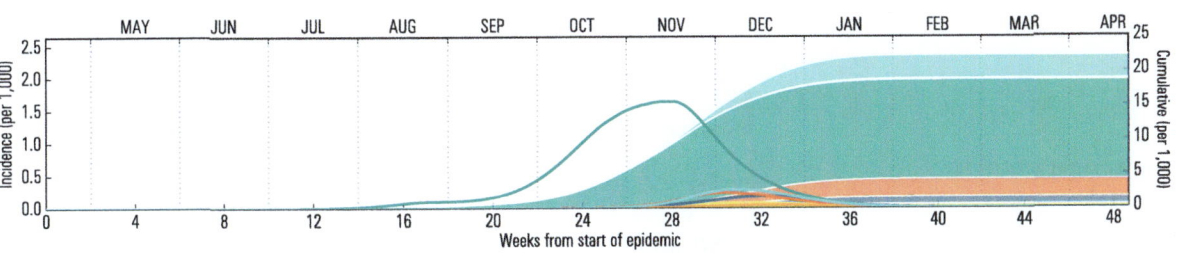

▲ CUMULATIVE INCIDENCE AND EPIDEMIC PEAKS

▼ ATTACK RATE COMPOSITION (ONE YEAR)

Americas 67%
South-East Asia 15%
Western Pacific 12%
Europe 4%
Africa 2%
<1%

▼ EPIDEMIC ACTIVITY

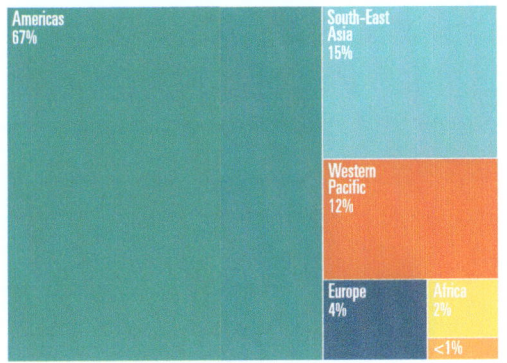

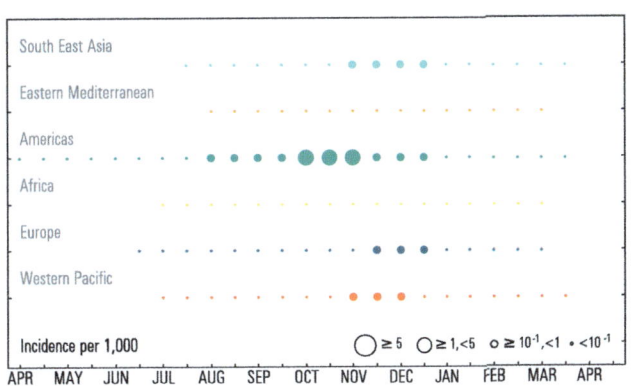

BUENOS AIRES, ARGENTINA

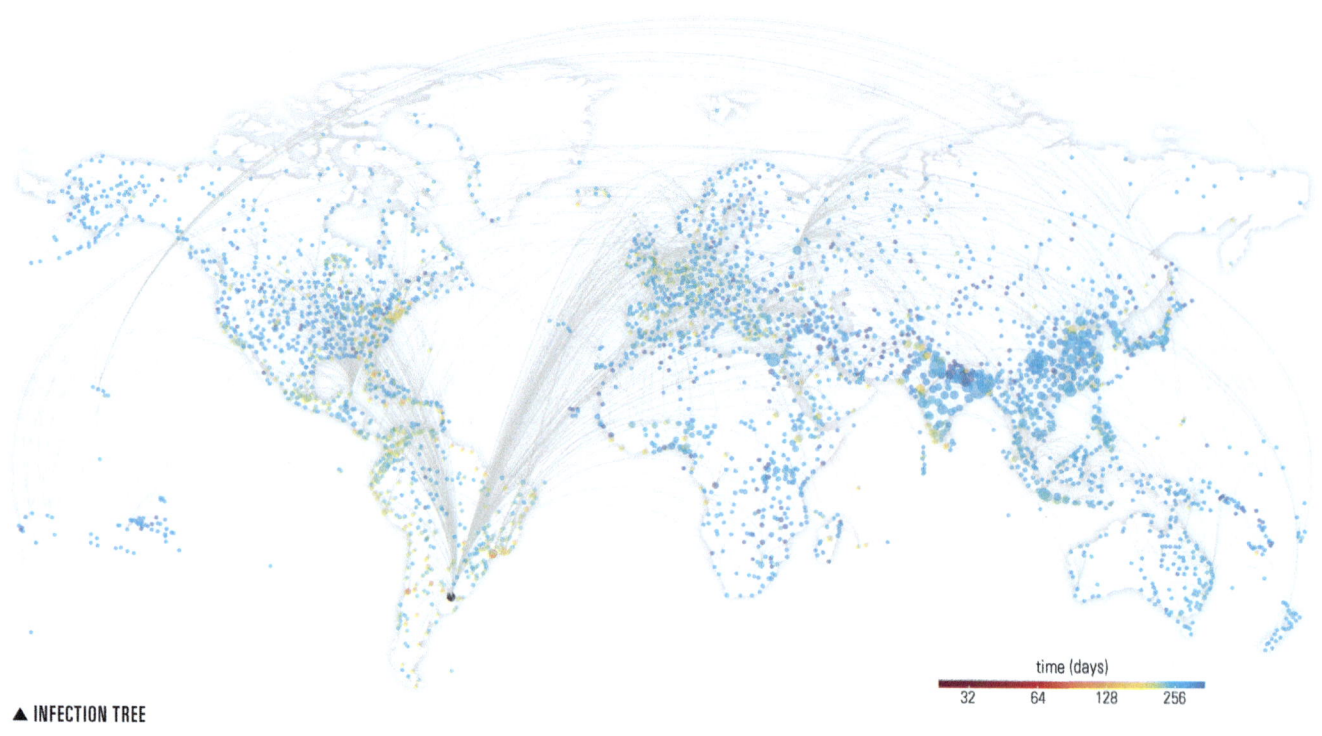

time (days)

32   64   128   256

▲ INFECTION TREE

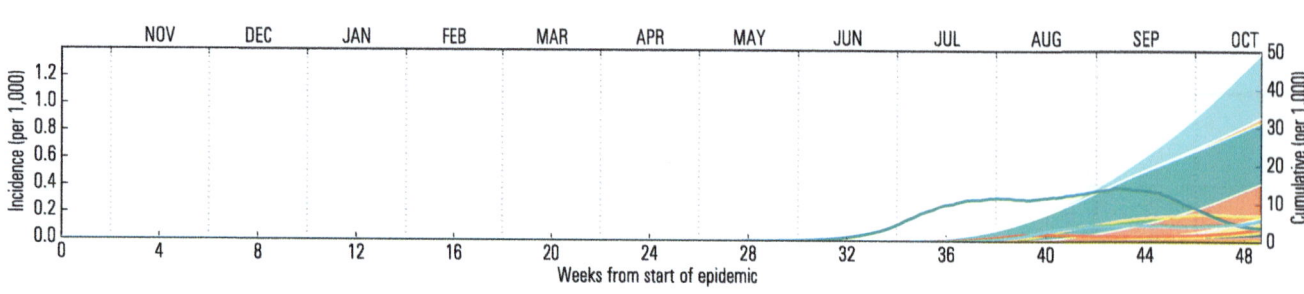

▲ CUMULATIVE INCIDENCE AND EPIDEMIC PEAKS

▼ AFFECTED COUNTRIES

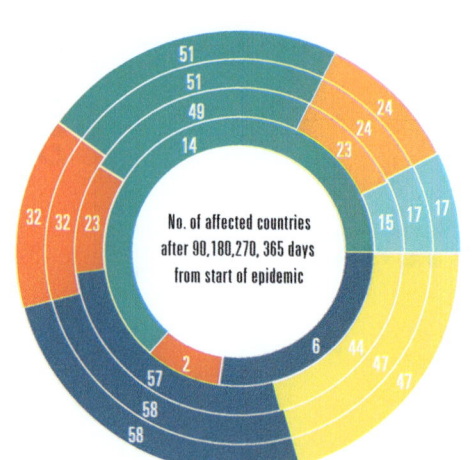

No. of affected countries
after 90,180,270, 365 days
from start of epidemic

▼ EPIDEMIC ACTIVITY

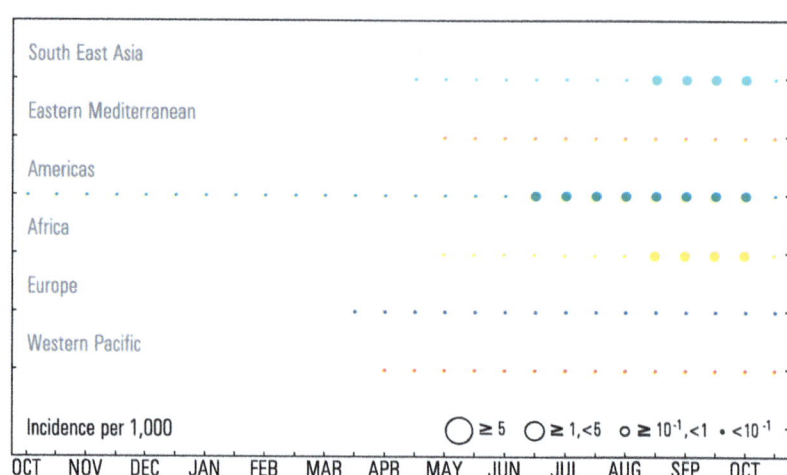

South East Asia

Eastern Mediterranean

Americas

Africa

Europe

Western Pacific

Incidence per 1,000   ◯ ≥ 5   ◯ ≥ 1,<5   ○ ≥ 10⁻¹,<1   • <10⁻¹

OCT   NOV   DEC   JAN   FEB   MAR   APR   MAY   JUN   JUL   AUG   SEP   OCT

WHO
REGIONS

■ Americas   ■ E. Mediterranean
■ Africa   ■ South-East Asia
■ Europe   ■ Western Pacific

INITIAL
CONDITIONS

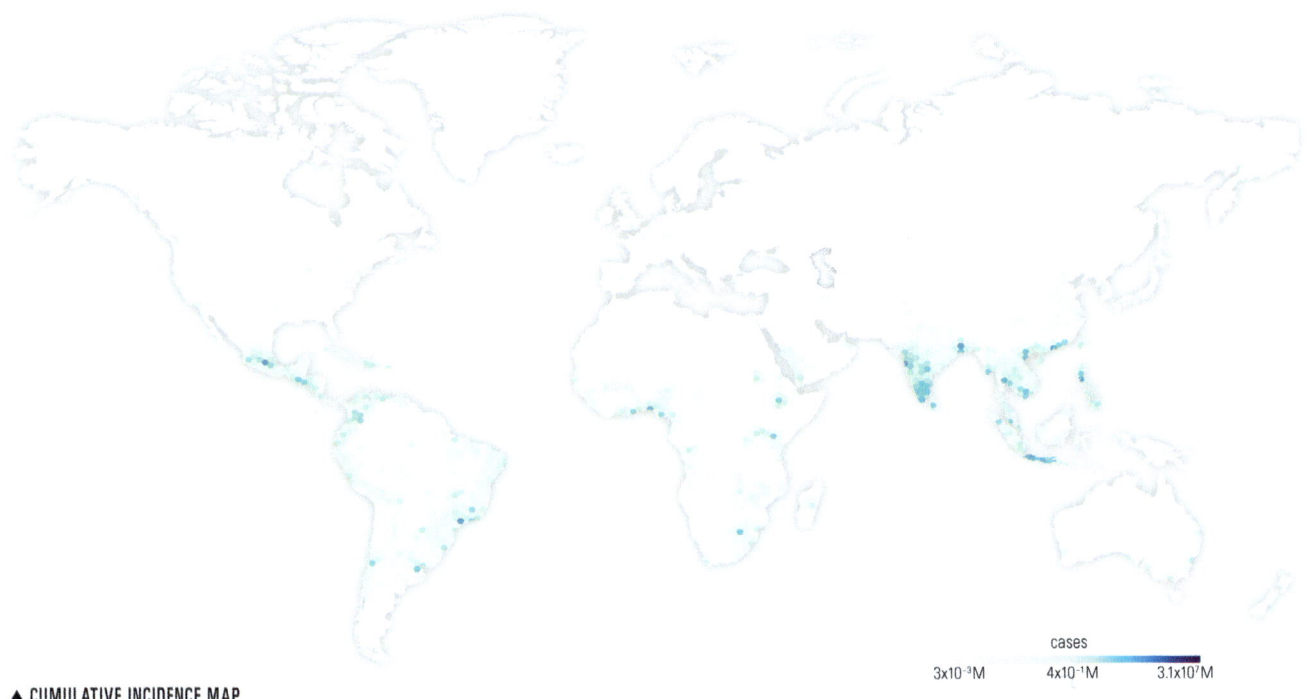

cases

3x10⁻³M   4x10⁻¹M   3.1x10⁷M

▲ CUMULATIVE INCIDENCE MAP

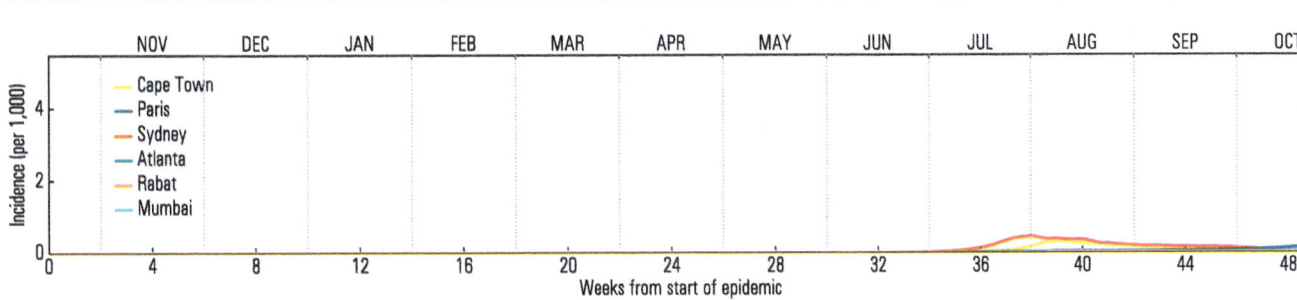

— Cape Town
— Paris
— Sydney
— Atlanta
— Rabat
— Mumbai

Incidence (per 1,000)

Weeks from start of epidemic

▲ URBAN LEVEL INCIDENCE

▼ RISK QUAD CHART

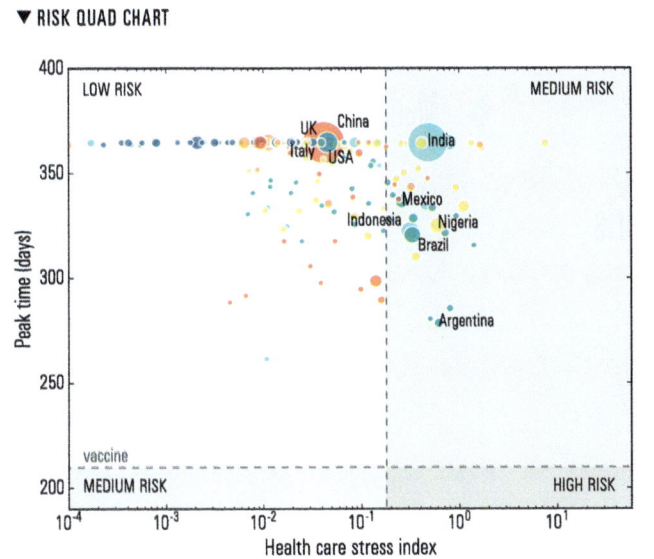

Peak time (days)

LOW RISK   MEDIUM RISK

UK   China
Italy USA   India

Mexico
Indonesia   Nigeria
            Brazil

Argentina

vaccine
MEDIUM RISK   HIGH RISK

Health care stress index

▼ ATTACK RATE COMPOSITION (ONE YEAR)

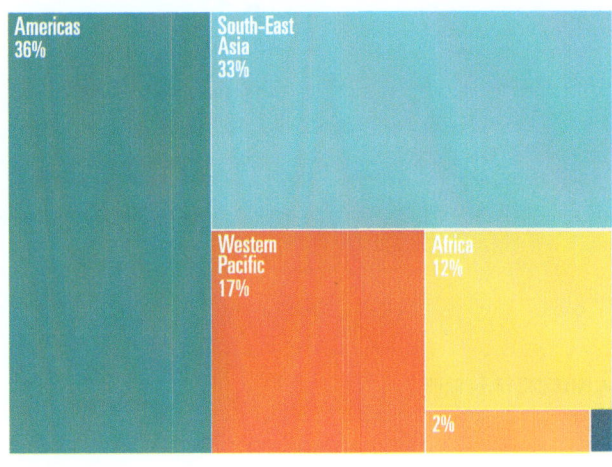

Americas
36%

South-East
Asia
33%

Western
Pacific
17%

Africa
12%

2%

BUENOS AIRES, ARGENTINA

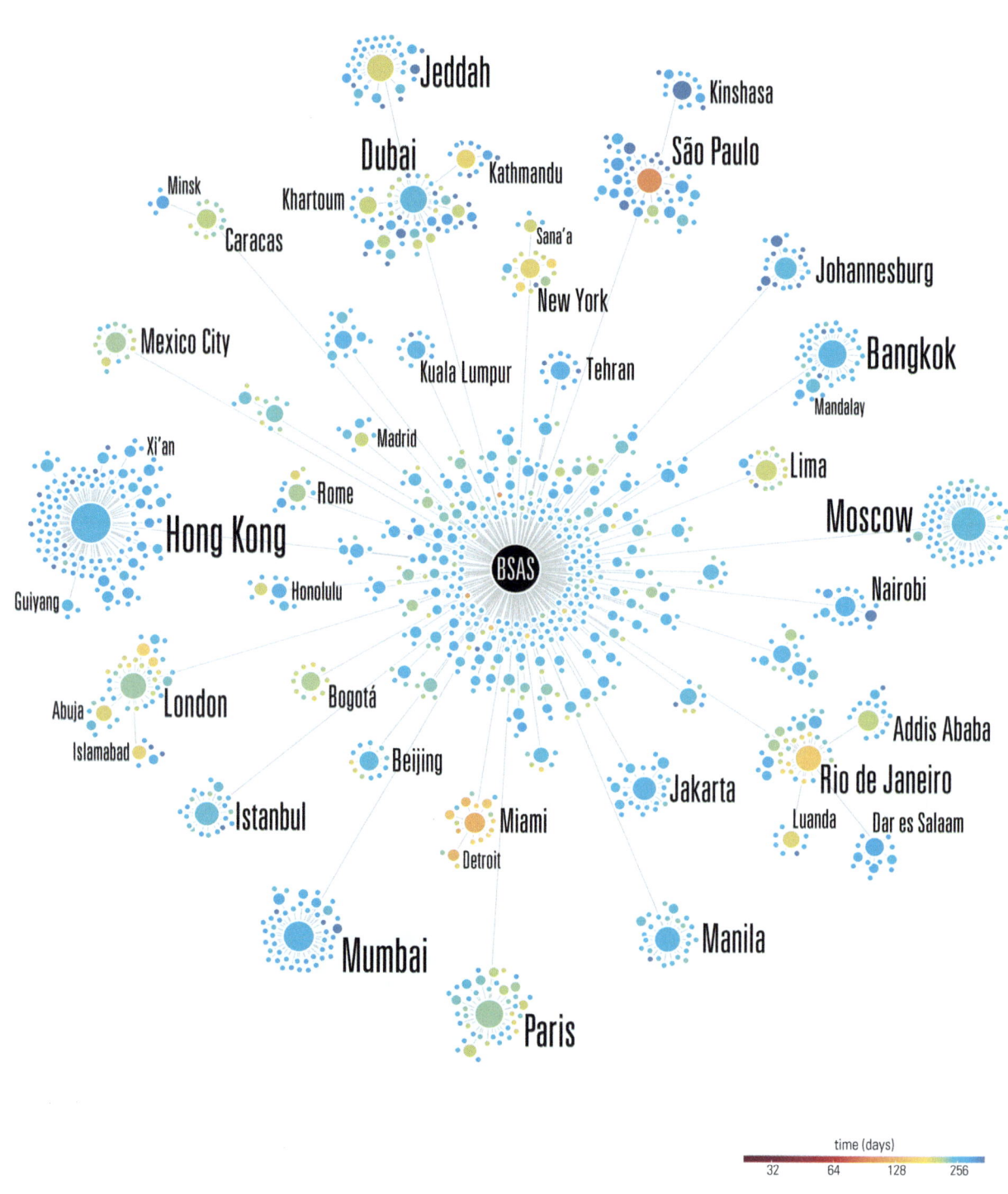

time (days)

32   64   128   256

▲ INFECTION TREE GRAPH

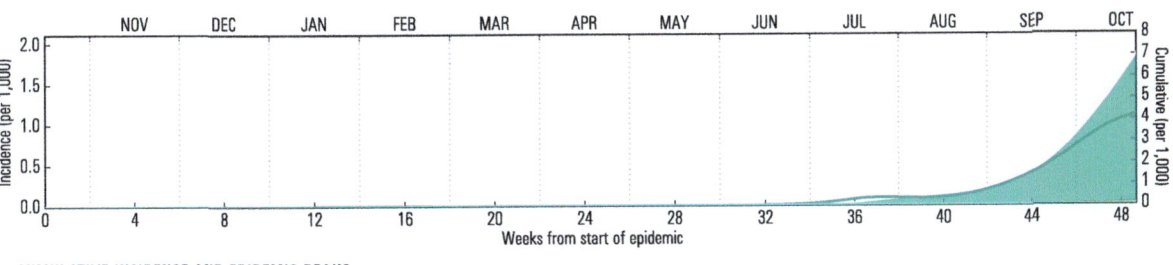

▲ CUMULATIVE INCIDENCE AND EPIDEMIC PEAKS

▼ ATTACK RATE COMPOSITION (ONE YEAR)

▼ EPIDEMIC ACTIVITY

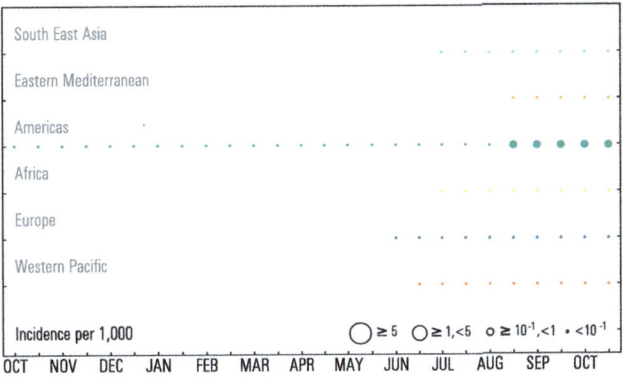

80% AIR TRAFFIC REDUCTION

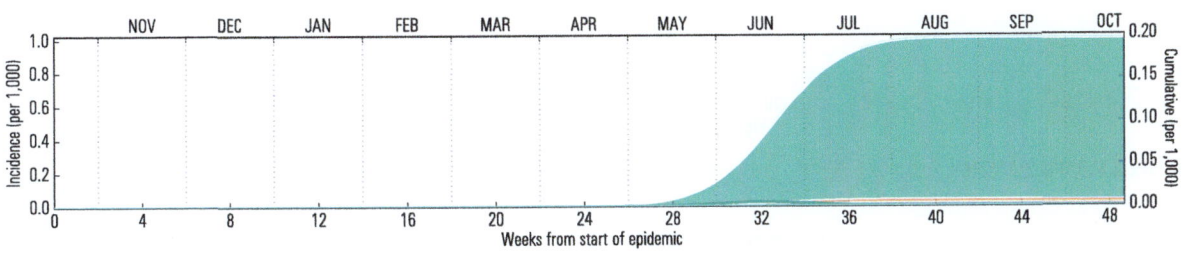

▲ CUMULATIVE INCIDENCE AND EPIDEMIC PEAKS

▼ ATTACK RATE COMPOSITION (ONE YEAR)

▼ EPIDEMIC ACTIVITY

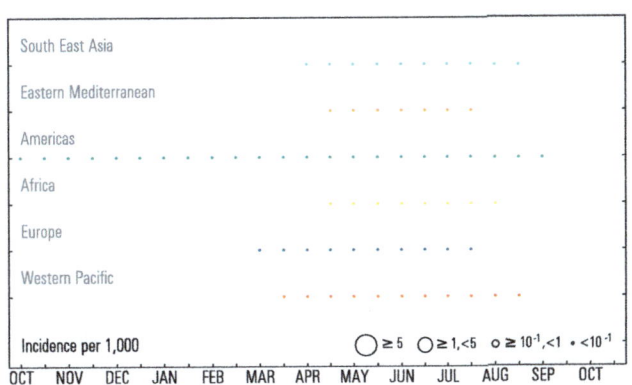

VACCINATIONS

BUENOS AIRES, ARGENTINA

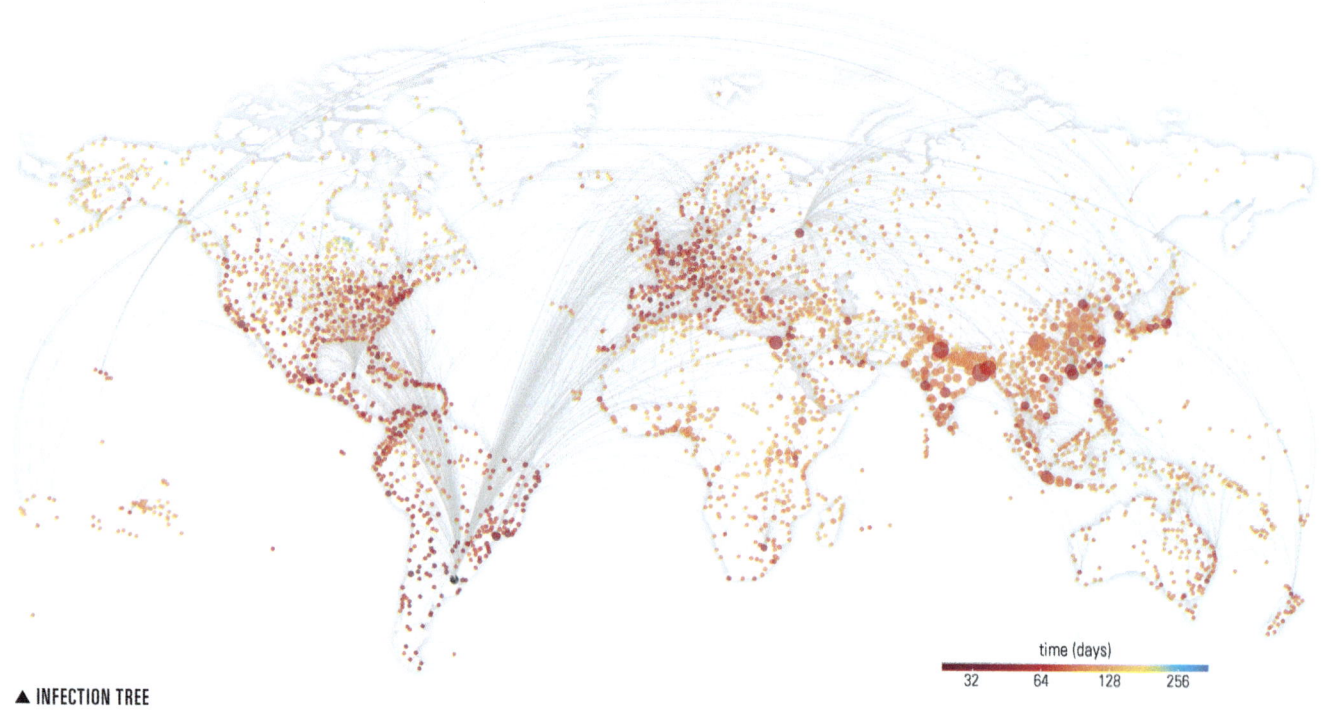

▲ INFECTION TREE

time (days)
32   64   128   256

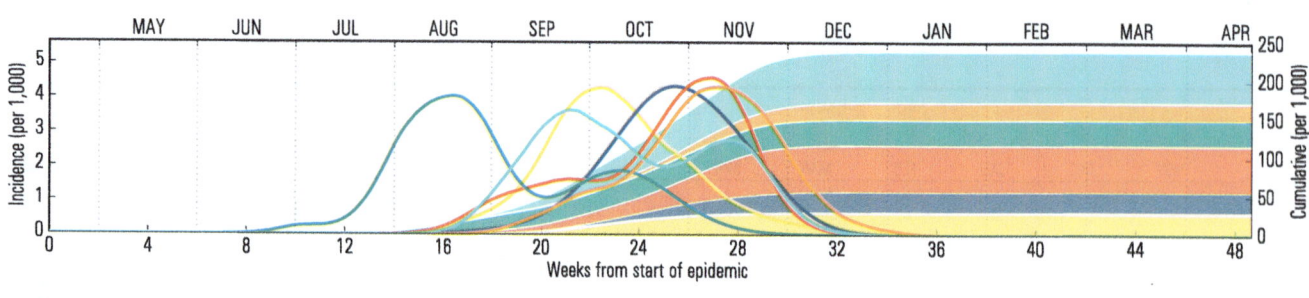

▲ CUMULATIVE INCIDENCE AND EPIDEMIC PEAKS

▼ AFFECTED COUNTRIES

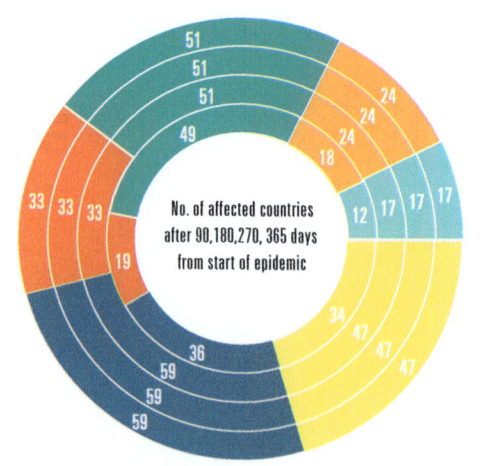

No. of affected countries after 90, 180, 270, 365 days from start of epidemic

▼ EPIDEMIC ACTIVITY

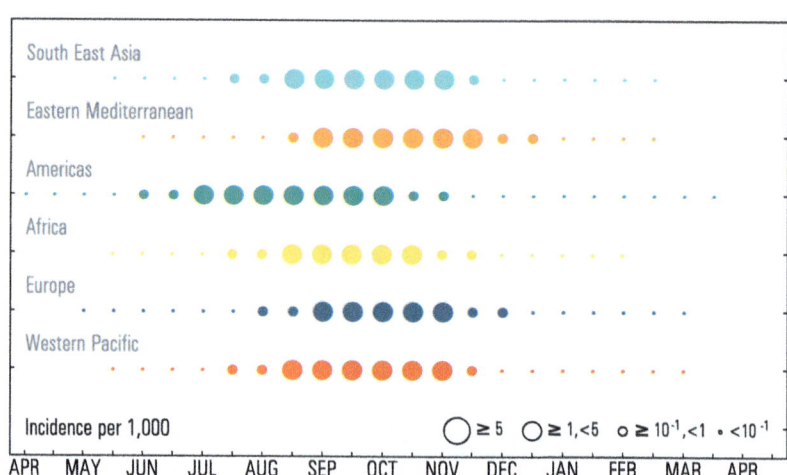

WHO
REGIONS

■ Americas ■ E. Mediterranean
■ Africa   ■ South-East Asia
■ Europe   ■ Western Pacific

INITIAL
CONDITIONS

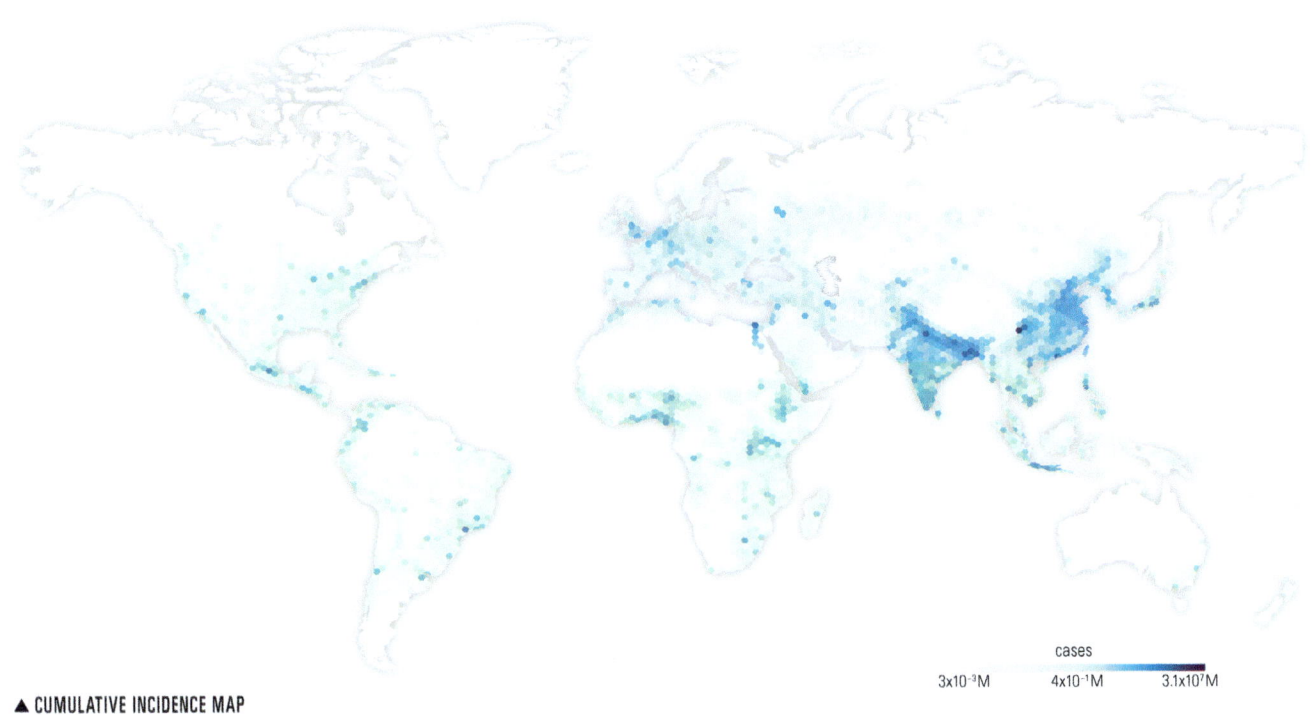

cases

3x10⁻³M   4x10⁻¹M   3.1x10⁷M

▲ CUMULATIVE INCIDENCE MAP

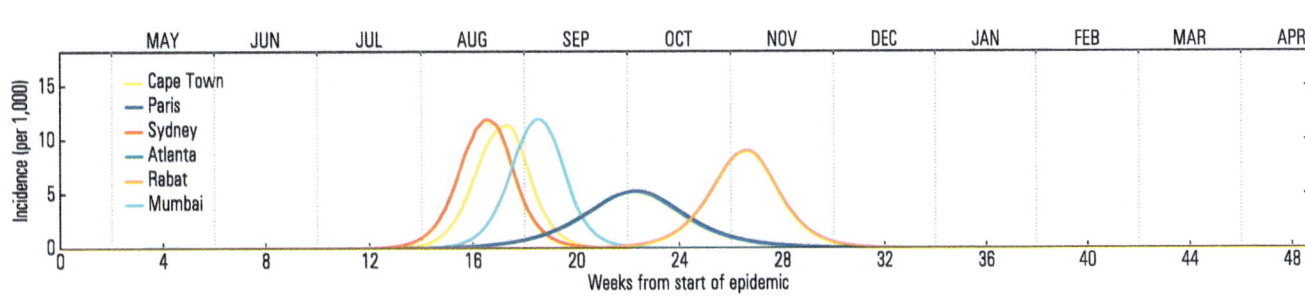

▲ URBAN LEVEL INCIDENCE

▼ RISK QUAD CHART

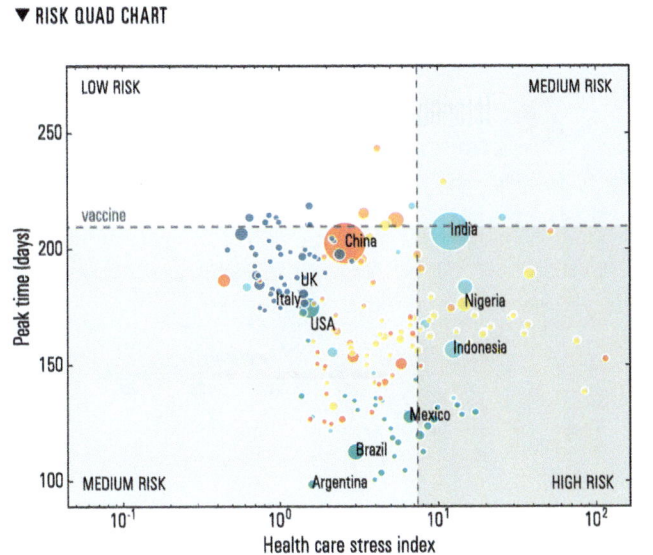

▼ ATTACK RATE COMPOSITION (ONE YEAR)

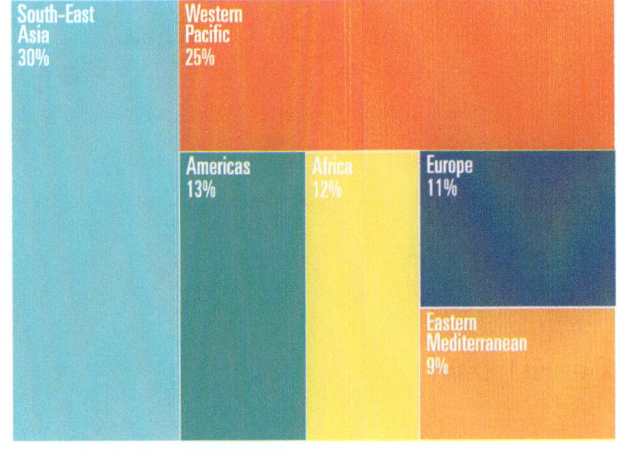

BUENOS AIRES, ARGENTINA

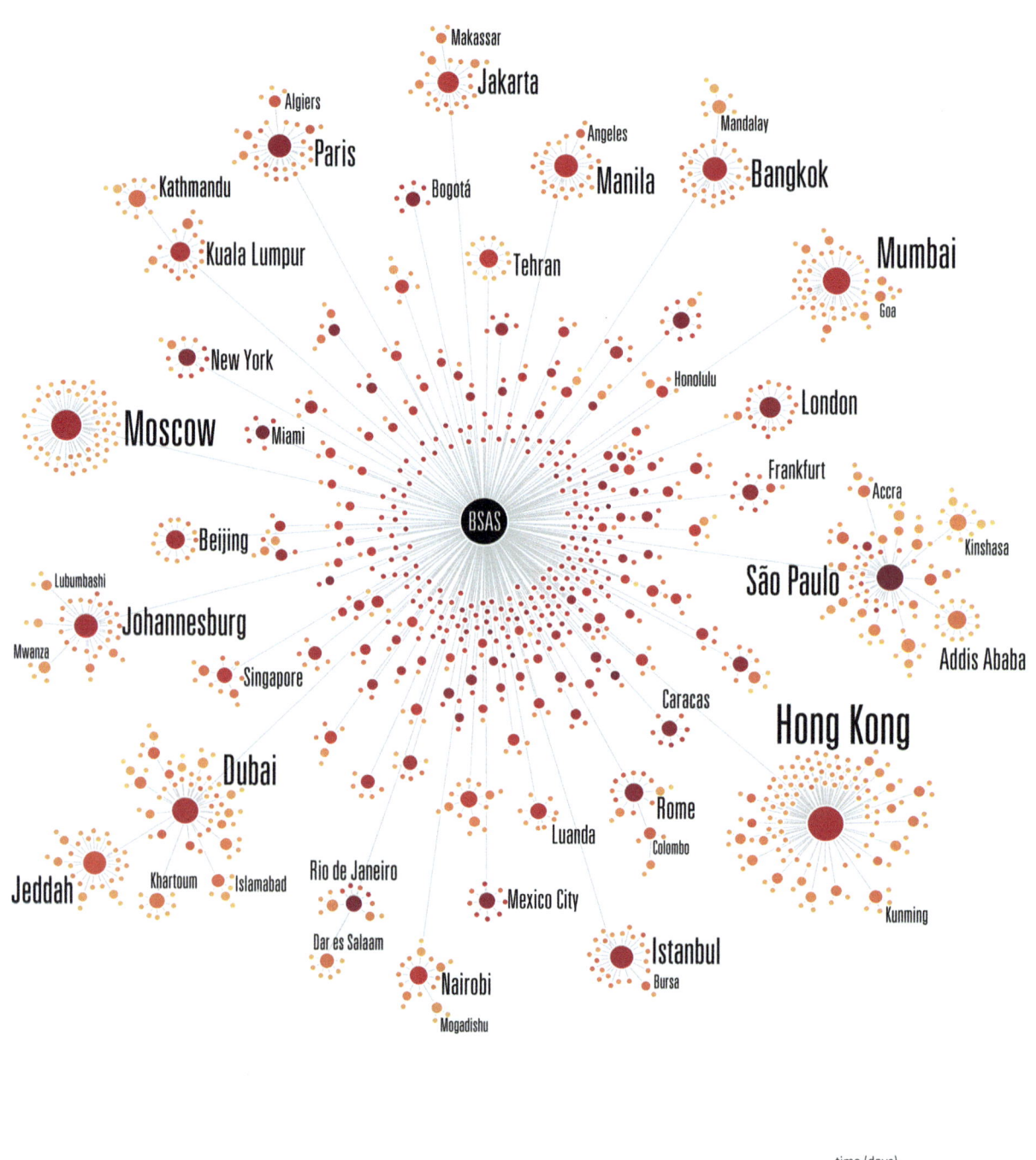

Makassar

Jakarta

Algiers

Paris

Mandalay

Angeles

Manila

Bangkok

Kathmandu

Bogotá

Mumbai

Kuala Lumpur

Goa

Tehran

New York

Honolulu

London

Moscow

Miami

Frankfurt

Accra

Beijing

BSAS

Kinshasa

São Paulo

Lubumbashi

Johannesburg

Addis Ababa

Mwanza

Singapore

Hong Kong

Caracas

Dubai

Rome

Luanda

Colombo

Kunming

Jeddah

Khartoum

Islamabad

Rio de Janeiro

Mexico City

Dar es Salaam

Istanbul

Nairobi

Bursa

Mogadishu

time (days)

32   64   128   256

▲ INFECTION TREE GRAPH

- Americas
- Africa
- Europe
- E. Mediterranean
- South-East Asia
- Western Pacific

INITIAL CONDITIONS

 APR  R₀ 2.0

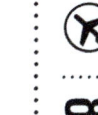 80% AIR TRAFFIC REDUCTION

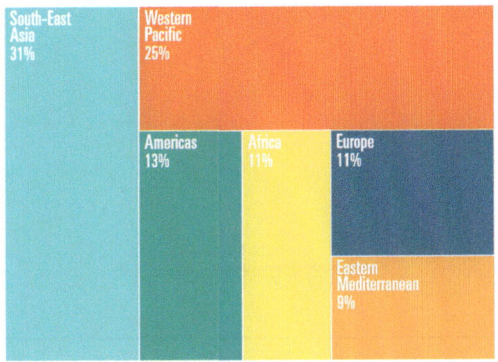

▲ CUMULATIVE INCIDENCE AND EPIDEMIC PEAKS

▼ ATTACK RATE COMPOSITION (ONE YEAR)

South-East Asia 31%
Western Pacific 25%
Americas 13%
Africa 11%
Europe 11%
Eastern Mediterranean 9%

▼ EPIDEMIC ACTIVITY

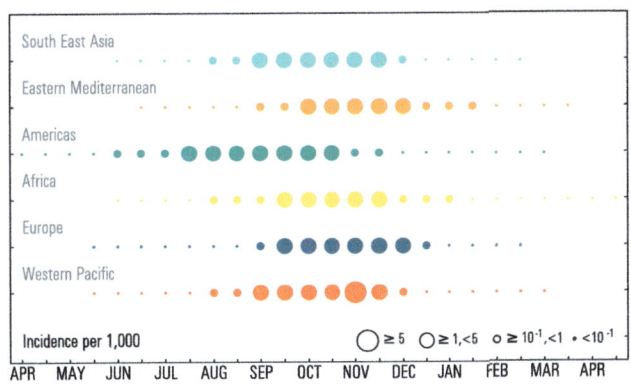

Incidence per 1,000   ◯ ≥ 5   ◯ ≥ 1,<5   ○ ≥ 10⁻¹,<1   • <10⁻¹

 VACCINATIONS

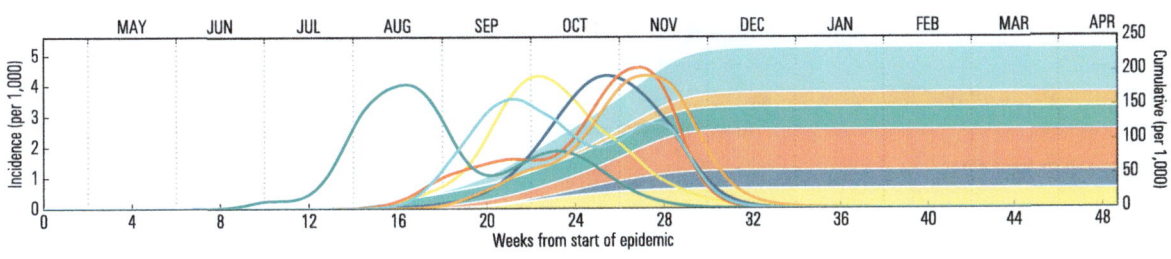

▲ CUMULATIVE INCIDENCE AND EPIDEMIC PEAKS

▼ ATTACK RATE COMPOSITION (ONE YEAR)

South-East Asia 30%
Western Pacific 25%
Americas 14%
Africa 12%
Europe 11%
Eastern Mediterranean 8%

▼ EPIDEMIC ACTIVITY

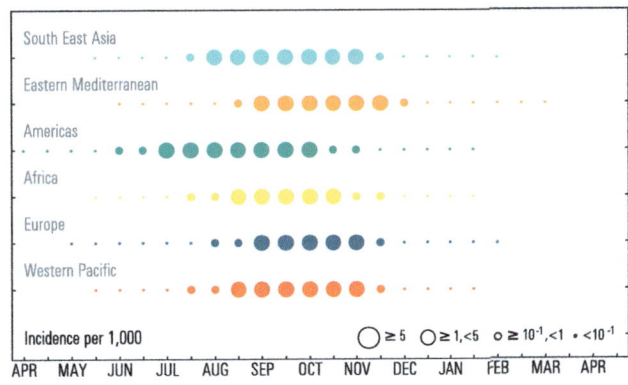

Incidence per 1,000   ◯ ≥ 5   ◯ ≥ 1,<5   ○ ≥ 10⁻¹,<1   • <10⁻¹

 ORIGIN CITY

BUENOS AIRES, ARGENTINA

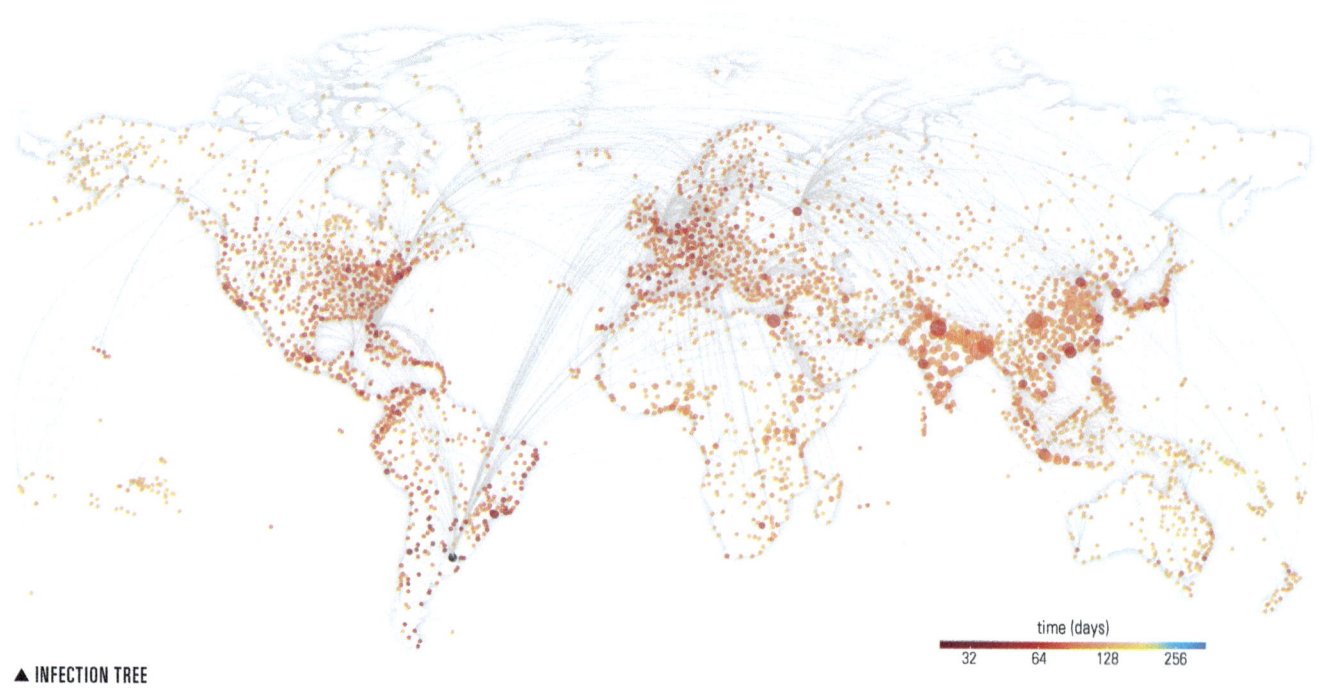

time (days)

32   64   128   256

▲ INFECTION TREE

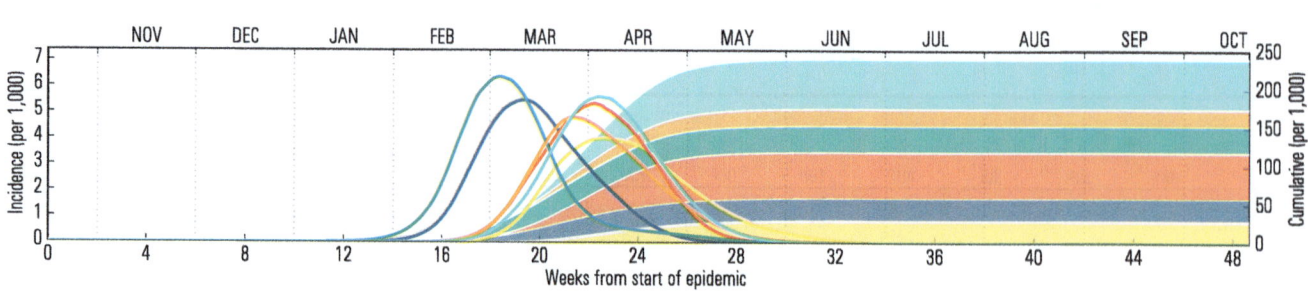

▲ CUMULATIVE INCIDENCE AND EPIDEMIC PEAKS

▼ AFFECTED COUNTRIES

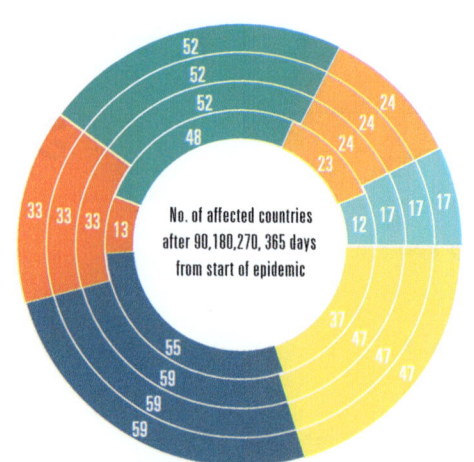

No. of affected countries
after 90,180,270, 365 days
from start of epidemic

▼ EPIDEMIC ACTIVITY

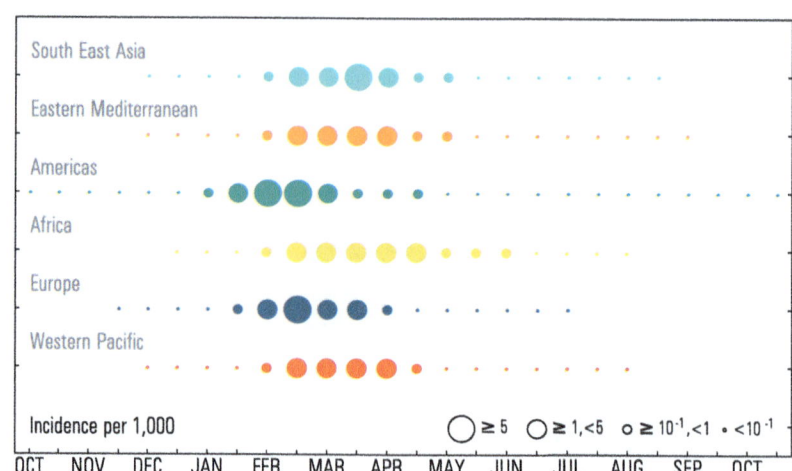

INITIAL CONDITIONS

OCT   2.0

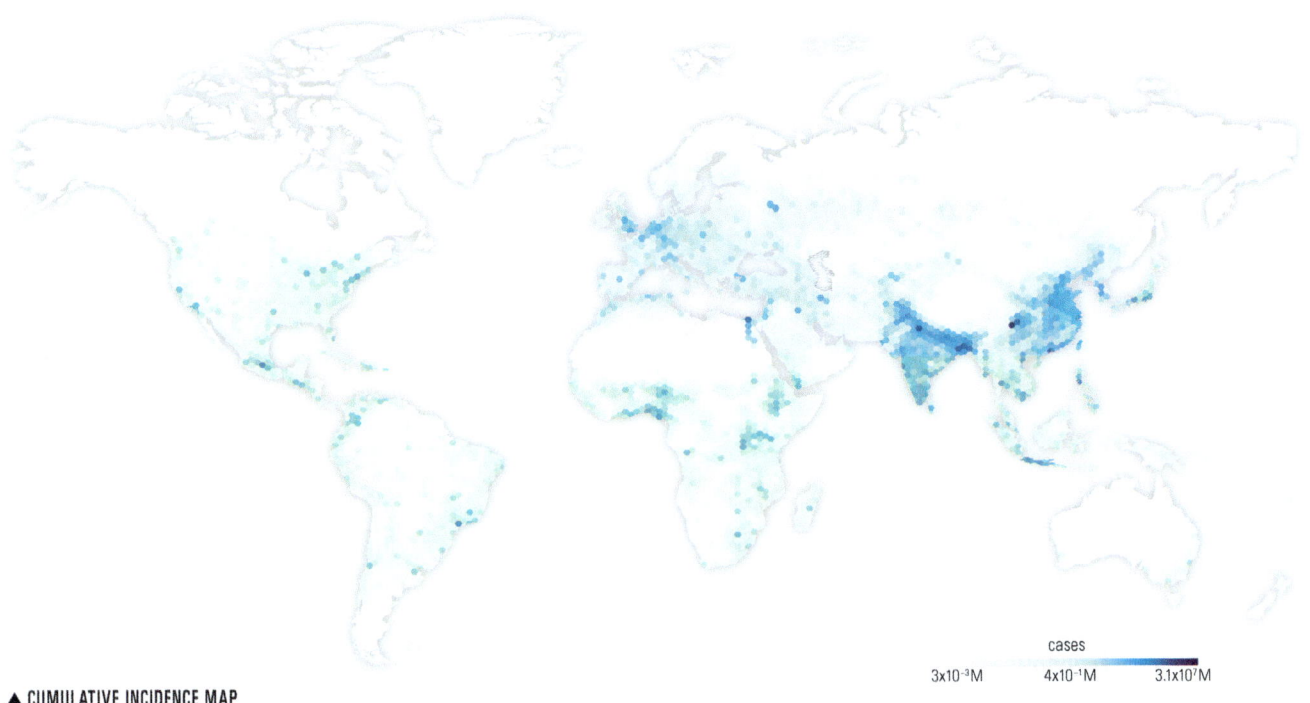

cases

3x10⁻³M   4x10⁻¹M   3.1x10⁷M

▲ CUMULATIVE INCIDENCE MAP

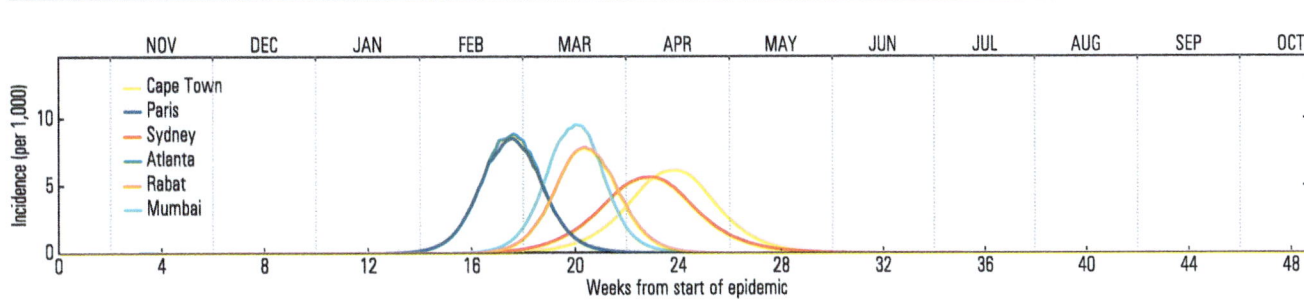

▲ URBAN LEVEL INCIDENCE

▼ RISK QUAD CHART

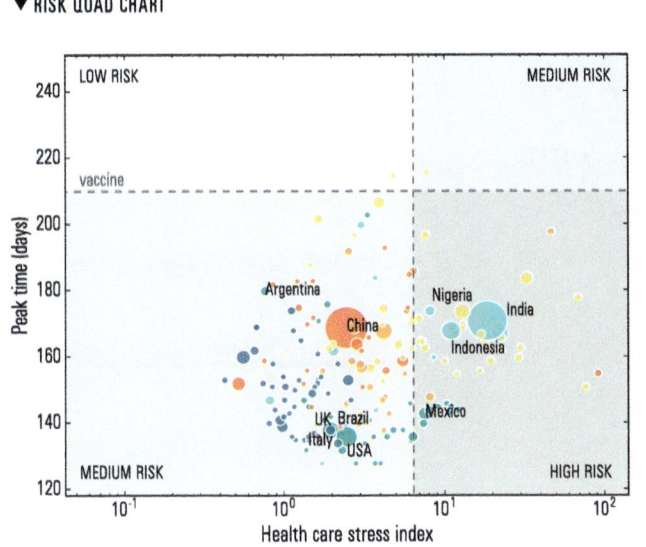

▼ ATTACK RATE COMPOSITION (ONE YEAR)

BUENOS AIRES, ARGENTINA

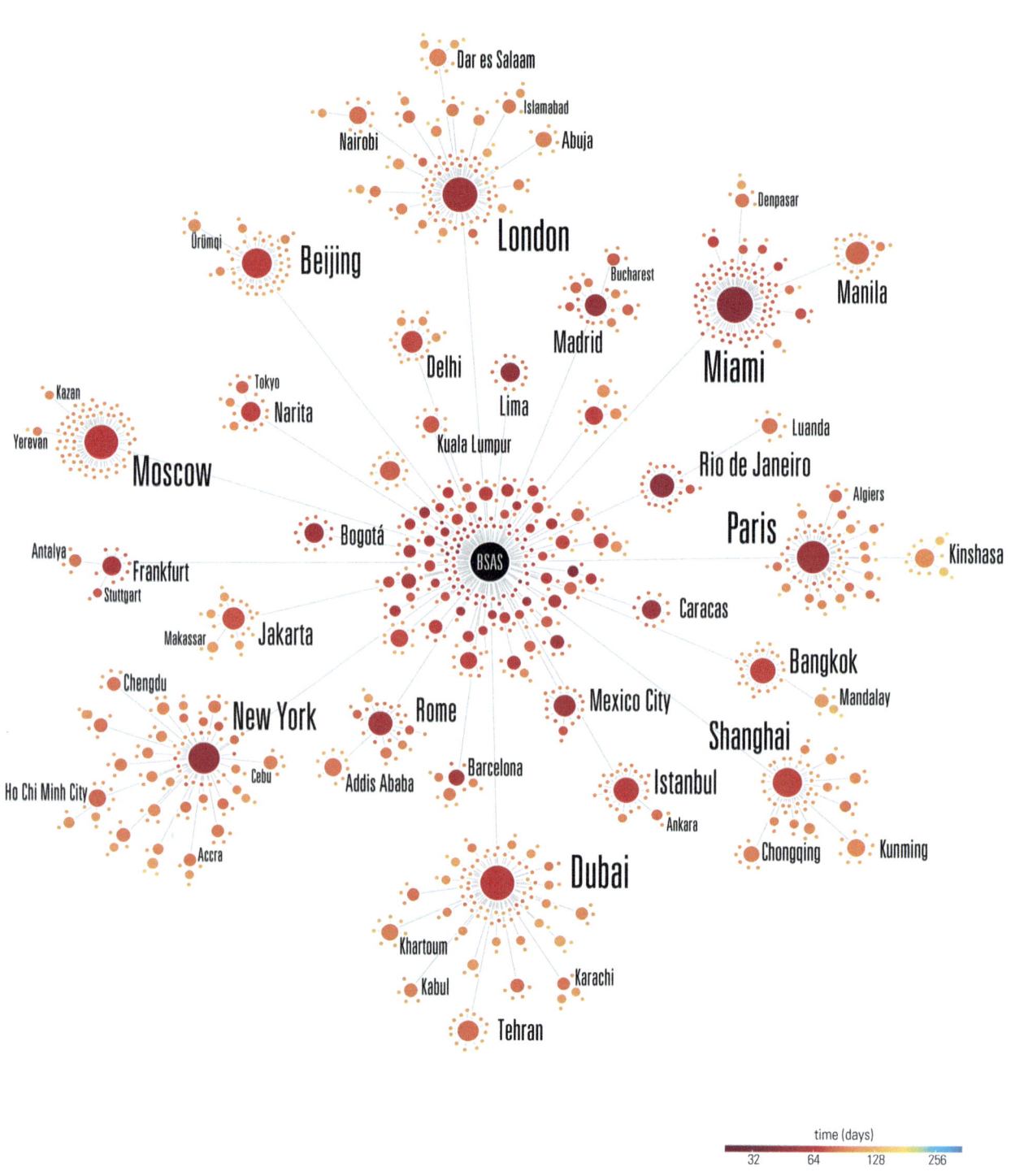

time (days)

32   64   128   256

▲ INFECTION TREE GRAPH

INITIAL CONDITIONS

 80% AIR TRAFFIC REDUCTION

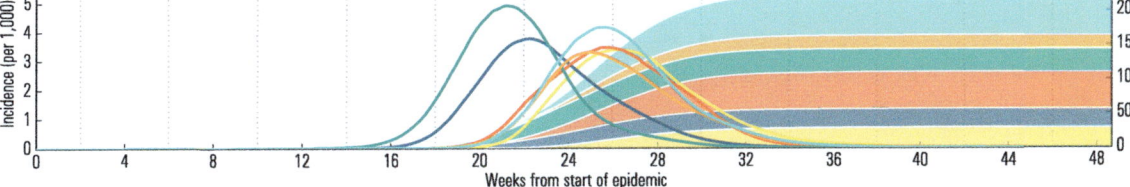

▲ CUMULATIVE INCIDENCE AND EPIDEMIC PEAKS

▼ ATTACK RATE COMPOSITION (ONE YEAR)

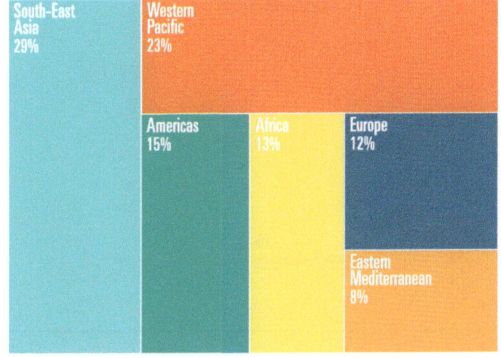

▼ EPIDEMIC ACTIVITY

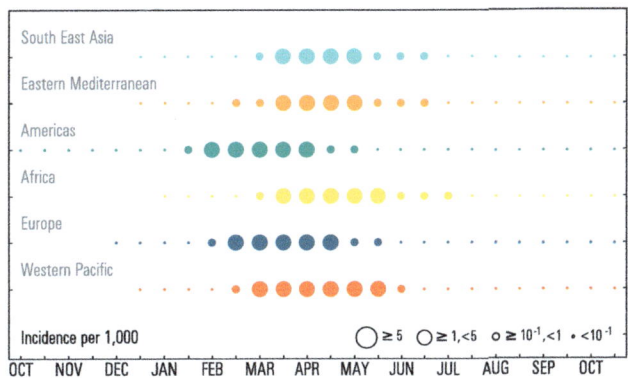

VACCINATIONS

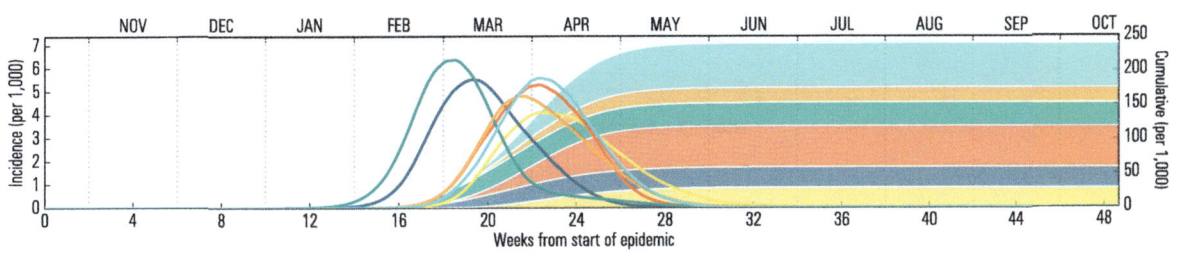

▲ CUMULATIVE INCIDENCE AND EPIDEMIC PEAKS

▼ ATTACK RATE COMPOSITION (ONE YEAR)

▼ EPIDEMIC ACTIVITY

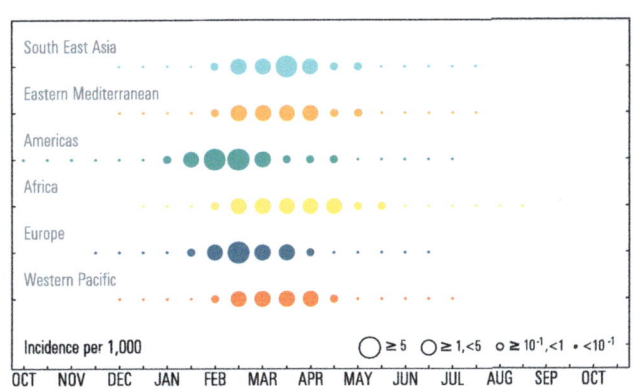

HANOI, VIETNAM

ORIGIN CITY

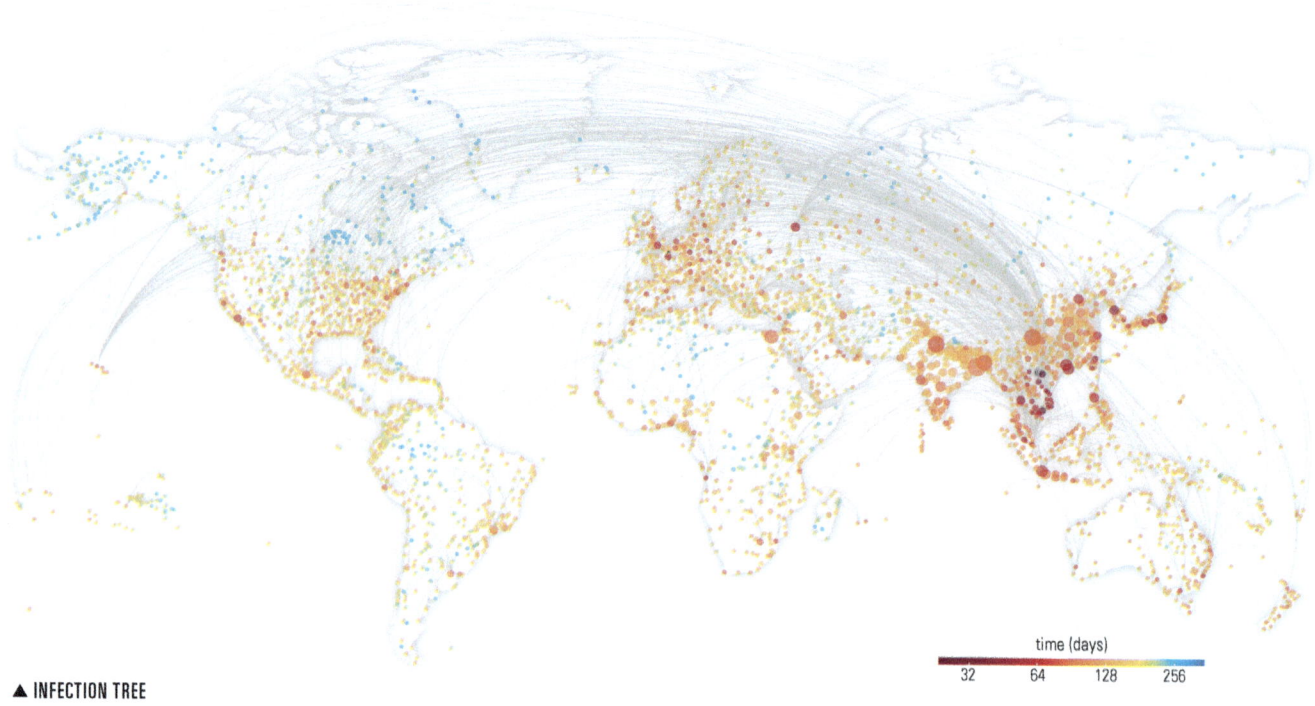

time (days)
32   64   128   256

▲ INFECTION TREE

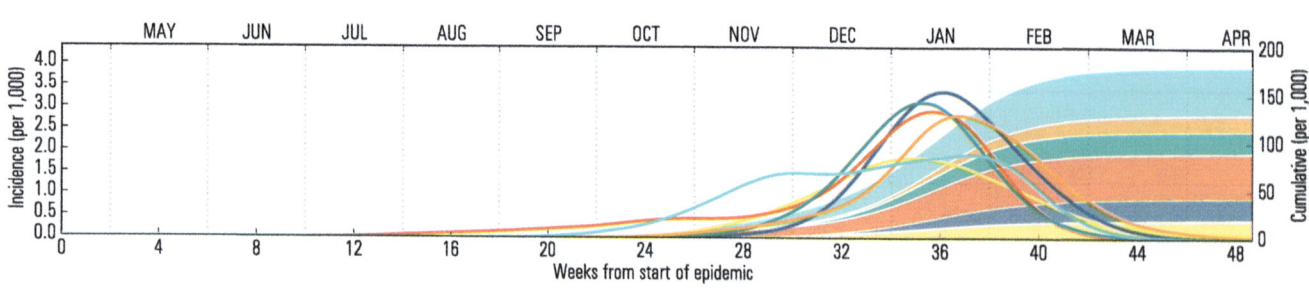

▲ CUMULATIVE INCIDENCE AND EPIDEMIC PEAKS

▼ AFFECTED COUNTRIES

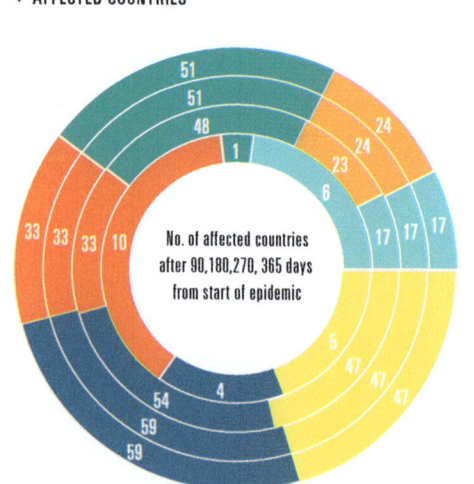

No. of affected countries
after 90,180,270, 365 days
from start of epidemic

▼ EPIDEMIC ACTIVITY

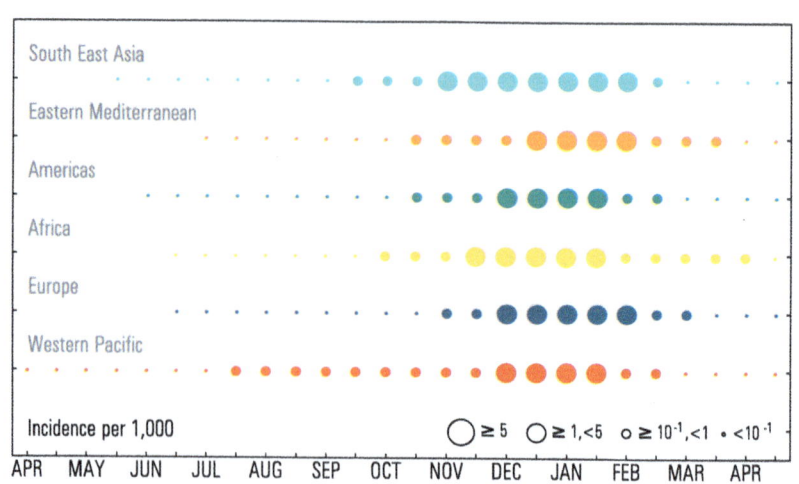

WHO REGIONS

- Americas
- Africa
- Europe
- E. Mediterranean
- South-East Asia
- Western Pacific

INITIAL CONDITIONS

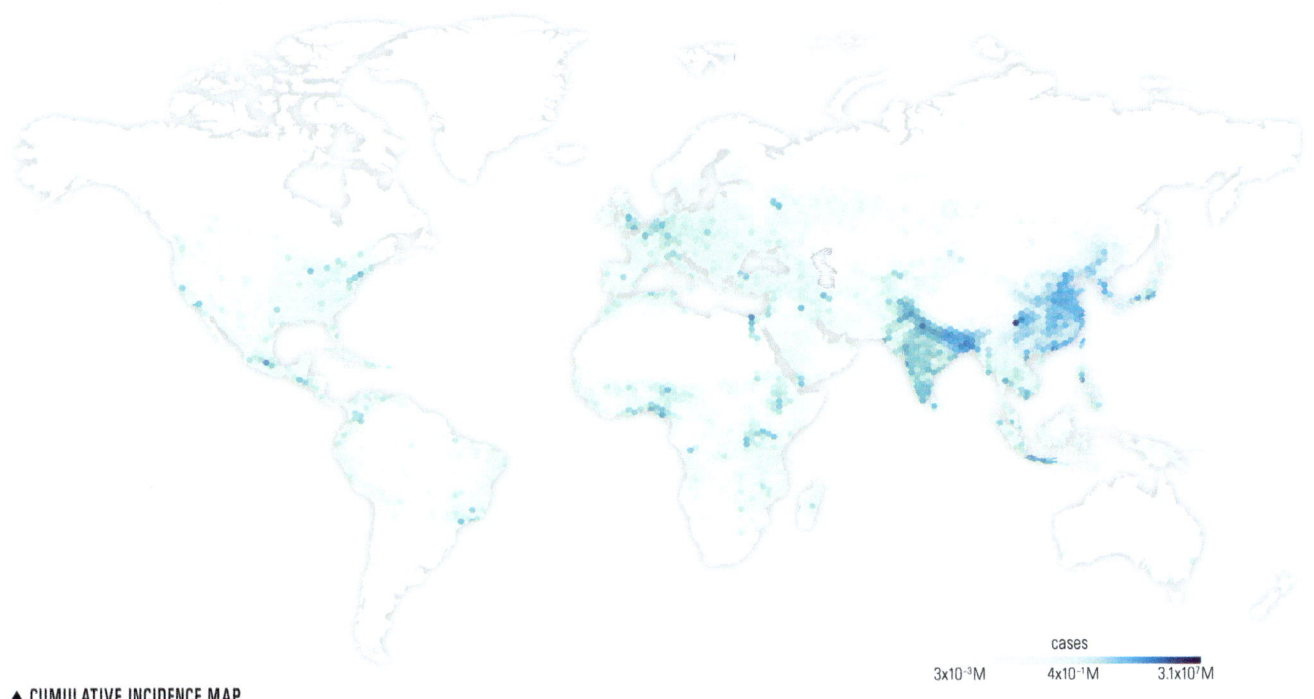

cases

3x10⁻³M   4x10⁻¹M   3.1x10⁷M

▲ CUMULATIVE INCIDENCE MAP

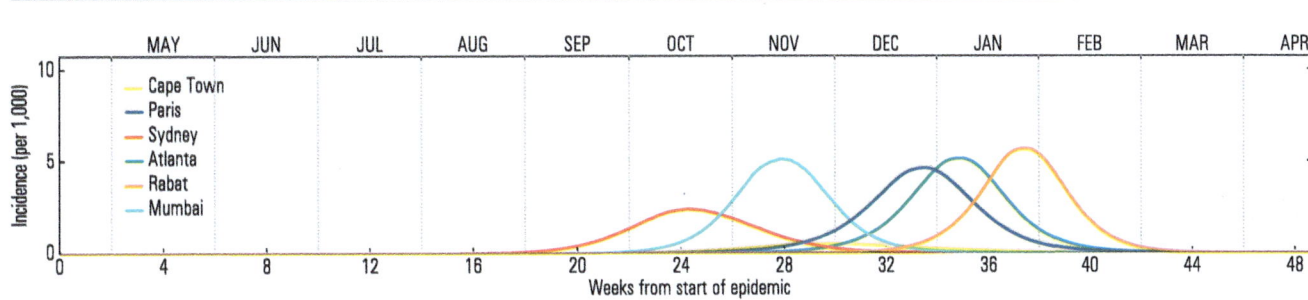

▲ URBAN LEVEL INCIDENCE

▼ RISK QUAD CHART

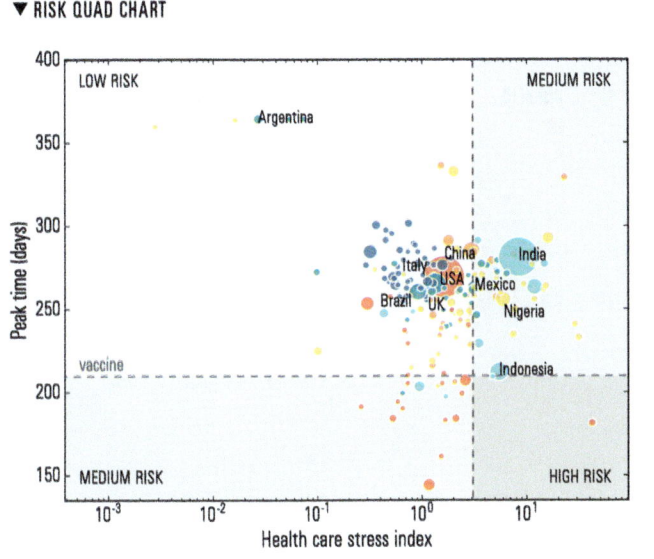

▼ ATTACK RATE COMPOSITION (ONE YEAR)

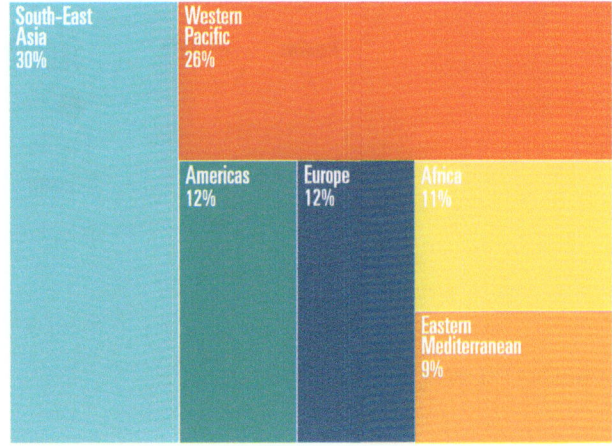

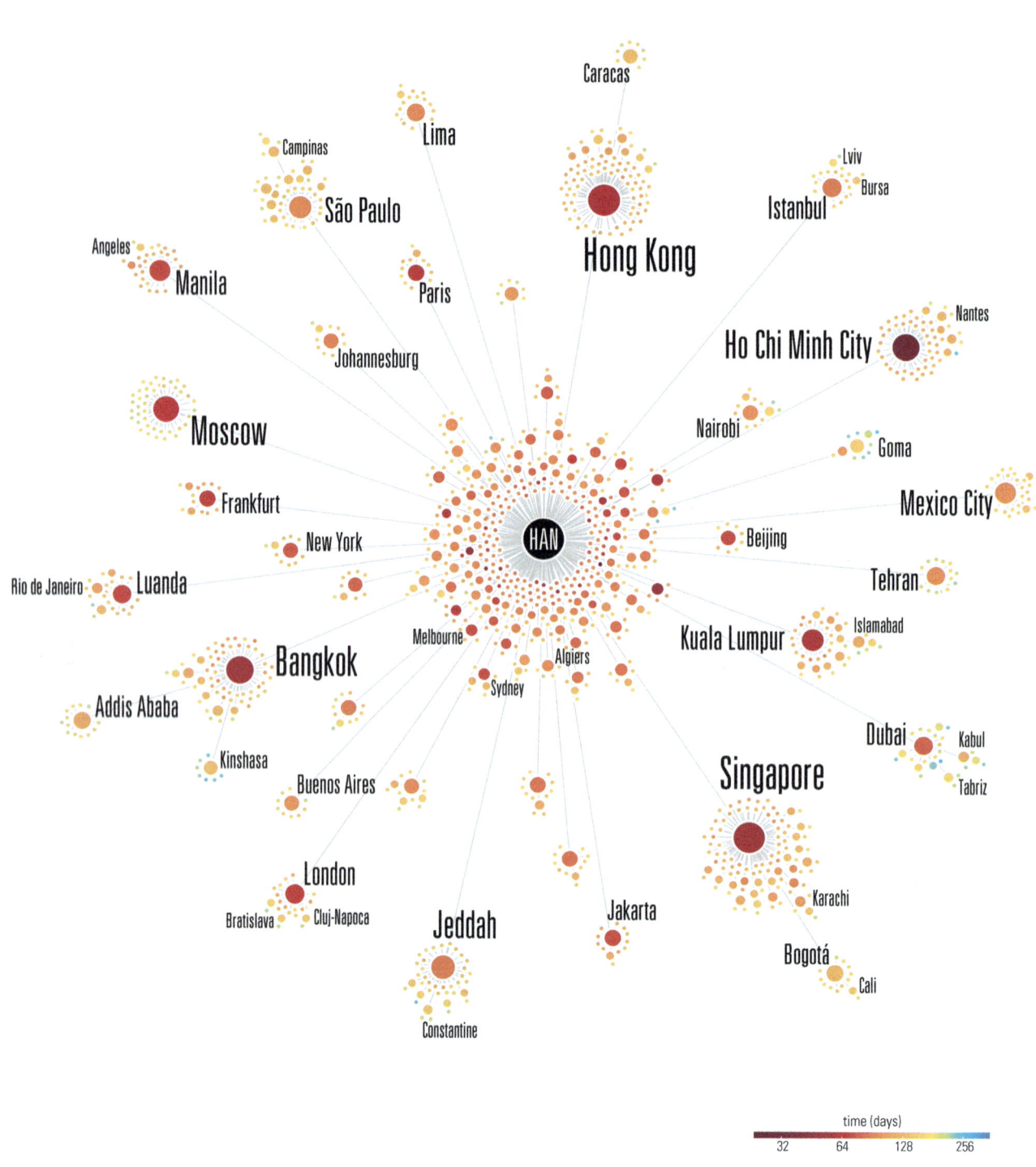

ORIGIN CITY : HANOI, VIETNAM

Caracas

Lima

Campinas

São Paulo

Hong Kong

Lviv

Bursa

Istanbul

Manila

Angeles

Paris

Ho Chi Minh City

Nantes

Johannesburg

Moscow

Nairobi

Goma

Frankfurt

Beijing

Mexico City

New York

HAN

Tehran

Rio de Janeiro

Luanda

Kuala Lumpur

Islamabad

Melbourne

Algiers

Bangkok

Sydney

Addis Ababa

Dubai

Kabul

Kinshasa

Singapore

Tabriz

Buenos Aires

London

Karachi

Bratislava

Cluj-Napoca

Jakarta

Bogotá

Jeddah

Cali

Constantine

time (days)

32   64   128   256

▲ INFECTION TREE GRAPH

■ Americas
□ Africa
■ Europe
■ E. Mediterranean
■ South-East Asia
■ Western Pacific

 APR
 R₀ 1.5

80% AIR TRAFFIC REDUCTION

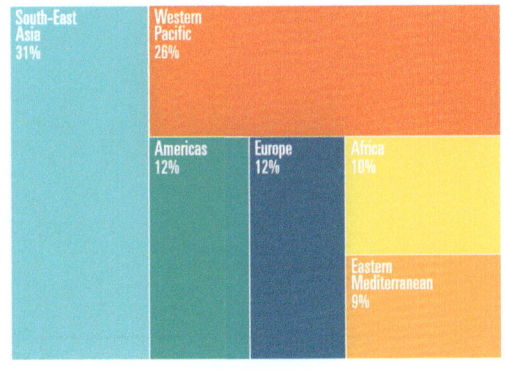

▲ CUMULATIVE INCIDENCE AND EPIDEMIC PEAKS

▼ ATTACK RATE COMPOSITION (ONE YEAR)

▼ EPIDEMIC ACTIVITY

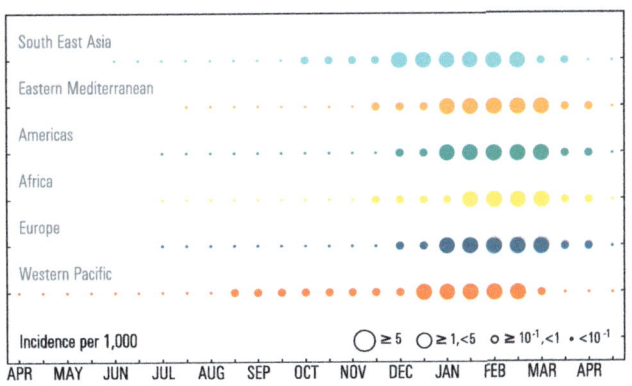

South-East Asia 31%
Western Pacific 26%
Americas 12%
Europe 12%
Africa 10%
Eastern Mediterranean 9%

Incidence per 1,000   ○ ≥ 5   ○ ≥ 1,<5   ○ ≥ 10⁻¹,<1   • <10⁻¹

VACCINATIONS

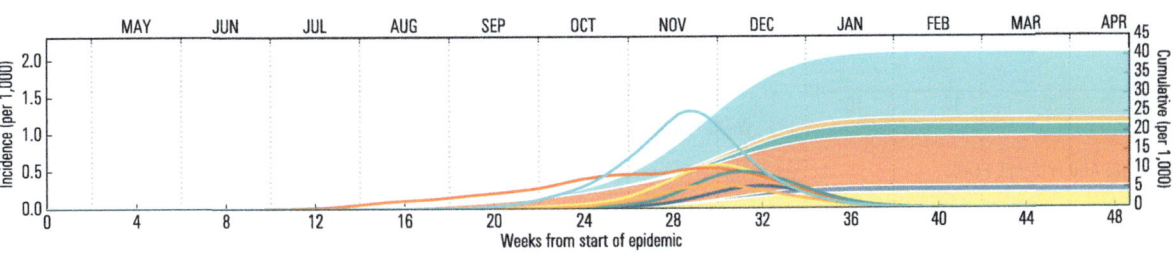

▲ CUMULATIVE INCIDENCE AND EPIDEMIC PEAKS

▼ ATTACK RATE COMPOSITION (ONE YEAR)

▼ EPIDEMIC ACTIVITY

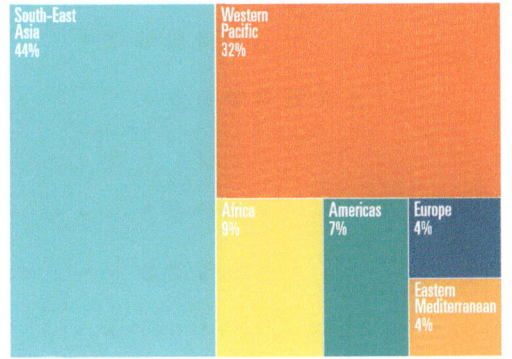

South-East Asia 44%
Western Pacific 32%
Africa 9%
Americas 7%
Europe 4%
Eastern Mediterranean 4%

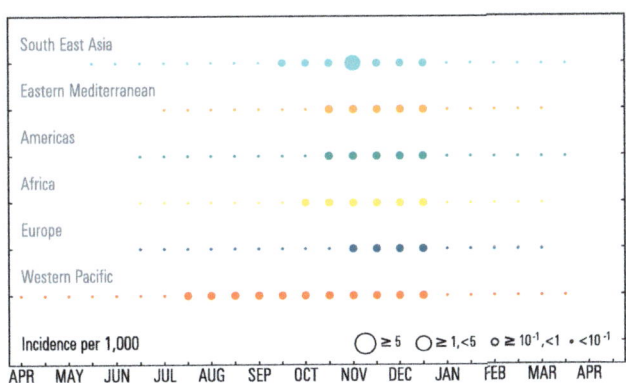

Incidence per 1,000   ○ ≥ 5   ○ ≥ 1,<5   ○ ≥ 10⁻¹,<1   • <10⁻¹

HANOI, VIETNAM

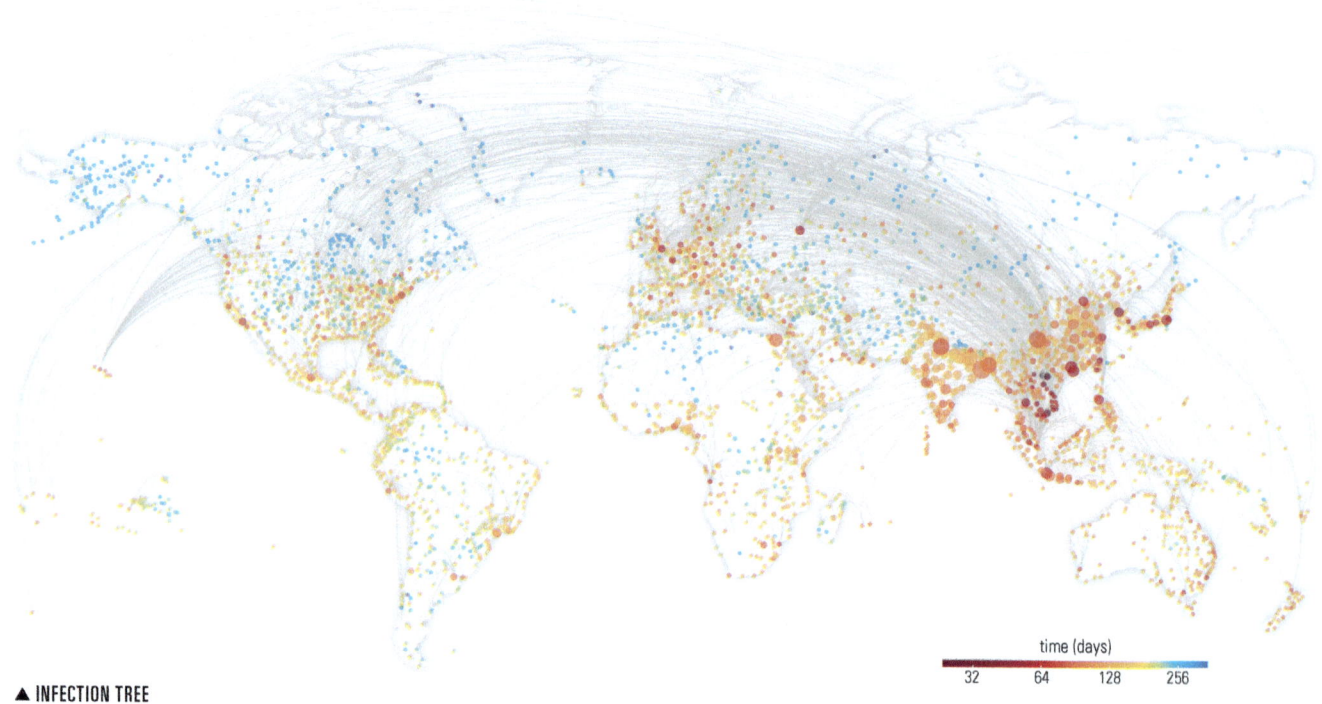

time (days)

32 64 128 256

▲ INFECTION TREE

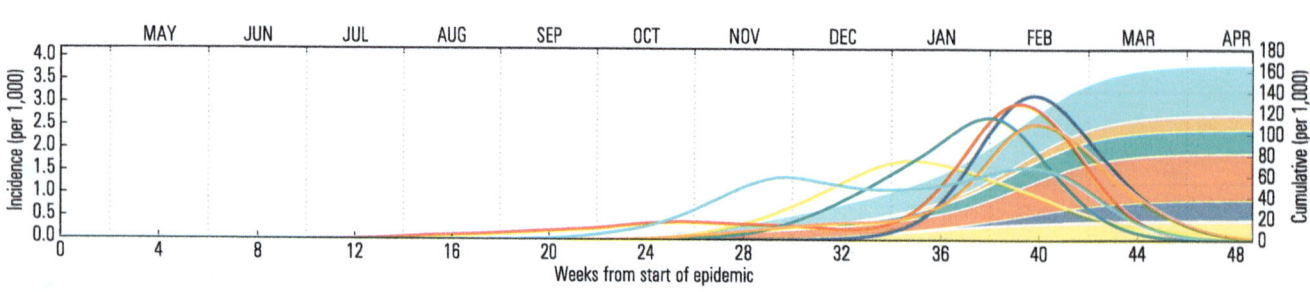

▲ CUMULATIVE INCIDENCE AND EPIDEMIC PEAKS

▼ AFFECTED COUNTRIES

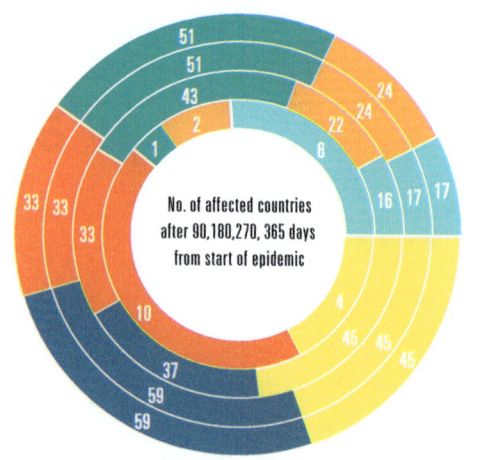

No. of affected countries
after 90, 180, 270, 365 days
from start of epidemic

▼ EPIDEMIC ACTIVITY

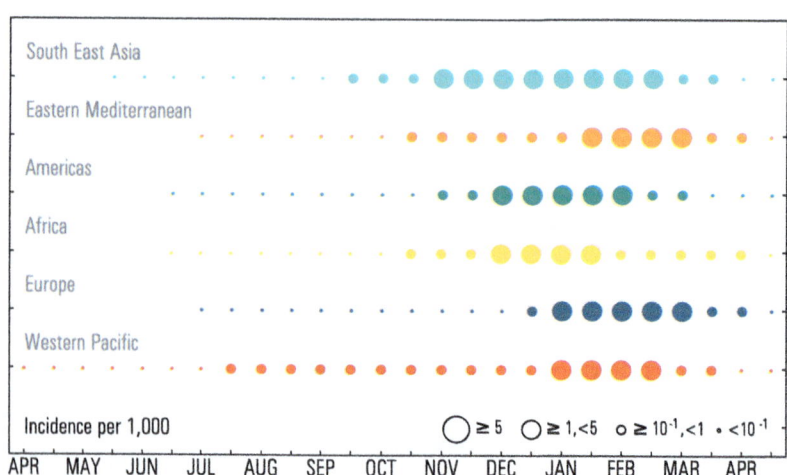

South East Asia

Eastern Mediterranean

Americas

Africa

Europe

Western Pacific

Incidence per 1,000   ○ ≥ 5   ○ ≥ 1,<5   ∘ ≥ 10⁻¹,<1   · <10⁻¹

APR MAY JUN JUL AUG SEP OCT NOV DEC JAN FEB MAR APR

■ Americas   ■ E. Mediterranean
■ Africa   ■ South-East Asia
■ Europe   ■ Western Pacific

APR | 1.5

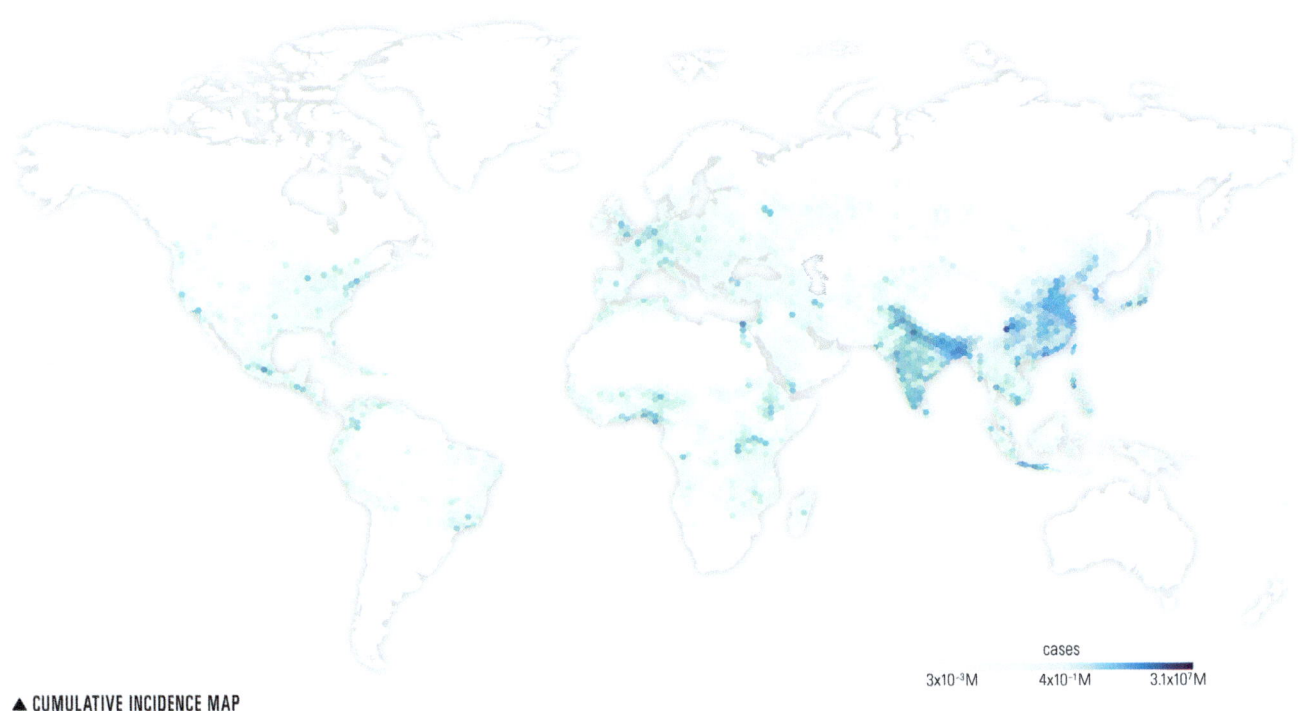

cases

3x10^{-3}M   4x10^{-1}M   3.1x10^{7}M

▲ CUMULATIVE INCIDENCE MAP

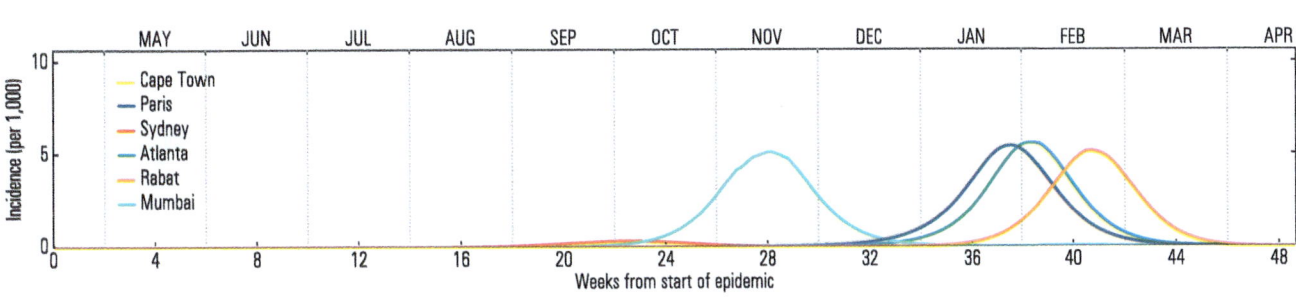

▲ URBAN LEVEL INCIDENCE

▼ RISK QUAD CHART

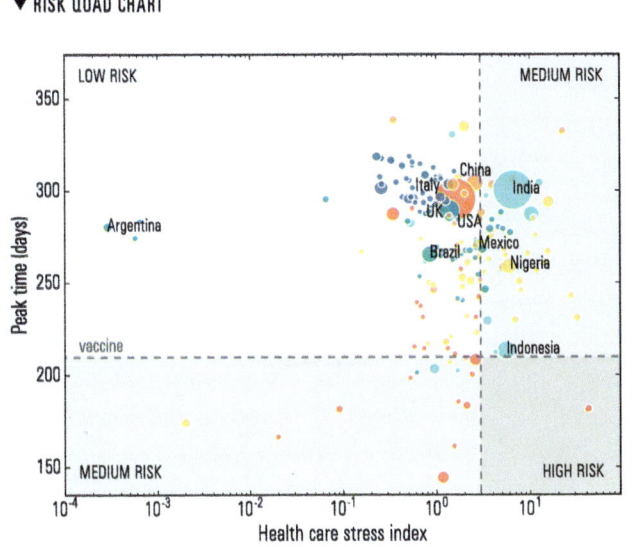

▼ ATTACK RATE COMPOSITION (ONE YEAR)

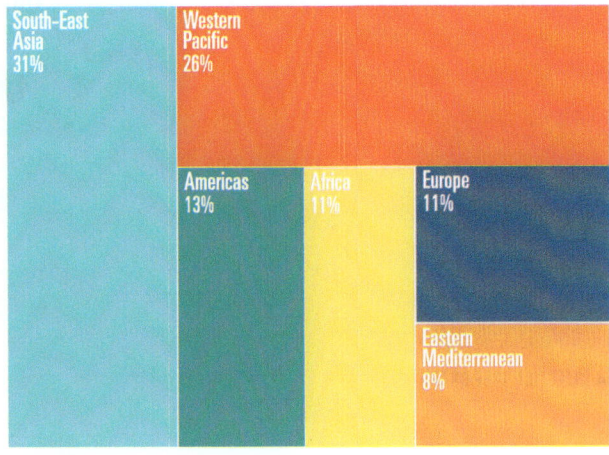

HANOI, VIETNAM

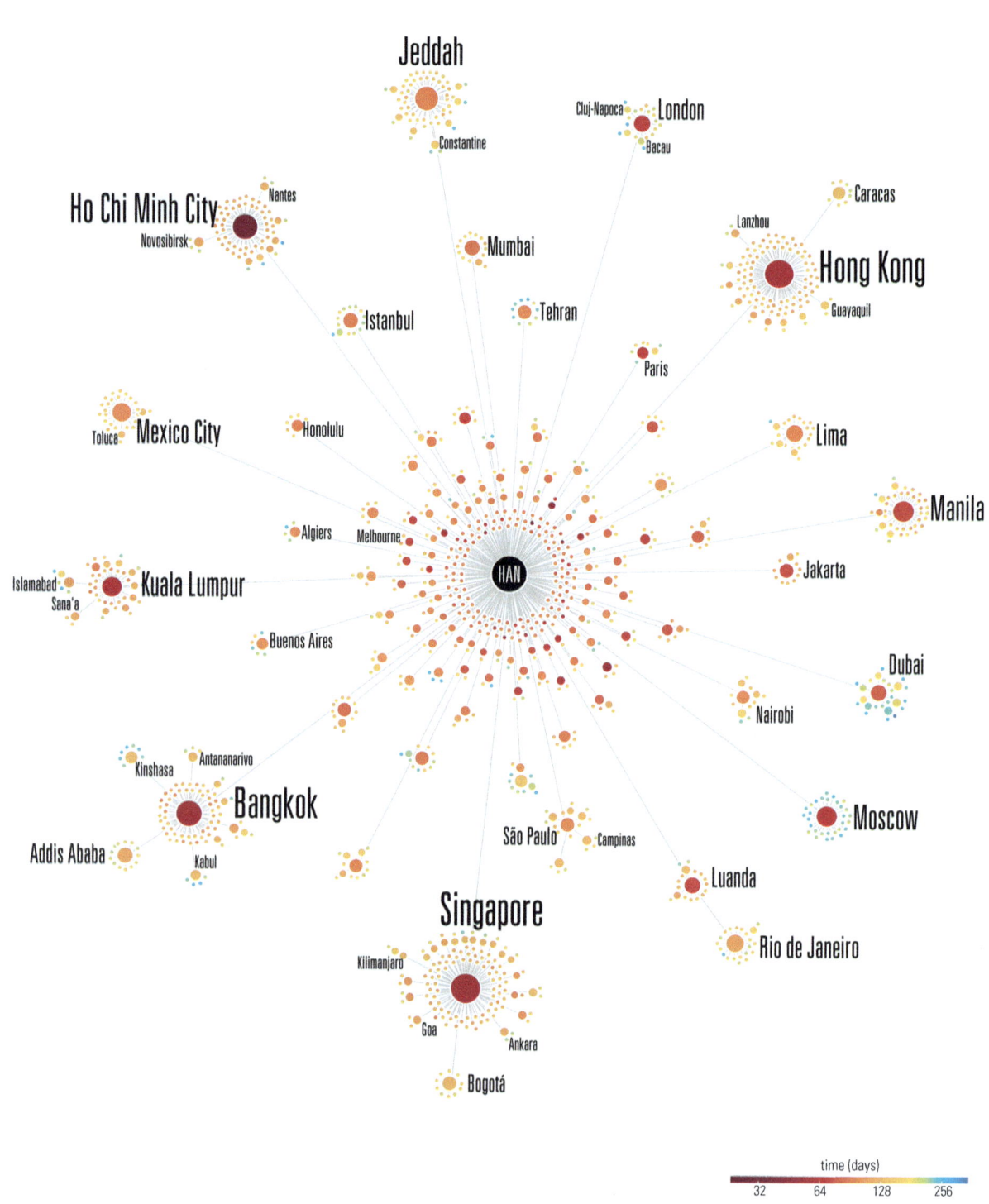

Jeddah

Cluj-Napoca London
Bacau

Constantine

Caracas
Lanzhou

Ho Chi Minh City Nantes

Mumbai Hong Kong

Novosibirsk

Guayaquil

Istanbul Tehran

Paris

Mexico City Honolulu Lima

Toluca

Manila

Algiers Melbourne

Jakarta

Kuala Lumpur

Islamabad

Sana'a Buenos Aires

Dubai

Nairobi

HAN

Antananarivo

Kinshasa

Bangkok Moscow

Addis Ababa São Paulo Campinas

Kabul Luanda

Singapore Rio de Janeiro

Kilimanjaro

Goa

Ankara

Bogotá

time (days)

32   64   128   256

▲ INFECTION TREE GRAPH

- Americas
- Africa
- Europe
- E. Mediterranean
- South-East Asia
- Western Pacific

INITIAL CONDITIONS

 APR
 R₀ 1.5

80% AIR TRAFFIC REDUCTION

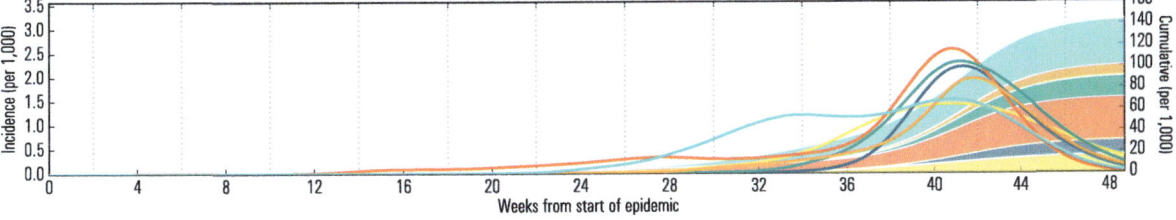

▲ CUMULATIVE INCIDENCE AND EPIDEMIC PEAKS

▼ ATTACK RATE COMPOSITION (ONE YEAR)

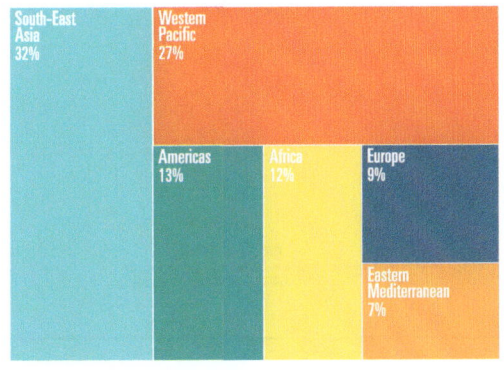

South-East Asia 32%
Western Pacific 27%
Americas 13%
Africa 12%
Europe 9%
Eastern Mediterranean 7%

▼ EPIDEMIC ACTIVITY

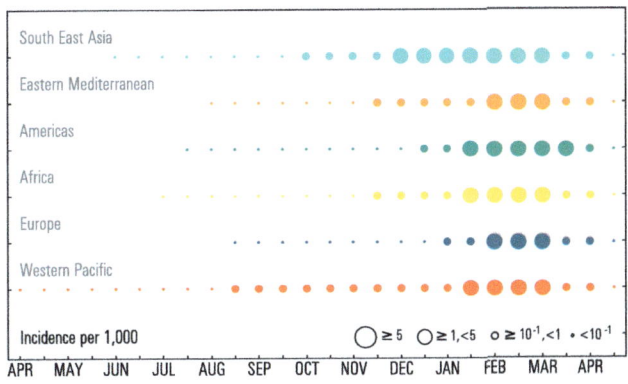

Incidence per 1,000   ⬤ ≥ 5   ◯ ≥ 1,<5   ○ ≥ 10⁻¹,<1   • <10⁻¹

VACCINATIONS

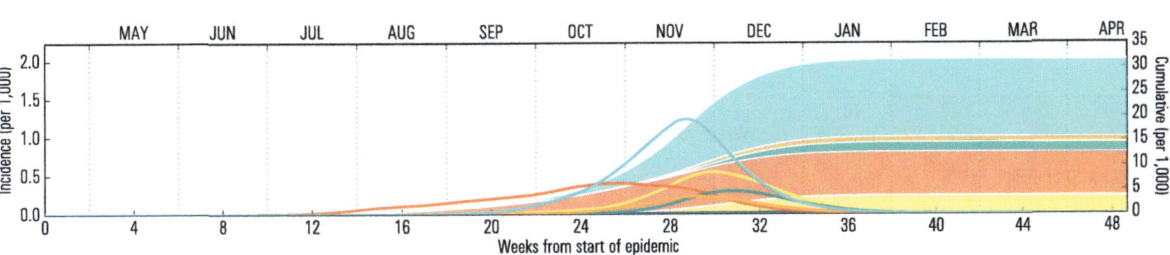

▲ CUMULATIVE INCIDENCE AND EPIDEMIC PEAKS

▼ ATTACK RATE COMPOSITION (ONE YEAR)

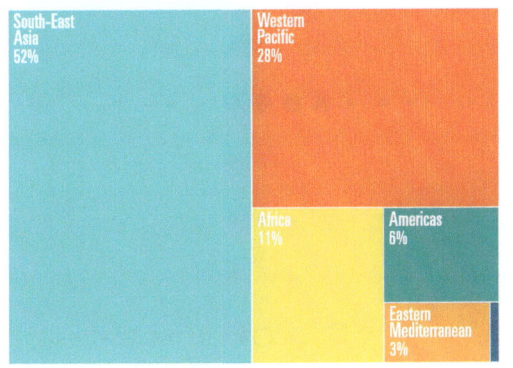

South-East Asia 52%
Western Pacific 28%
Africa 11%
Americas 6%
Eastern Mediterranean 3%

▼ EPIDEMIC ACTIVITY

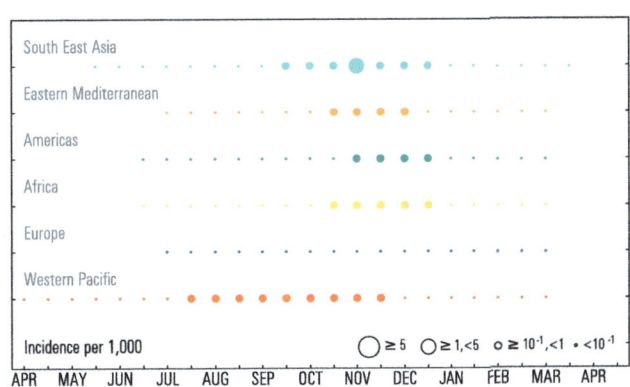

Incidence per 1,000   ⬤ ≥ 5   ◯ ≥ 1,<5   ○ ≥ 10⁻¹,<1   • <10⁻¹

HANOI, VIETNAM

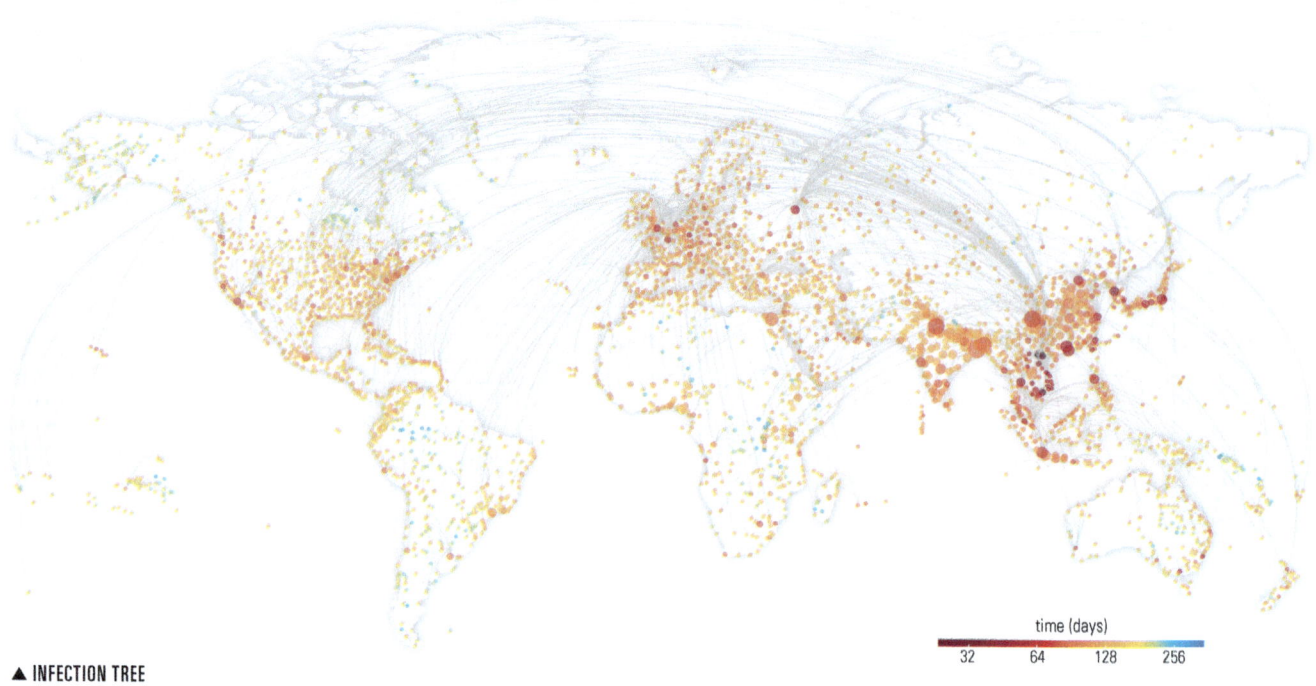

time (days)

32   64   128   256

▲ INFECTION TREE

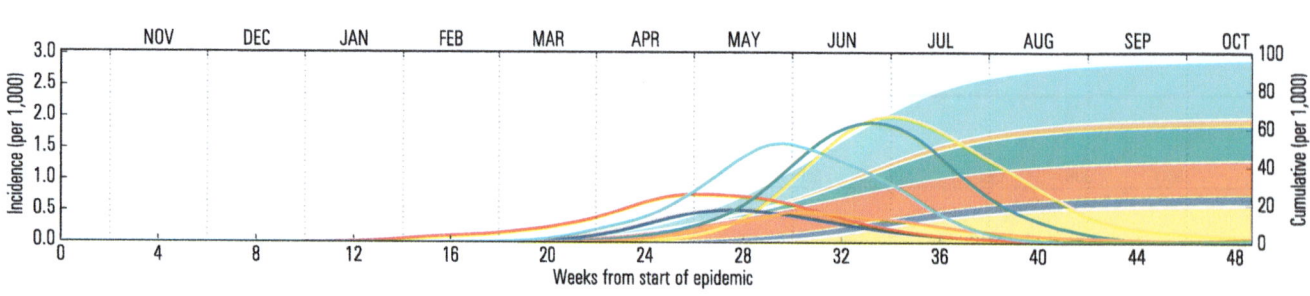

▲ CUMULATIVE INCIDENCE AND EPIDEMIC PEAKS

▼ AFFECTED COUNTRIES

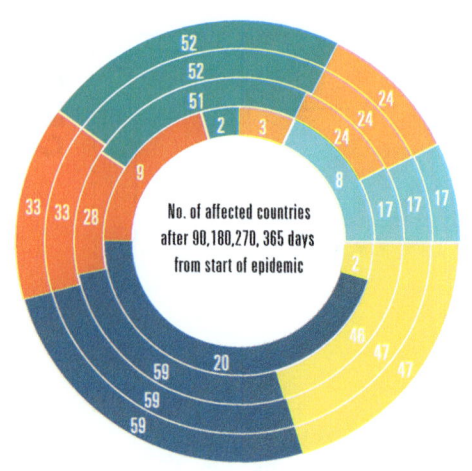

No. of affected countries after 90, 180, 270, 365 days from start of epidemic

▼ EPIDEMIC ACTIVITY

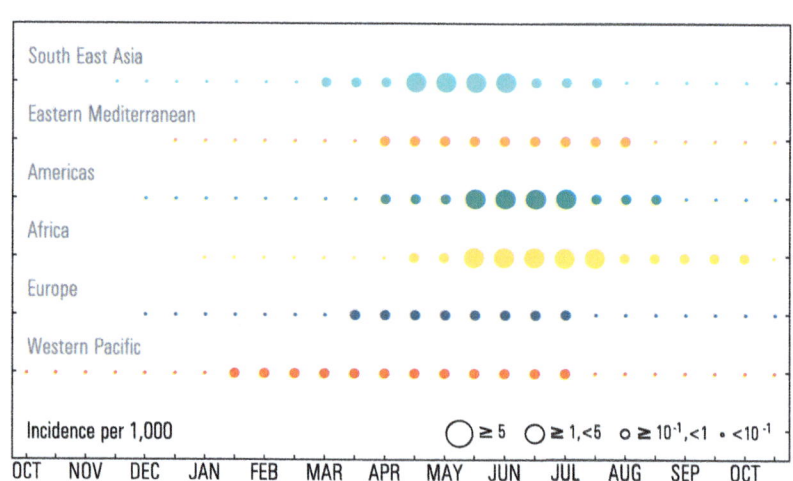

South East Asia

Eastern Mediterranean

Americas

Africa

Europe

Western Pacific

Incidence per 1,000   ◯ ≥ 5   ◯ ≥ 1,<5   ○ ≥ 10⁻¹,<1   · <10⁻¹

OCT NOV DEC JAN FEB MAR APR MAY JUN JUL AUG SEP OCT

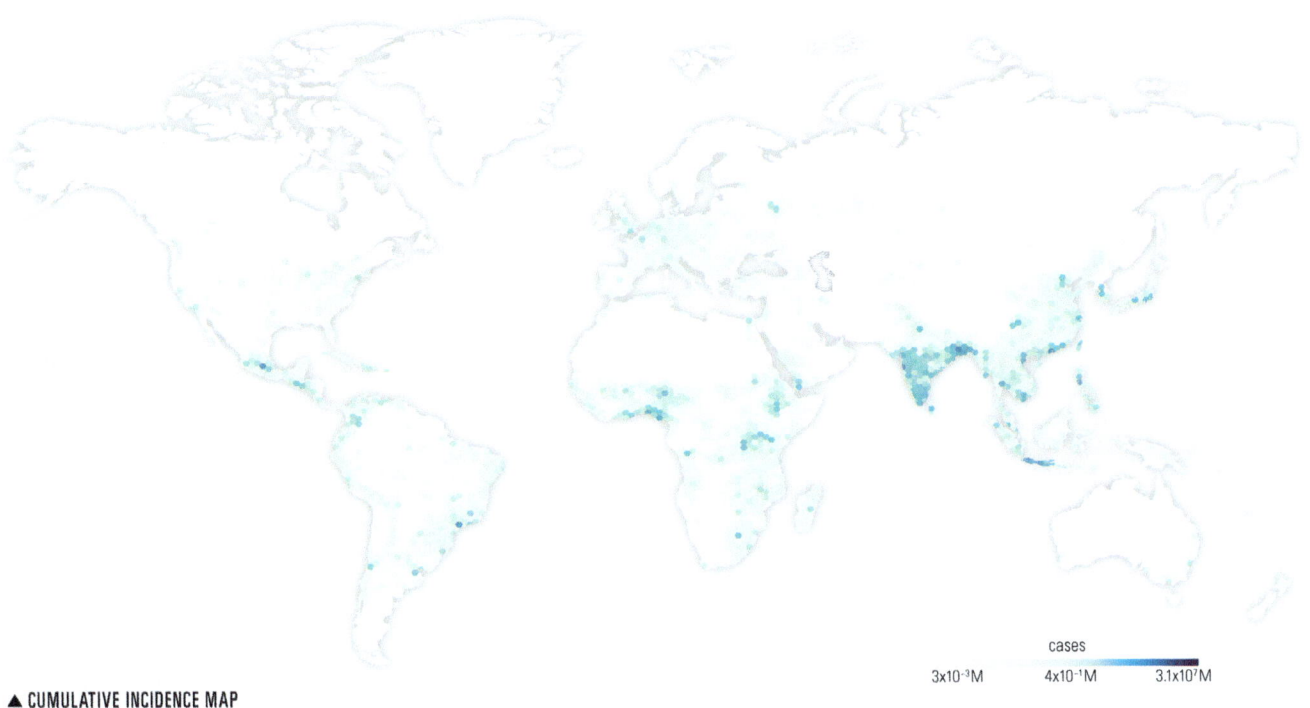

cases

3x10⁻³M   4x10⁻¹M   3.1x10⁷M

▲ CUMULATIVE INCIDENCE MAP

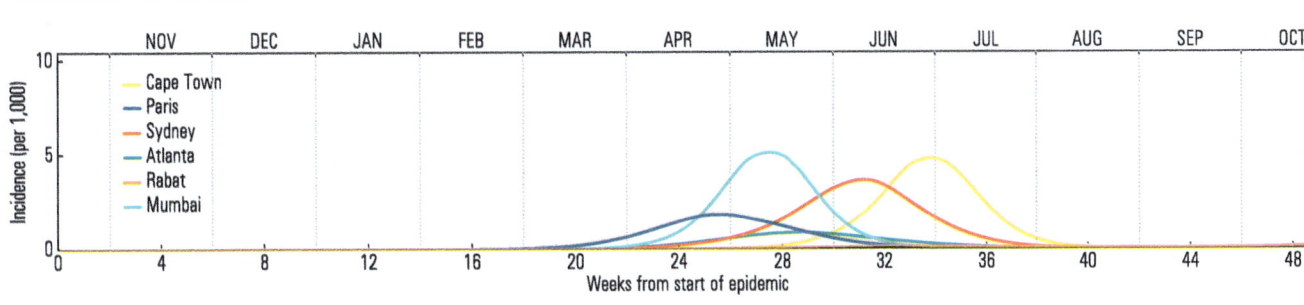

▲ URBAN LEVEL INCIDENCE

▼ RISK QUAD CHART

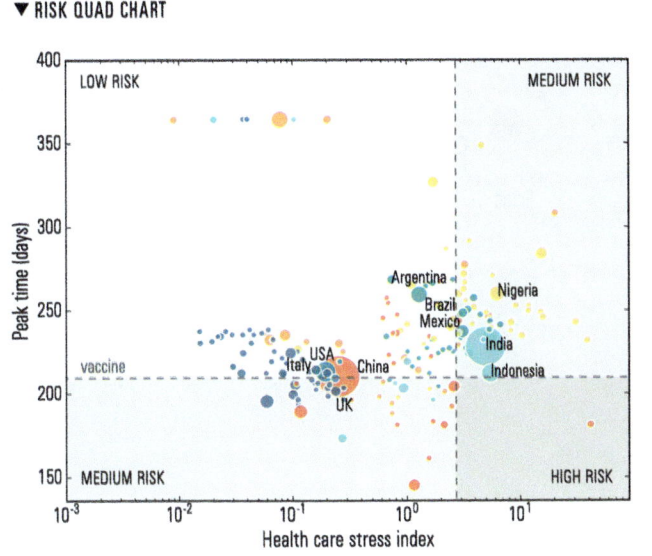

▼ ATTACK RATE COMPOSITION (ONE YEAR)

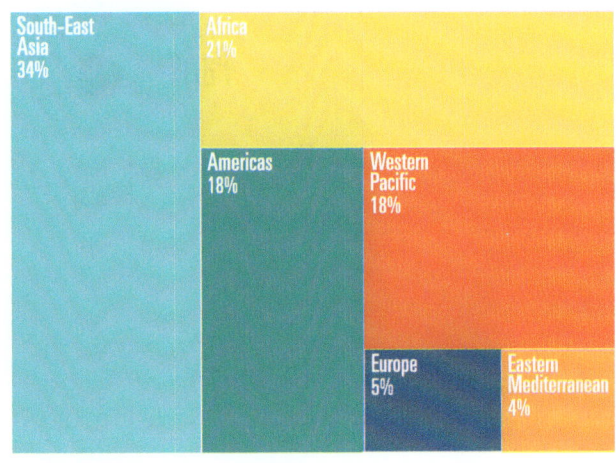

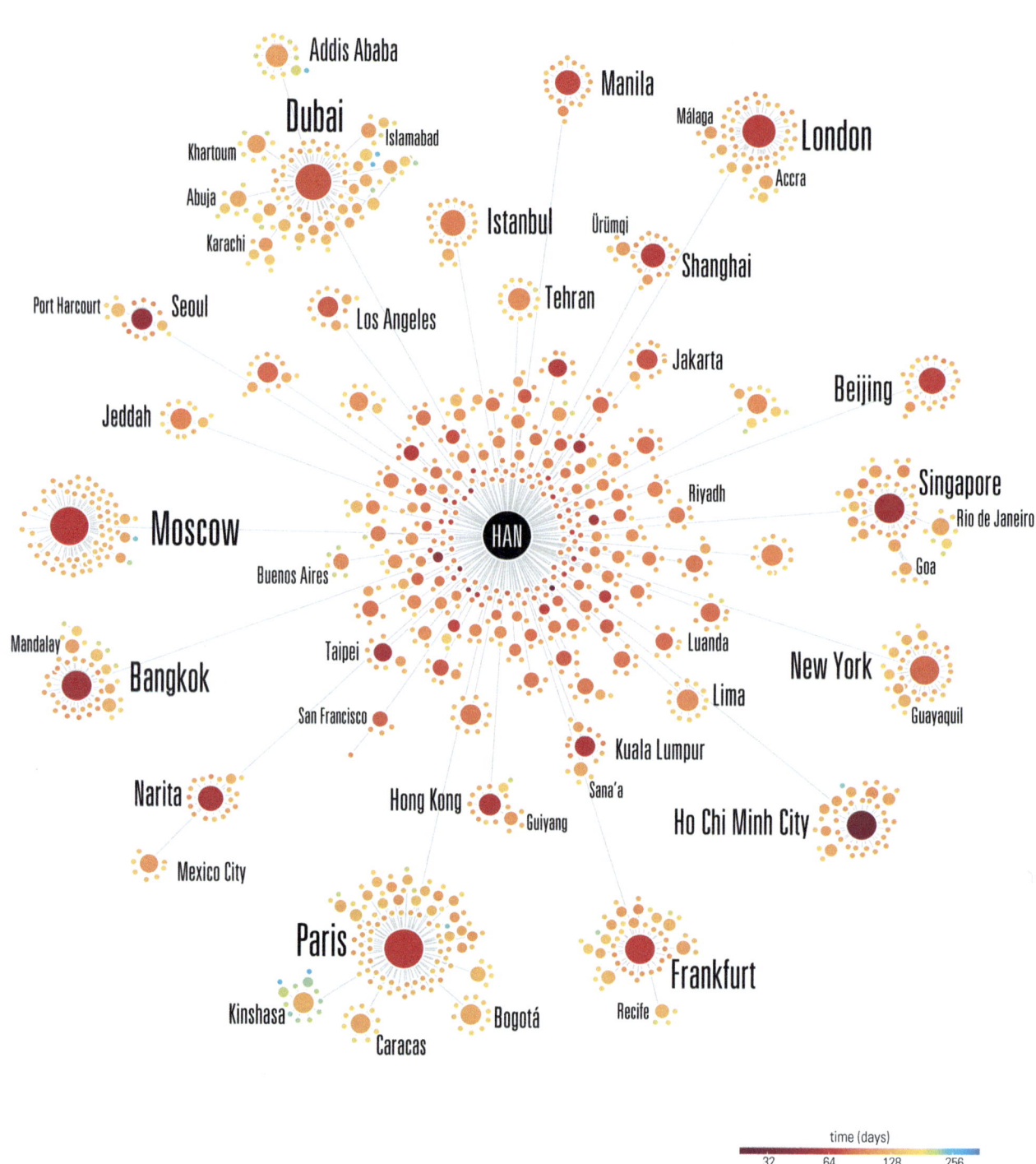

Addis Ababa

Dubai

Manila

Málaga

London

Khartoum

Islamabad

Accra

Abuja

Istanbul

Ürümqi

Karachi

Shanghai

Tehran

Port Harcourt

Seoul

Los Angeles

Jakarta

Beijing

Jeddah

Riyadh

Singapore

Rio de Janeiro

Moscow

HAN

Goa

Buenos Aires

Mandalay

Luanda

New York

Bangkok

Taipei

Lima

Guayaquil

San Francisco

Kuala Lumpur

Narita

Hong Kong

Sana'a

Guiyang

Ho Chi Minh City

Mexico City

Paris

Frankfurt

Kinshasa

Bogotá

Recife

Caracas

time (days)

32   64   128   256

▲ INFECTION TREE GRAPH

■ Americas ■ E. Mediterranean
■ Africa ■ South-East Asia
■ Europe ■ Western Pacific

 OCT  R₀ 1.5

 80% AIR TRAFFIC REDUCTION

▲ CUMULATIVE INCIDENCE AND EPIDEMIC PEAKS

▼ ATTACK RATE COMPOSITION (ONE YEAR)

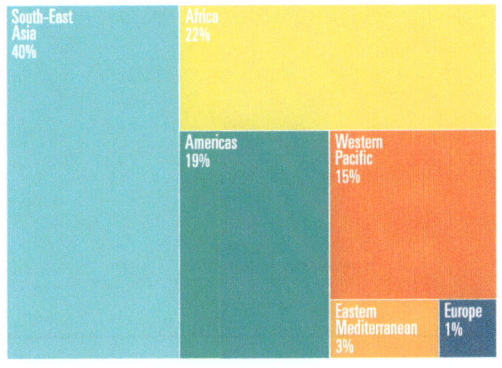

▼ EPIDEMIC ACTIVITY

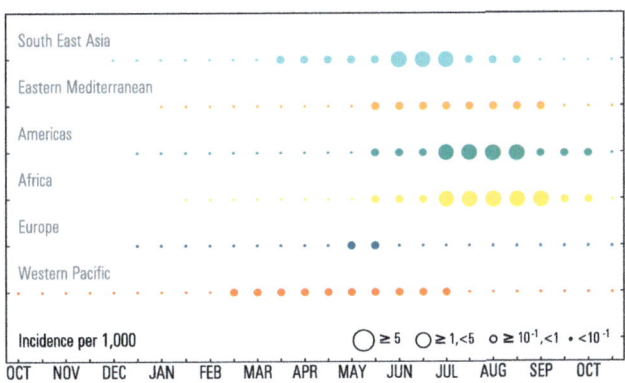

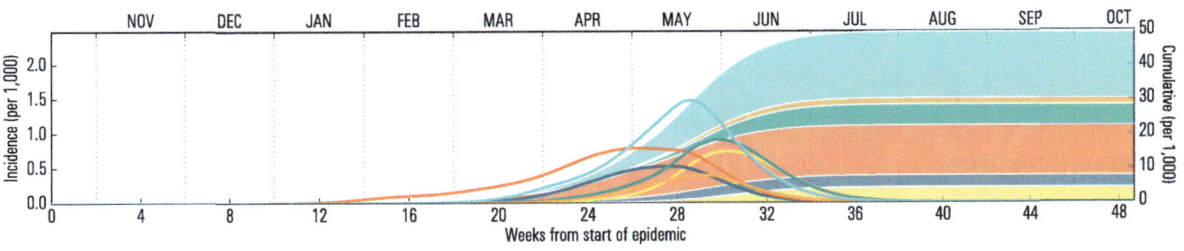

 VACCINATIONS

▲ CUMULATIVE INCIDENCE AND EPIDEMIC PEAKS

▼ ATTACK RATE COMPOSITION (ONE YEAR)

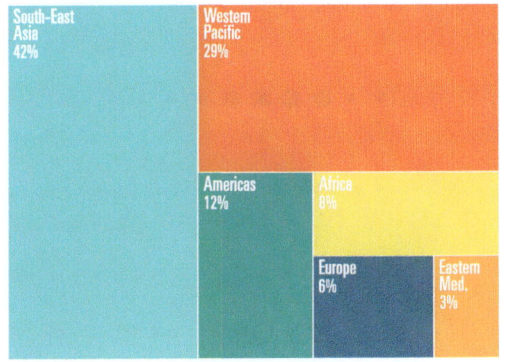

▼ EPIDEMIC ACTIVITY

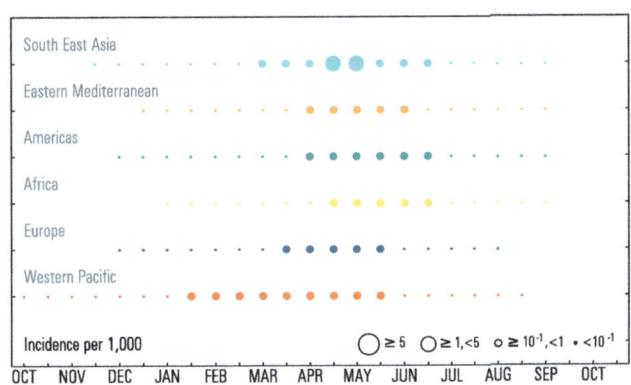

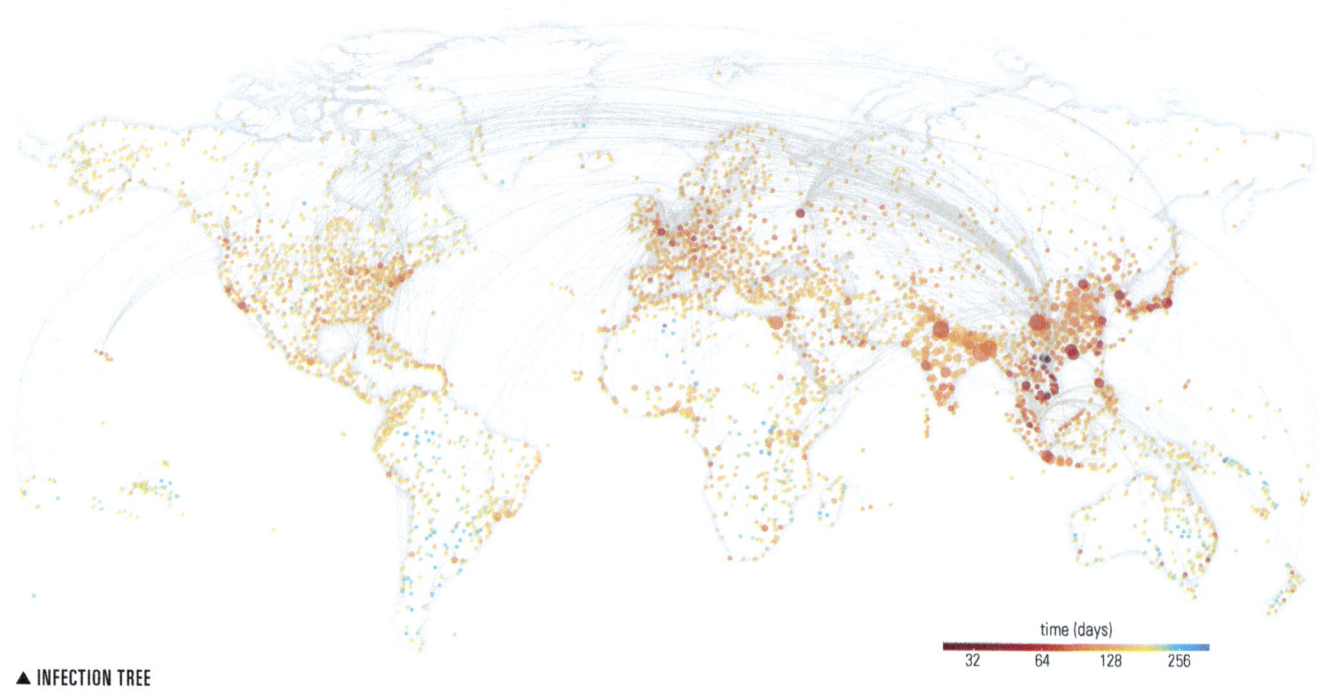

▲ INFECTION TREE

time (days)
32   64   128   256

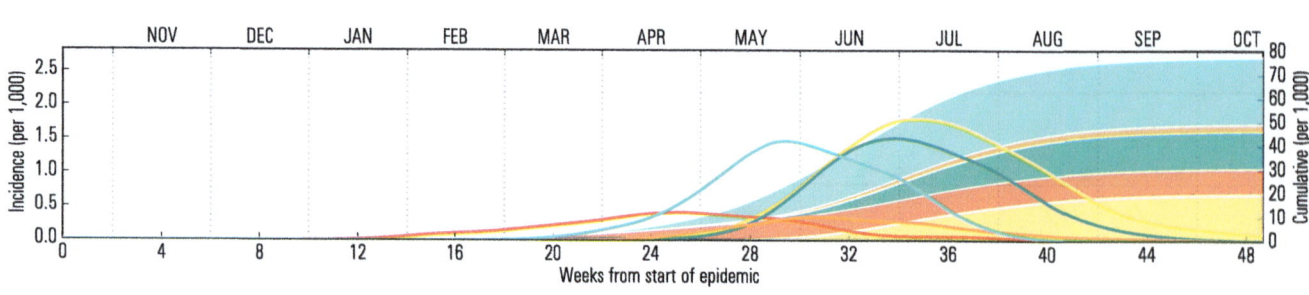

▲ CUMULATIVE INCIDENCE AND EPIDEMIC PEAKS

▼ AFFECTED COUNTRIES

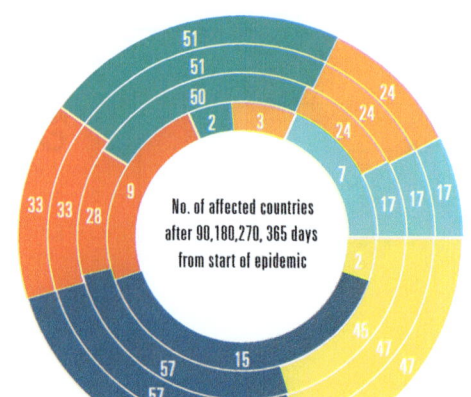

No. of affected countries after 90,180,270, 365 days from start of epidemic

▼ EPIDEMIC ACTIVITY

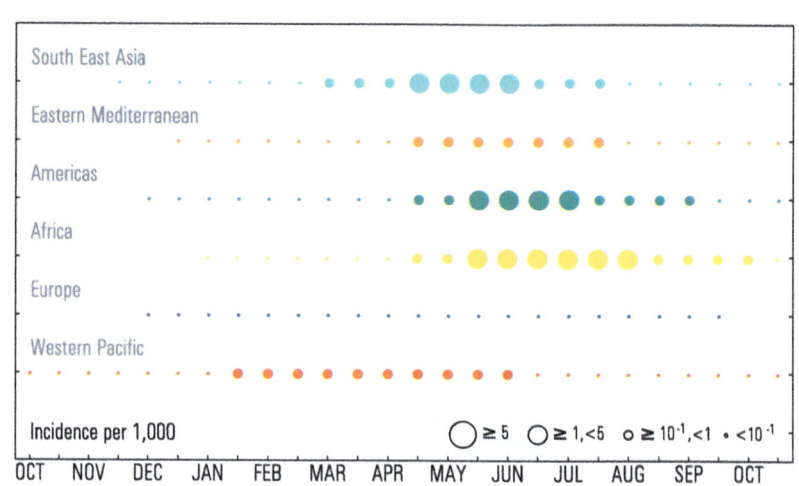

South East Asia

Eastern Mediterranean

Americas

Africa

Europe

Western Pacific

Incidence per 1,000

◯ ≥ 5   ◯ ≥ 1,<5   ○ ≥ 10⁻¹,<1   · <10⁻¹

OCT NOV DEC JAN FEB MAR APR MAY JUN JUL AUG SEP OCT

INITIAL CONDITIONS   OCT   1.5

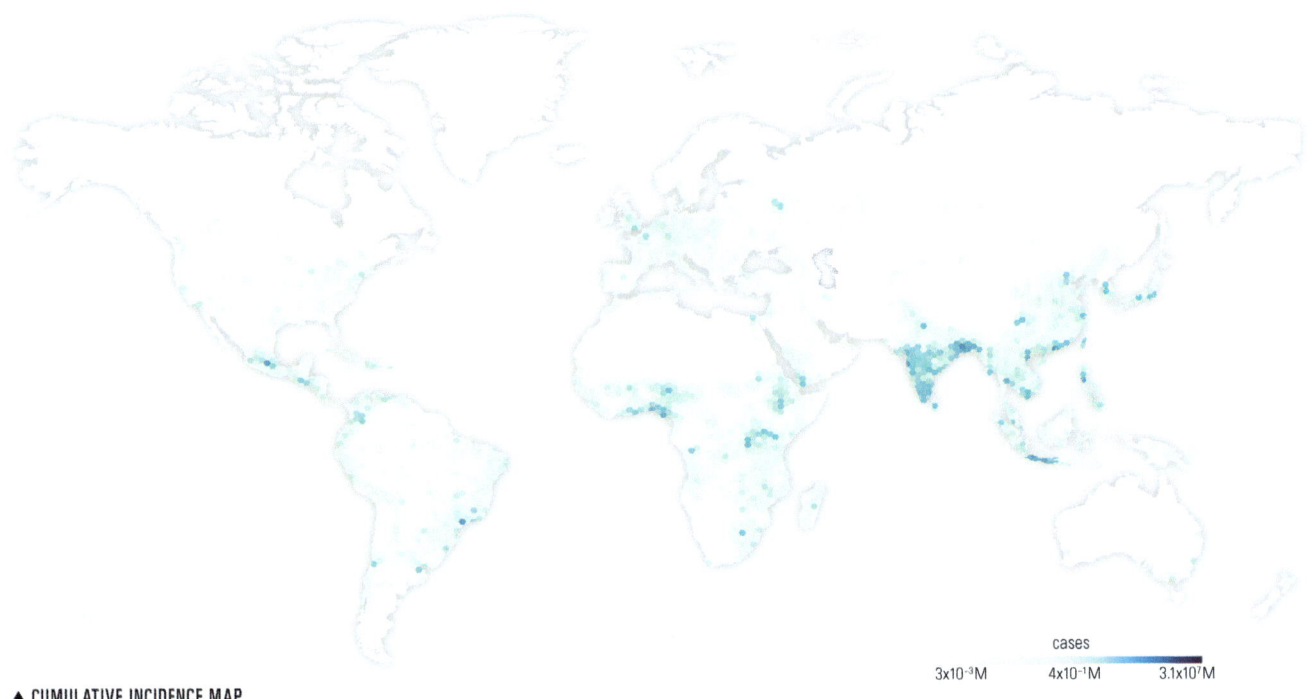

cases
3x10⁻³M   4x10⁻¹M   3.1x10⁷M

▲ CUMULATIVE INCIDENCE MAP

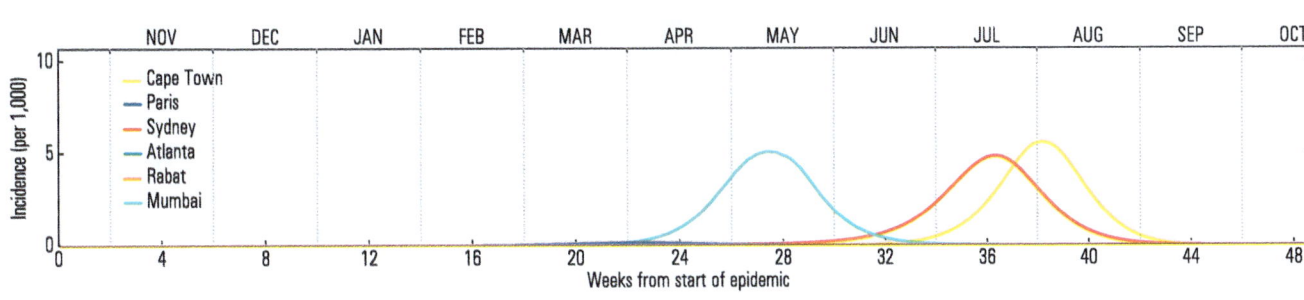

▲ URBAN LEVEL INCIDENCE

▼ RISK QUAD CHART

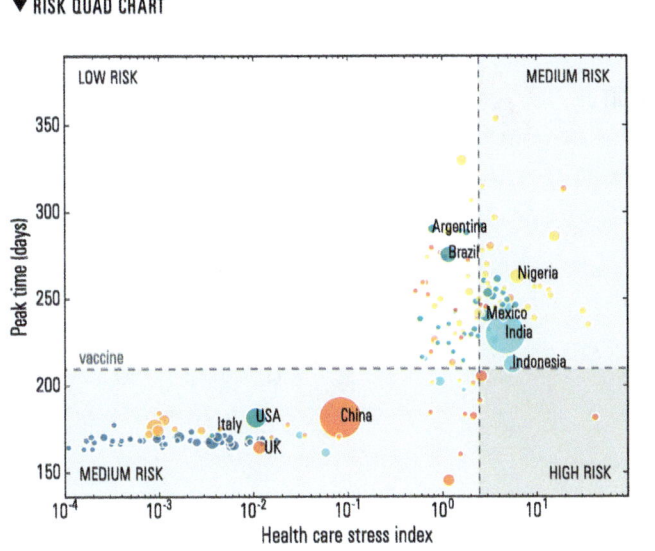

▼ ATTACK RATE COMPOSITION (ONE YEAR)

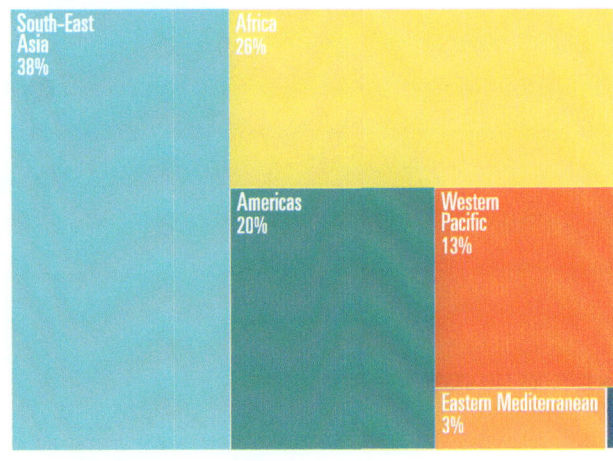

HANOI, VIETNAM

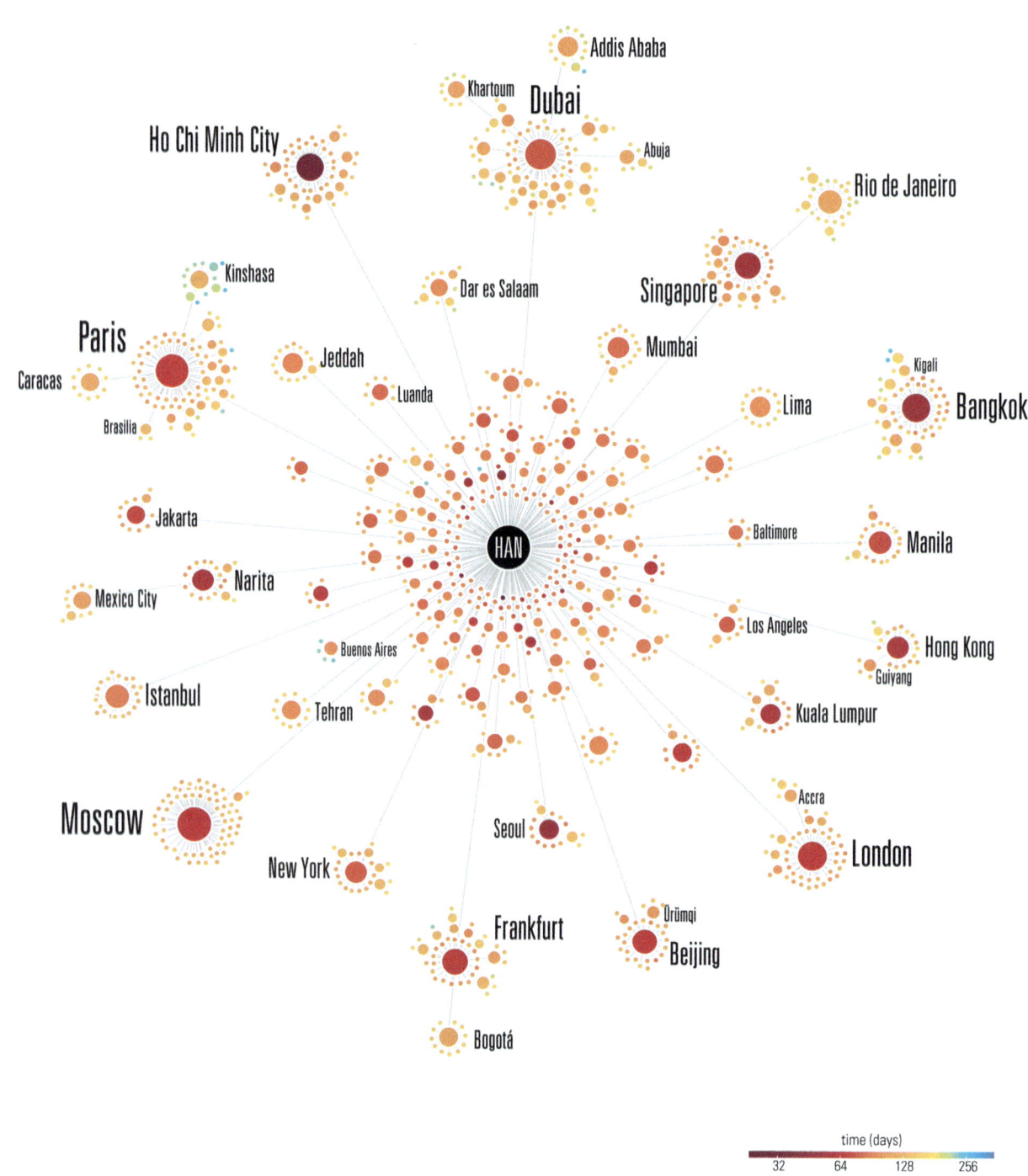

Ho Chi Minh City

Addis Ababa

Khartoum

Dubai

Abuja

Rio de Janeiro

Kinshasa

Dar es Salaam

Singapore

Paris

Jeddah

Mumbai

Caracas

Luanda

Kigali

Brasilia

Lima

Bangkok

Jakarta

Baltimore

Manila

Mexico City

Narita

HAN

Los Angeles

Hong Kong

Buenos Aires

Guiyang

Istanbul

Kuala Lumpur

Tehran

Accra

Moscow

Seoul

London

New York

Orümqi

Frankfurt

Beijing

Bogotá

time (days)

32   64   128   256

▲ INFECTION TREE GRAPH

■ Americas
■ Africa
■ Europe
■ E. Mediterranean
■ South-East Asia
■ Western Pacific

OCT

R₀ 1.5

80% AIR TRAFFIC REDUCTION

NOV   DEC   JAN   FEB   MAR   APR   MAY   JUN   JUL   AUG   SEP   OCT

Incidence (per 1,000)

Cumulative (per 1,000)

Weeks from start of epidemic

▲ CUMULATIVE INCIDENCE AND EPIDEMIC PEAKS

▼ ATTACK RATE COMPOSITION (ONE YEAR)

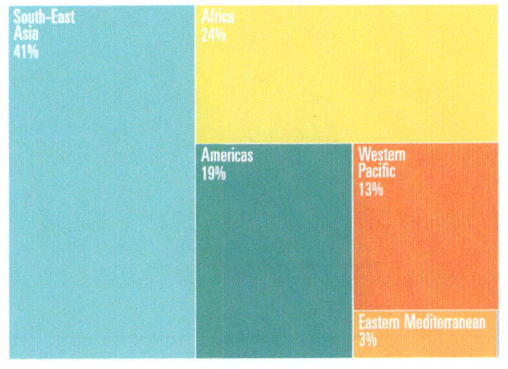

South-East Asia 41%
Africa 24%
Americas 19%
Western Pacific 13%
Eastern Mediterranean 3%

▼ EPIDEMIC ACTIVITY

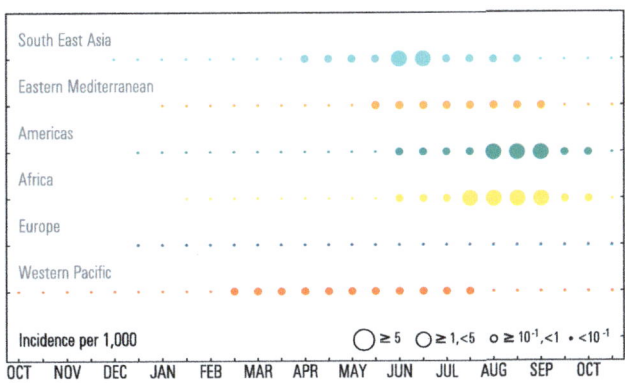

South East Asia
Eastern Mediterranean
Americas
Africa
Europe
Western Pacific

Incidence per 1,000   ◯ ≥ 5   ◯ ≥ 1,<5   ○ ≥ 10⁻¹,<1   • <10⁻¹

OCT   NOV   DEC   JAN   FEB   MAR   APR   MAY   JUN   JUL   AUG   SEP   OCT

VACCINATIONS

NOV   DEC   JAN   FEB   MAR   APR   MAY   JUN   JUL   AUG   SEP   OCT

Incidence (per 1,000)

Cumulative (per 1,000)

Weeks from start of epidemic

▲ CUMULATIVE INCIDENCE AND EPIDEMIC PEAKS

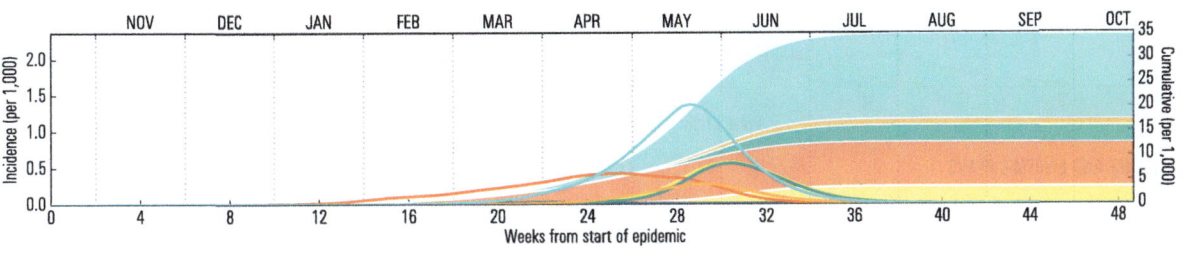

▼ ATTACK RATE COMPOSITION (ONE YEAR)

South-East Asia 53%
Western Pacific 25%
Africa 10%
Americas 9%
Eastern Med. 3%

▼ EPIDEMIC ACTIVITY

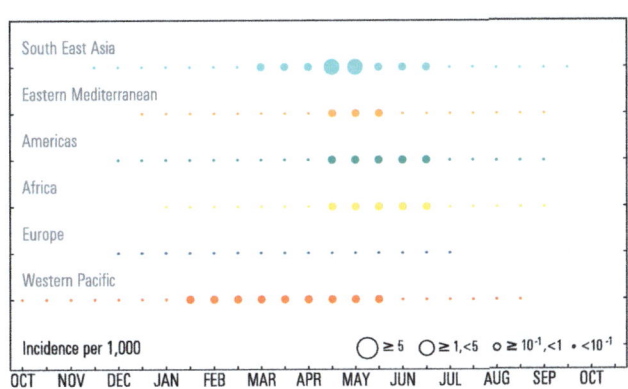

South East Asia
Eastern Mediterranean
Americas
Africa
Europe
Western Pacific

Incidence per 1,000   ◯ ≥ 5   ◯ ≥ 1,<5   ○ ≥ 10⁻¹,<1   • <10⁻¹

OCT   NOV   DEC   JAN   FEB   MAR   APR   MAY   JUN   JUL   AUG   SEP   OCT

HANOI, VIETNAM

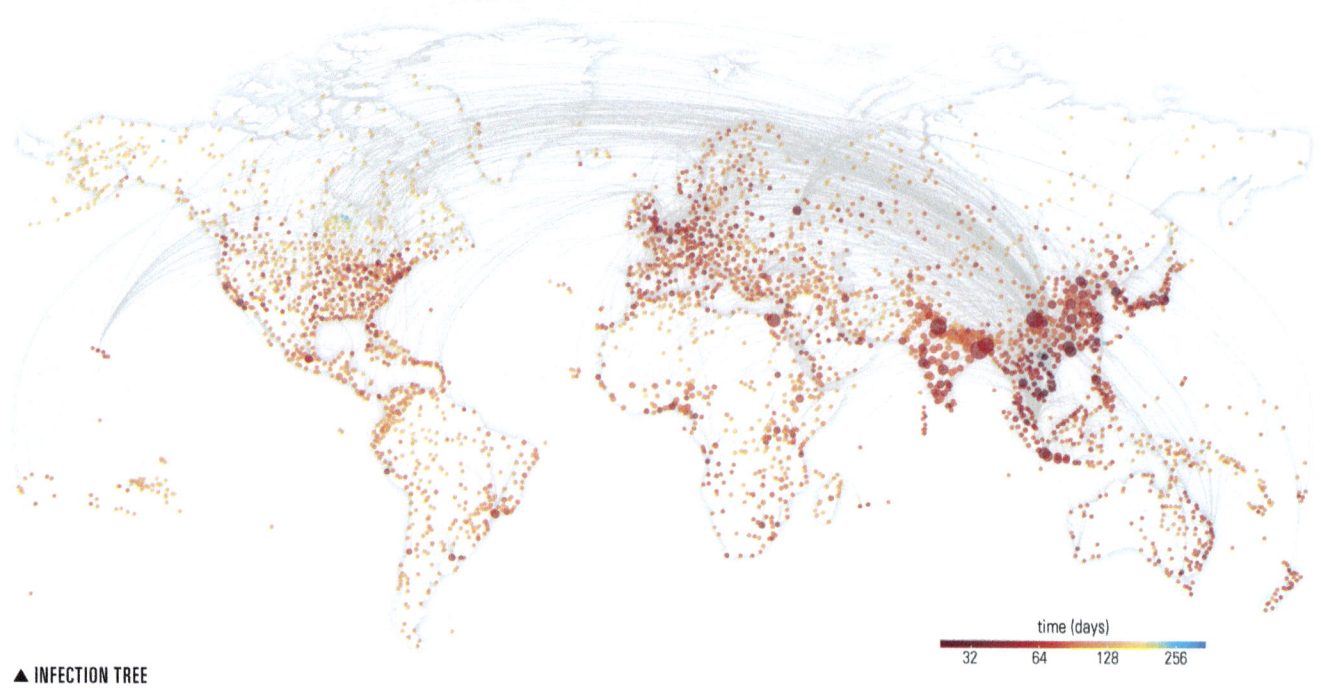

▲ INFECTION TREE

time (days)
32   64   128   256

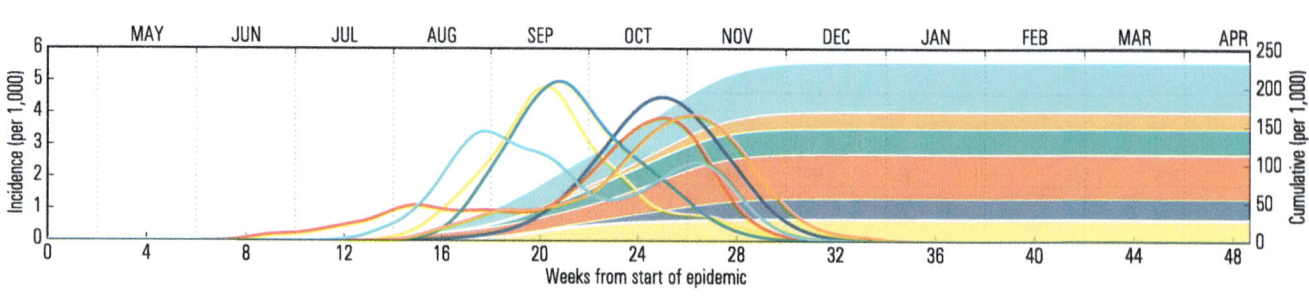

▲ CUMULATIVE INCIDENCE AND EPIDEMIC PEAKS

▼ AFFECTED COUNTRIES

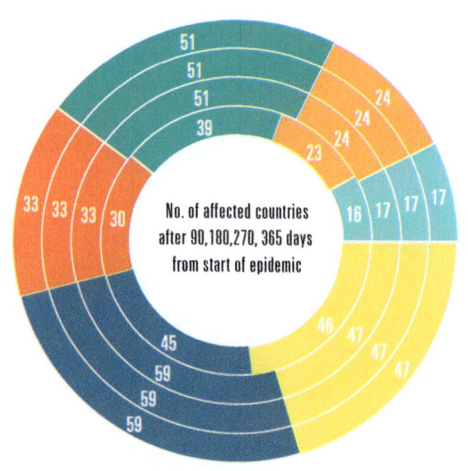

No. of affected countries after 90,180,270,365 days from start of epidemic

▼ EPIDEMIC ACTIVITY

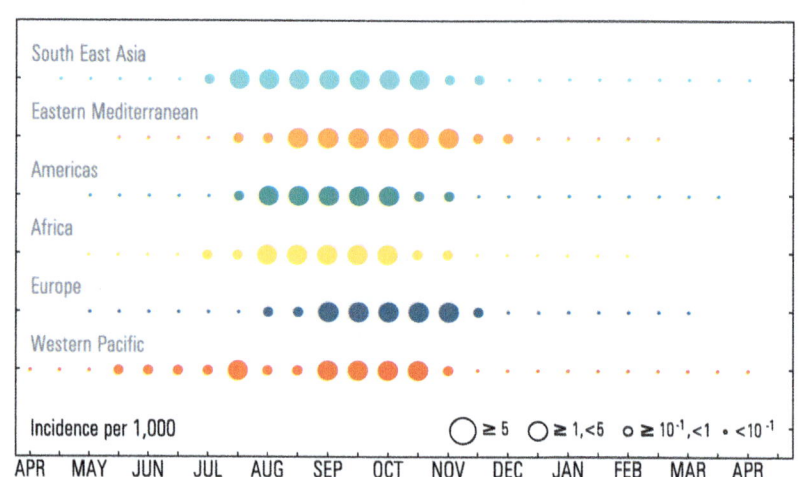

South East Asia

Eastern Mediterranean

Americas

Africa

Europe

Western Pacific

Incidence per 1,000

○ ≥ 5   ○ ≥ 1,<5   ○ ≥ 10⁻¹,<1   · <10⁻¹

APR MAY JUN JUL AUG SEP OCT NOV DEC JAN FEB MAR APR

■ Americas   ■ E. Mediterranean
■ Africa   ■ South-East Asia
■ Europe   ■ Western Pacific

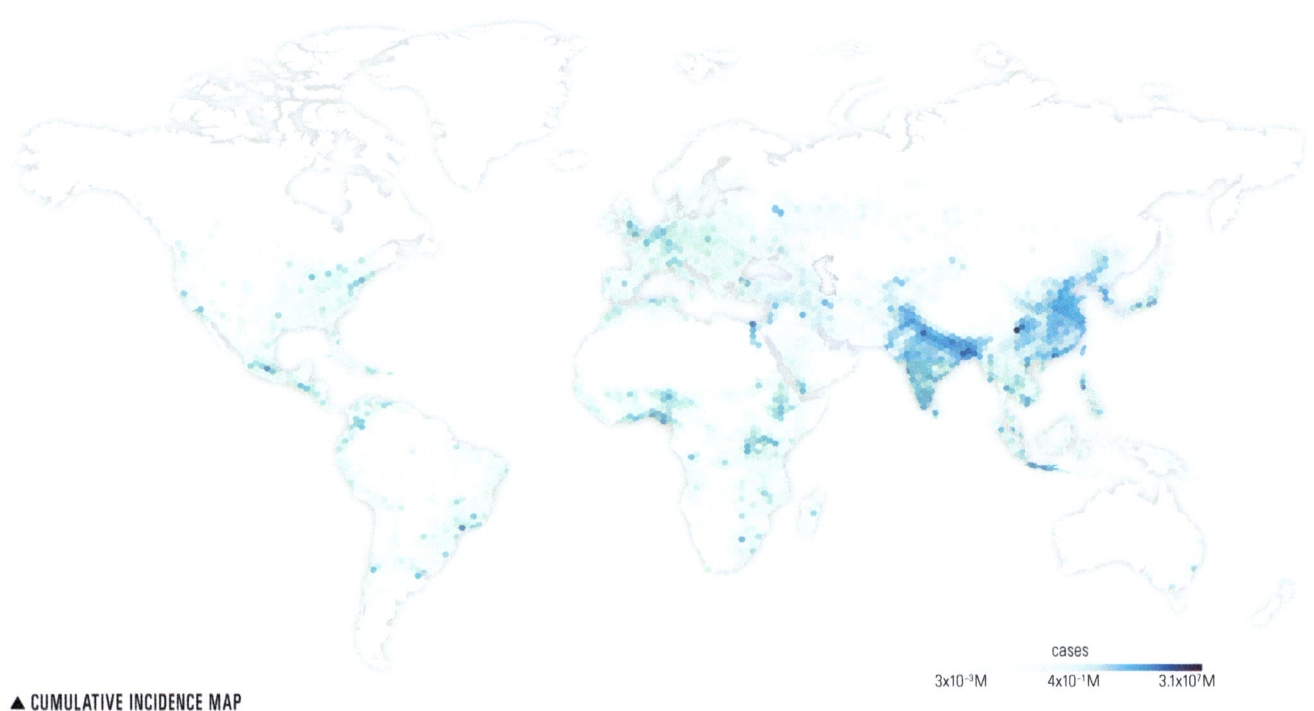

cases

3x10⁻³M   4x10⁻¹M   3.1x10⁷M

▲ CUMULATIVE INCIDENCE MAP

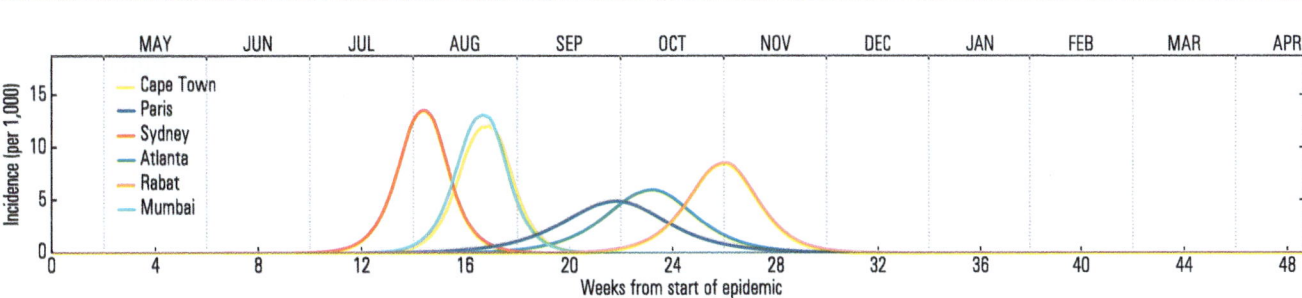

▲ URBAN LEVEL INCIDENCE

▼ RISK QUAD CHART

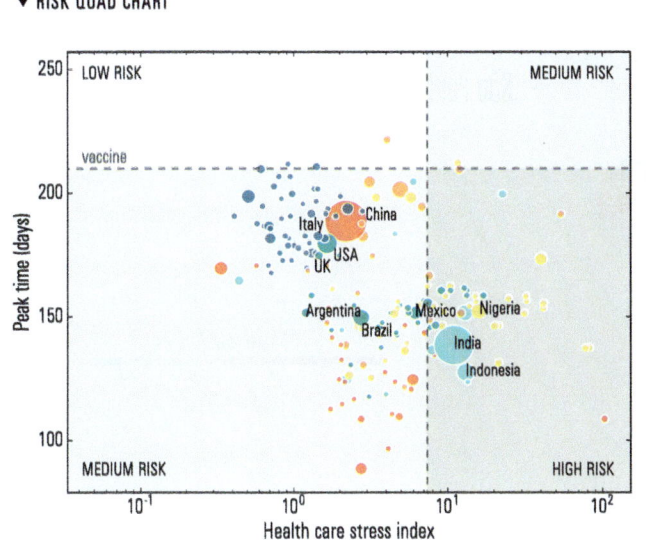

▼ ATTACK RATE COMPOSITION (ONE YEAR)

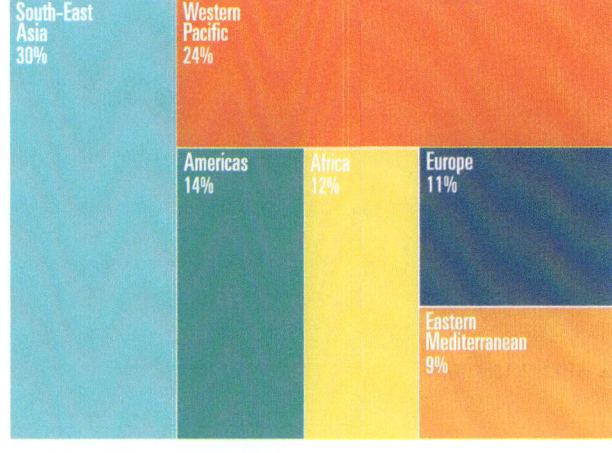

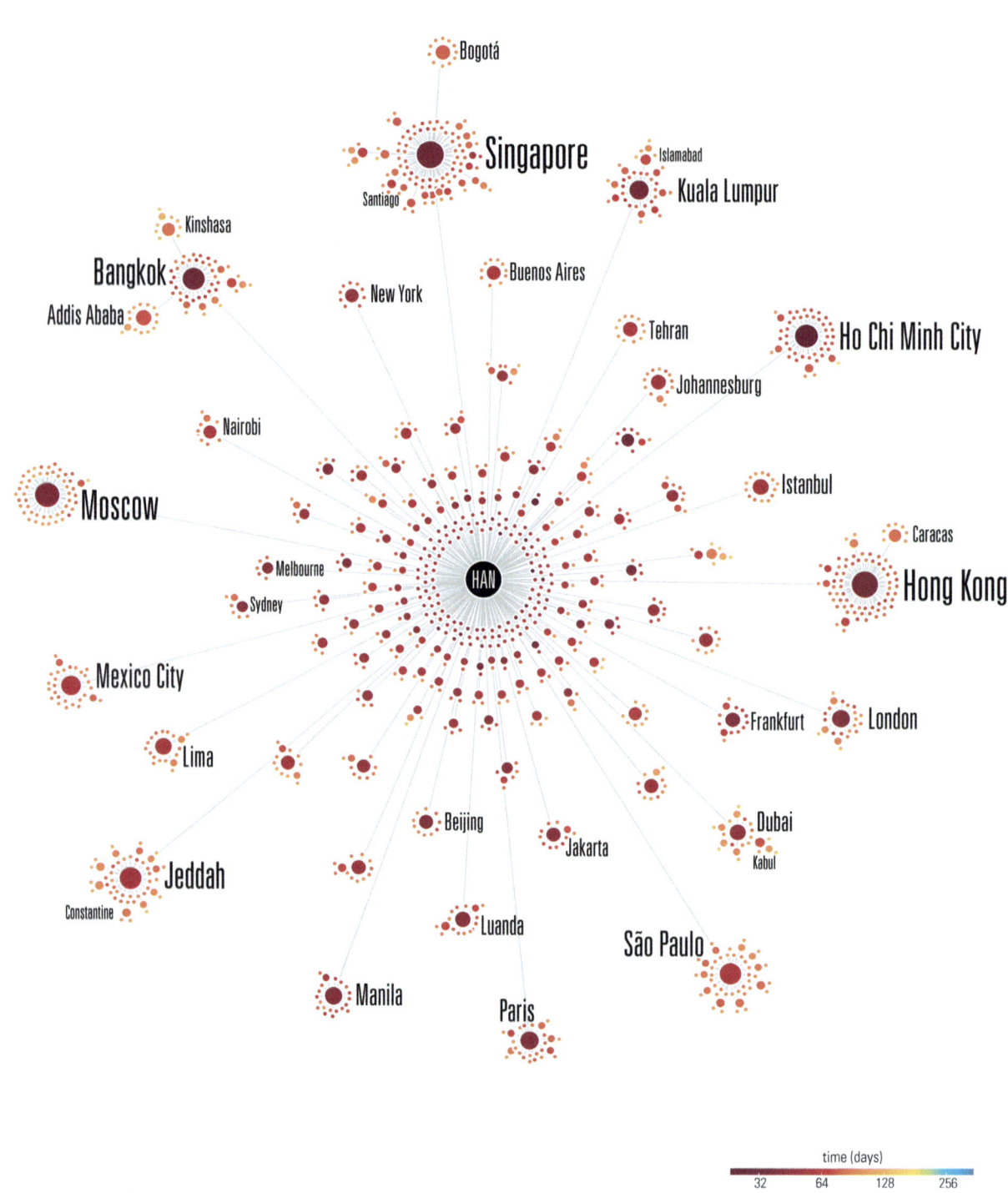

Bogotá

Singapore

Islamabad

Kuala Lumpur

Santiago

Kinshasa

Bangkok

Buenos Aires

New York

Addis Ababa

Tehran

Ho Chi Minh City

Johannesburg

Nairobi

Istanbul

Moscow

Caracas

Melbourne

HAN

Hong Kong

Sydney

Mexico City

Frankfurt

London

Lima

Dubai

Jeddah

Beijing

Jakarta

Kabul

Constantine

Luanda

São Paulo

Manila

Paris

time (days)

32   64   128   256

▲ INFECTION TREE GRAPH

■ Americas   ■ E. Mediterranean
■ Africa   ■ South-East Asia
■ Europe   ■ Western Pacific

INITIAL CONDITIONS

 APR    2.0 | 

80% AIR TRAFFIC REDUCTION

▲ CUMULATIVE INCIDENCE AND EPIDEMIC PEAKS

▼ ATTACK RATE COMPOSITION (ONE YEAR)

▼ EPIDEMIC ACTIVITY

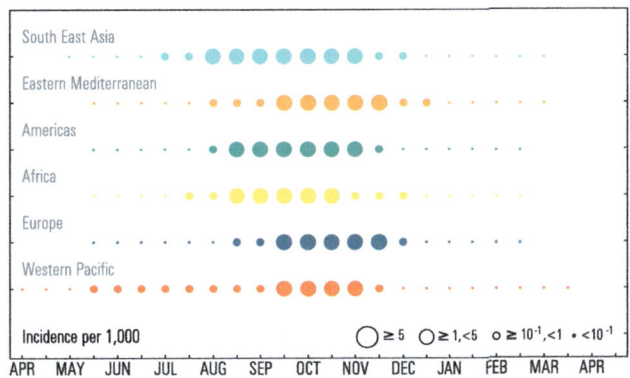

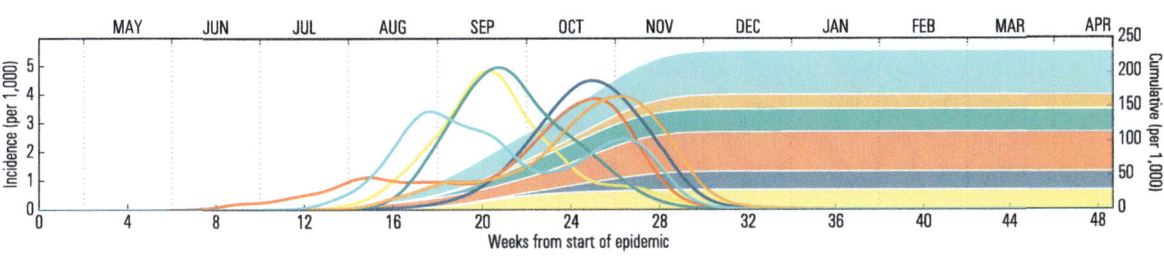

Incidence per 1,000   ○ ≥ 5   ○ ≥ 1,<5   ○ ≥ 10⁻¹,<1   • <10⁻¹

VACCINATIONS

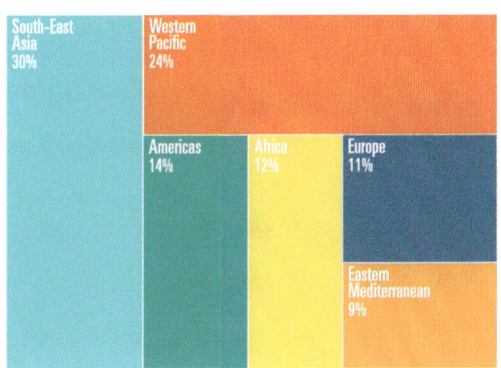

▲ CUMULATIVE INCIDENCE AND EPIDEMIC PEAKS

▼ ATTACK RATE COMPOSITION (ONE YEAR)

▼ EPIDEMIC ACTIVITY

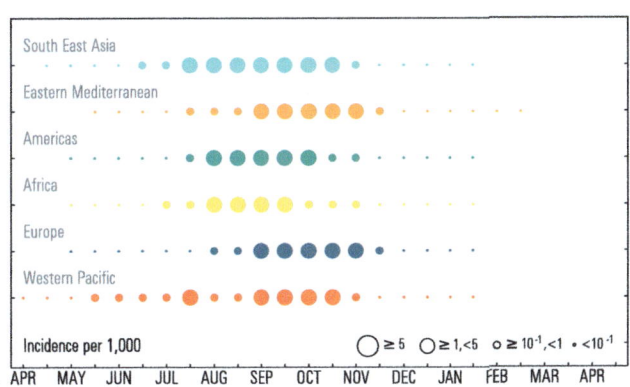

Incidence per 1,000   ○ ≥ 5   ○ ≥ 1,<5   ○ ≥ 10⁻¹,<1   • <10⁻¹

HANOI, VIETNAM

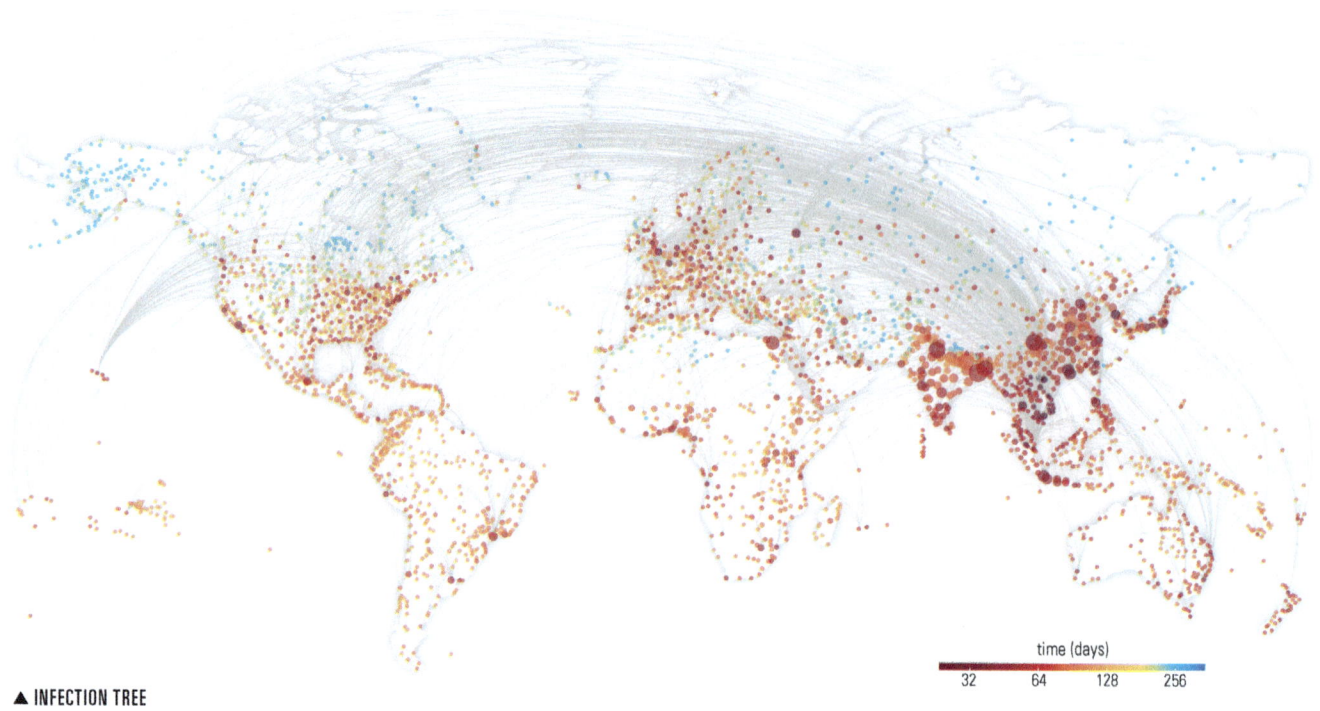

time (days)

32   64   128   256

▲ INFECTION TREE

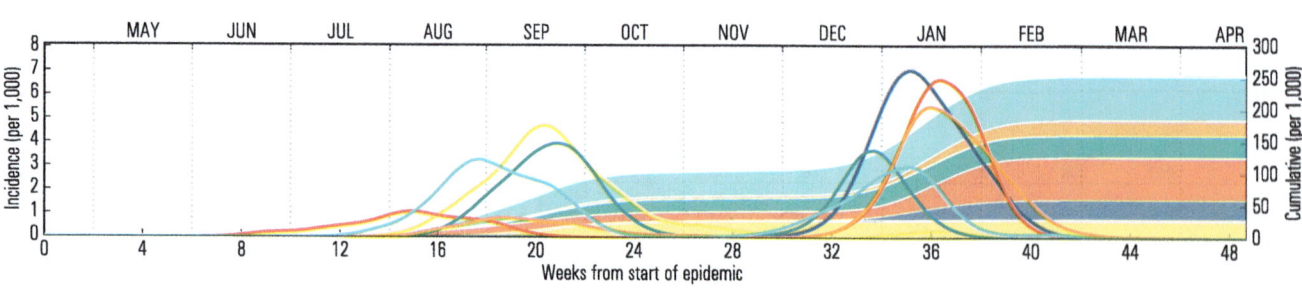

▲ CUMULATIVE INCIDENCE AND EPIDEMIC PEAKS

▼ AFFECTED COUNTRIES

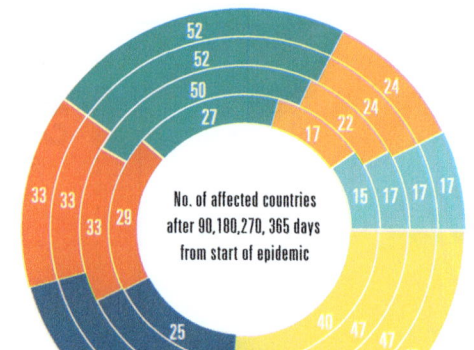

No. of affected countries
after 90,180,270, 365 days
from start of epidemic

▼ EPIDEMIC ACTIVITY

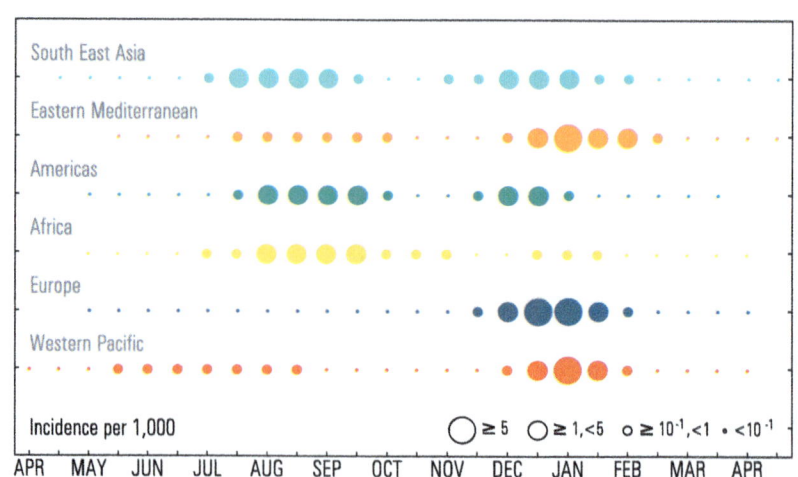

South East Asia

Eastern Mediterranean

Americas

Africa

Europe

Western Pacific

Incidence per 1,000

◯ ≥ 5   ◯ ≥ 1,<5   ○ ≥ 10⁻¹,<1   • < 10⁻¹

APR MAY JUN JUL AUG SEP OCT NOV DEC JAN FEB MAR APR

- Americas
- Africa
- Europe
- E. Mediterranean
- South-East Asia
- Western Pacific

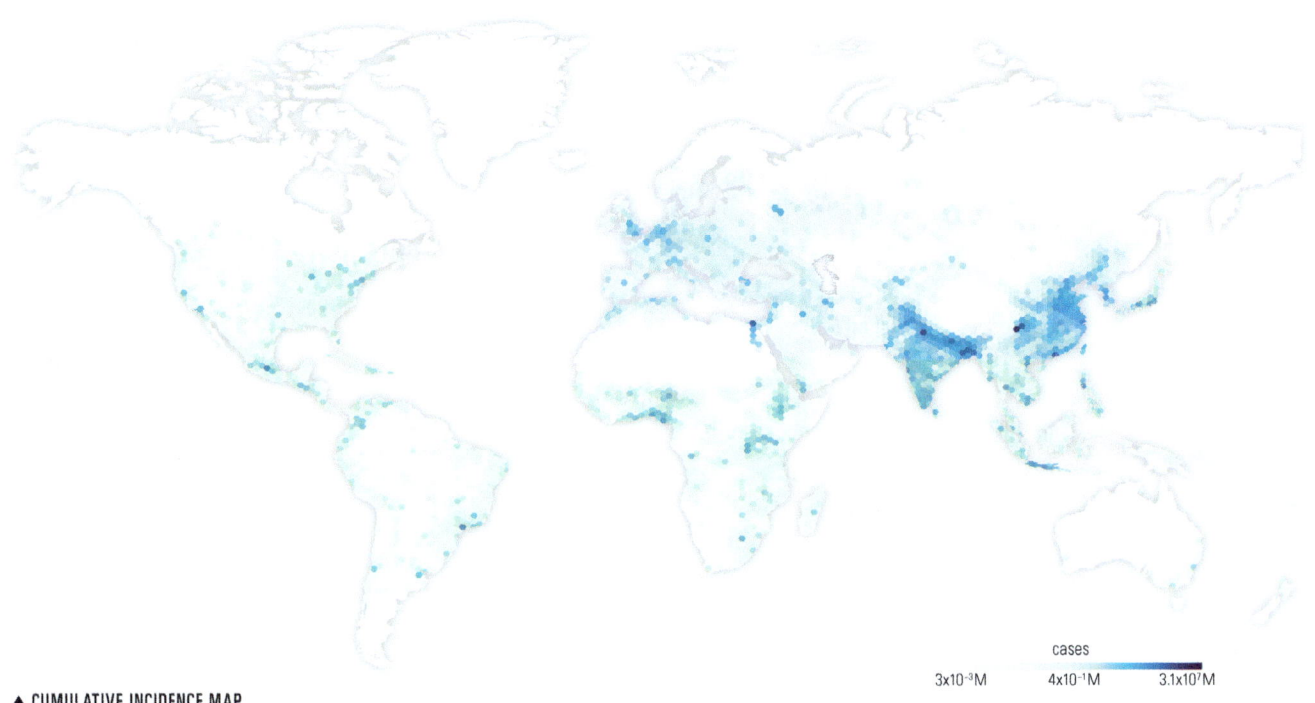

cases

3x10⁻³M   4x10⁻¹M   3.1x10⁷M

▲ CUMULATIVE INCIDENCE MAP

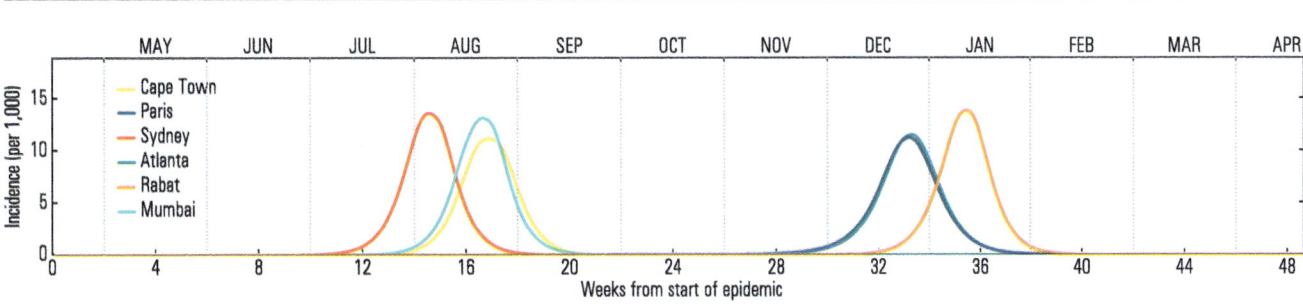

▲ URBAN LEVEL INCIDENCE

▼ RISK QUAD CHART

▼ ATTACK RATE COMPOSITION (ONE YEAR)

South-East Asia 30%
Western Pacific 25%
Americas 13%
Europe 12%
Africa 11%
Eastern Mediterranean 9%

HANOI, VIETNAM

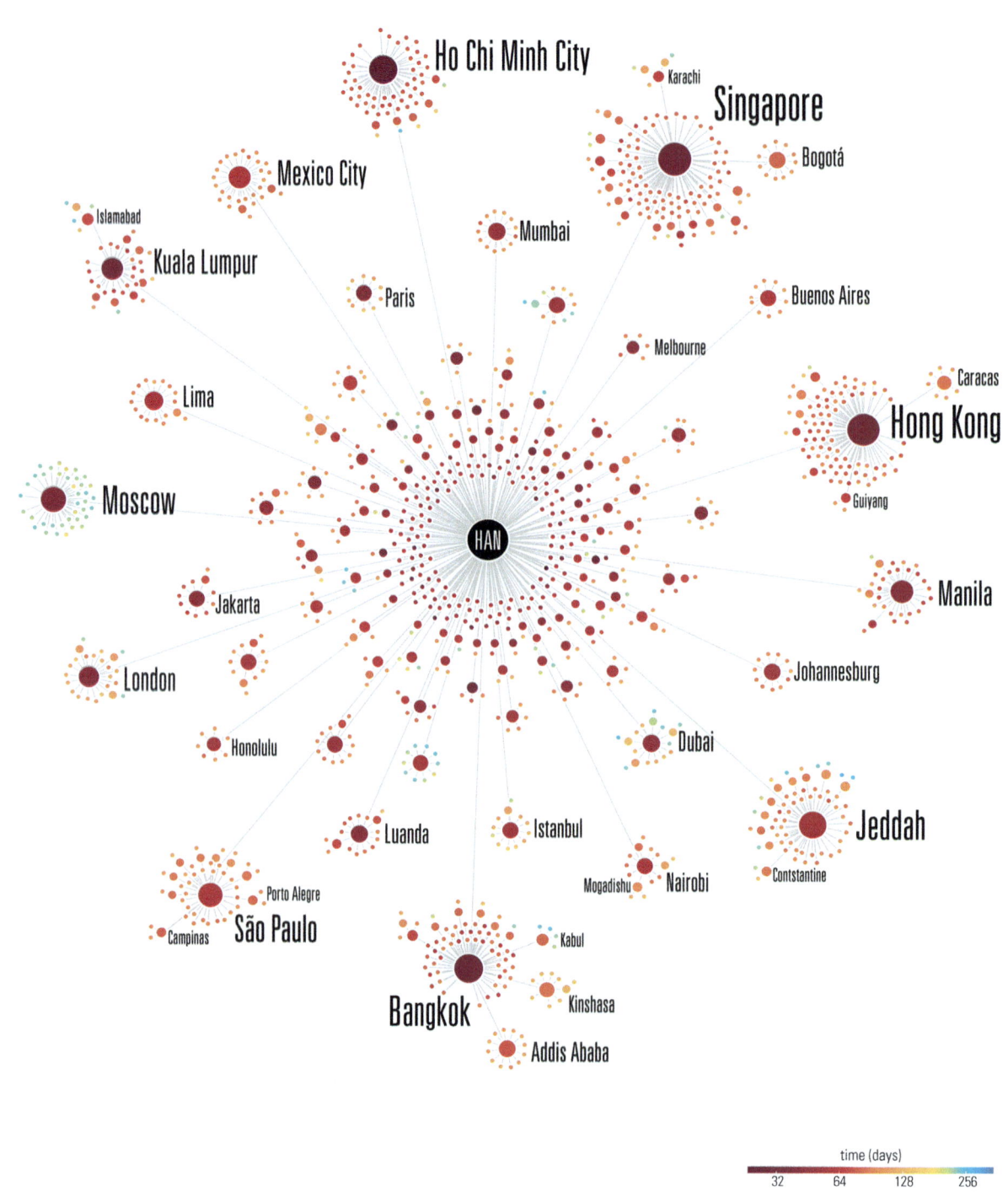

Ho Chi Minh City

Karachi

Singapore

Mexico City

Bogotá

Islamabad

Mumbai

Kuala Lumpur

Paris

Buenos Aires

Melbourne

Caracas

Lima

Hong Kong

Moscow

HAN

Guiyang

Jakarta

Manila

London

Johannesburg

Honolulu

Dubai

Luanda

Istanbul

Jeddah

Mogadishu

Nairobi

Contstantine

Porto Alegre

São Paulo

Kabul

Campinas

Kinshasa

Bangkok

Addis Ababa

time (days)

| 32 | 64 | 128 | 256 |

▲ INFECTION TREE GRAPH

80% AIR TRAFFIC REDUCTION

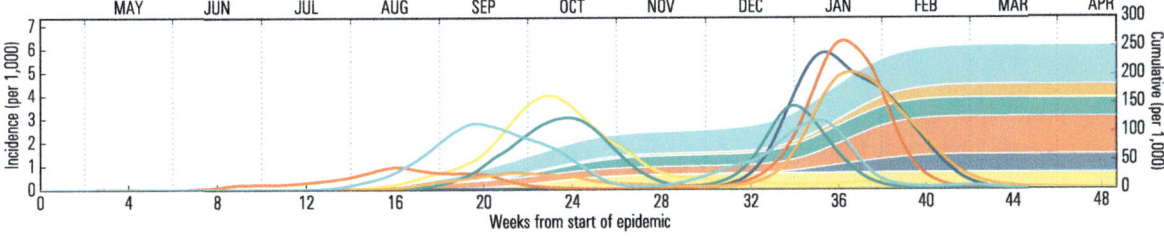

▲ CUMULATIVE INCIDENCE AND EPIDEMIC PEAKS

▼ ATTACK RATE COMPOSITION (ONE YEAR)

▼ EPIDEMIC ACTIVITY

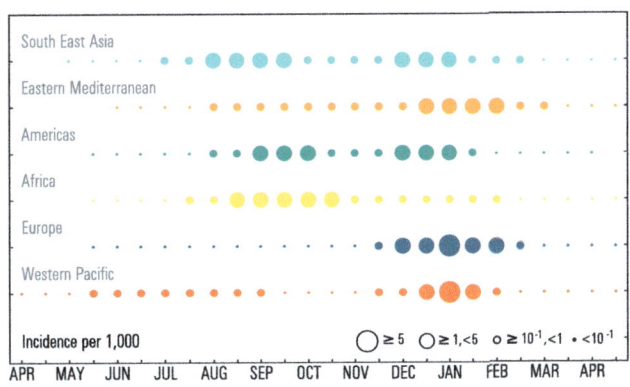

VACCINATIONS

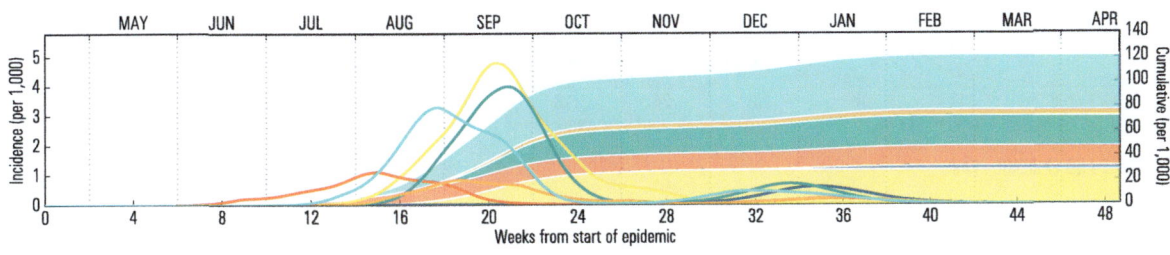

▲ CUMULATIVE INCIDENCE AND EPIDEMIC PEAKS

▼ ATTACK RATE COMPOSITION (ONE YEAR)

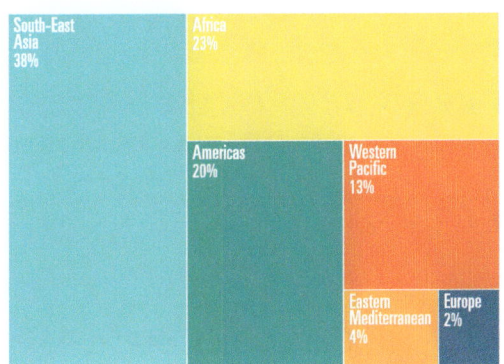

▼ EPIDEMIC ACTIVITY

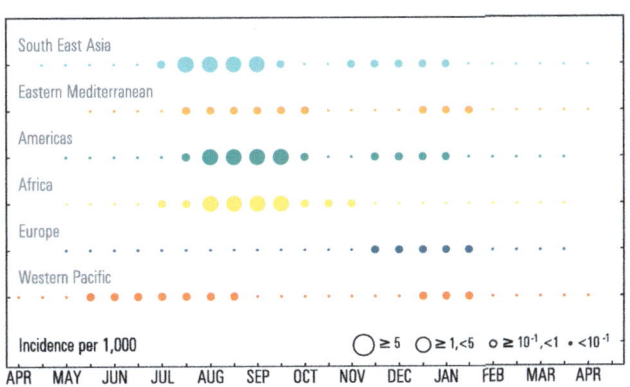

HANOI, VIETNAM

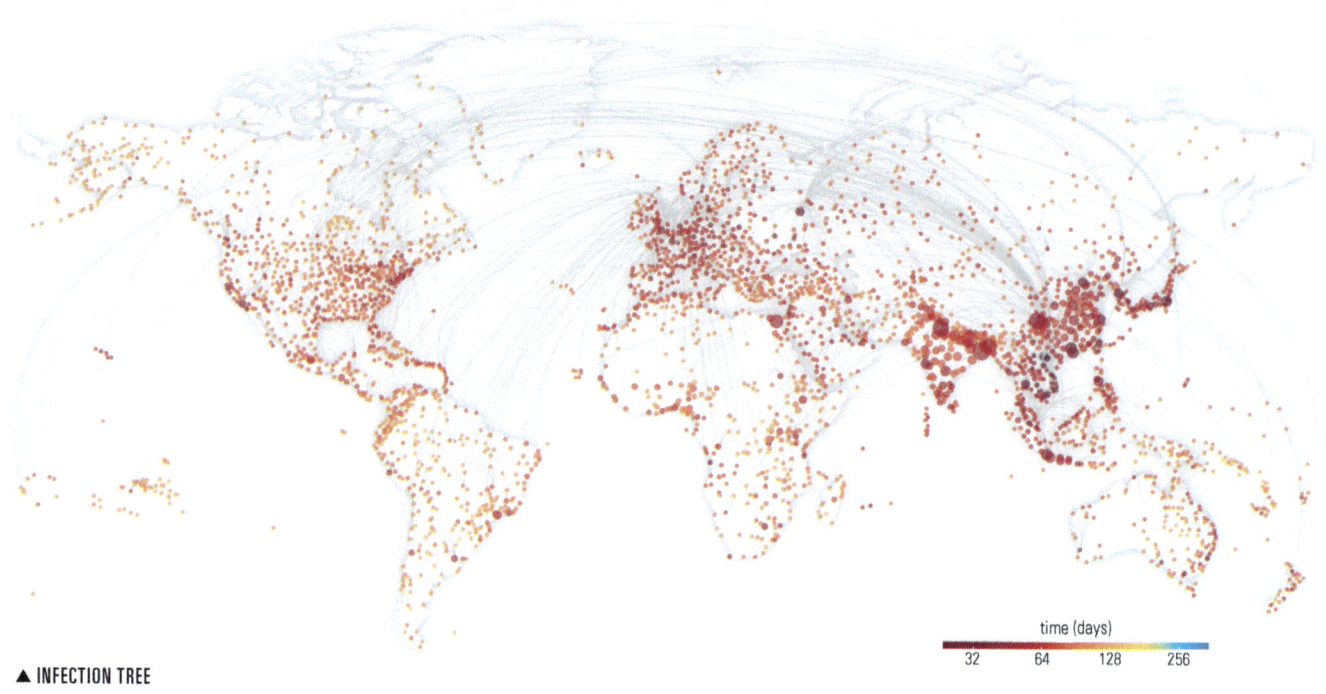

time (days)
32   64   128   256

▲ INFECTION TREE

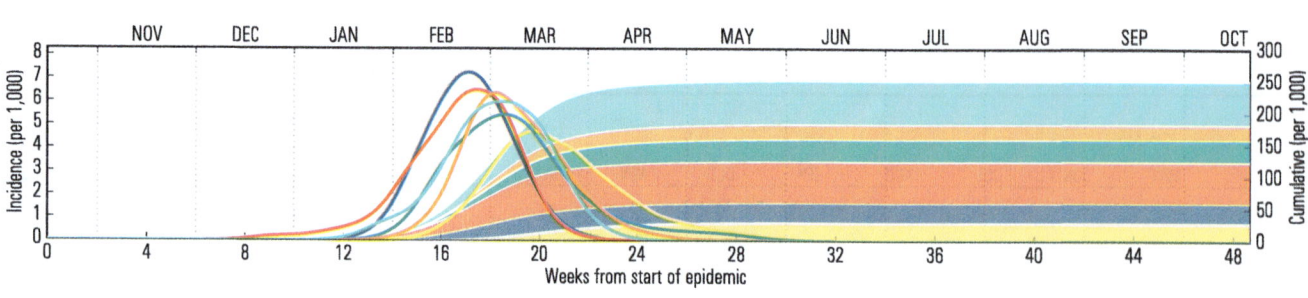

▲ CUMULATIVE INCIDENCE AND EPIDEMIC PEAKS

▼ AFFECTED COUNTRIES

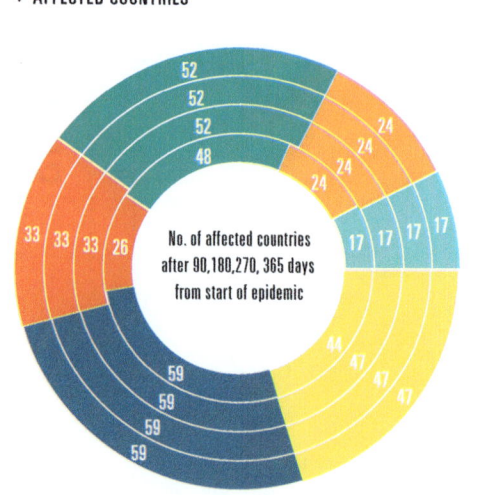

No. of affected countries after 90,180,270, 365 days from start of epidemic

52 · 52 · 52 · 48 · 24 · 24 · 24 · 24 · 17 · 17 · 17 · 17 · 33 · 33 · 33 · 26 · 59 · 59 · 59 · 59 · 44 · 41 · 41 · 41

▼ EPIDEMIC ACTIVITY

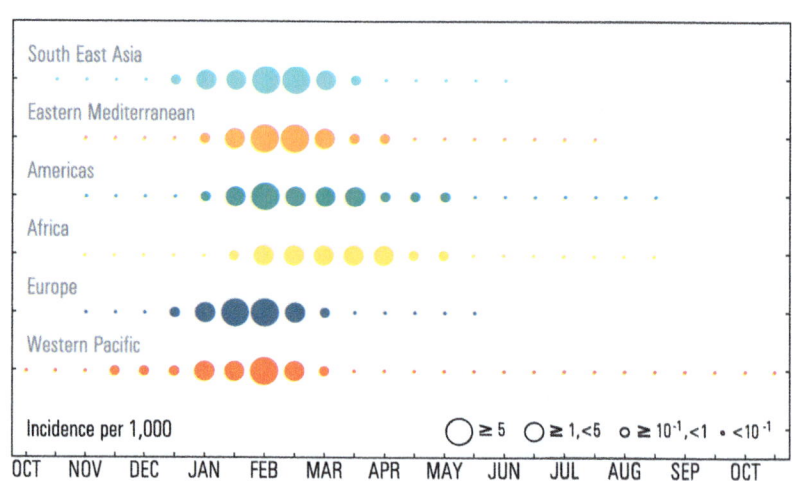

South East Asia

Eastern Mediterranean

Americas

Africa

Europe

Western Pacific

Incidence per 1,000

○ ≥ 5   ○ ≥ 1,<5   ○ ≥ 10⁻¹,<1   · <10⁻¹

OCT   NOV   DEC   JAN   FEB   MAR   APR   MAY   JUN   JUL   AUG   SEP   OCT

WHO REGIONS
- Americas
- Africa
- Europe
- E. Mediterranean
- South-East Asia
- Western Pacific

INITIAL CONDITIONS

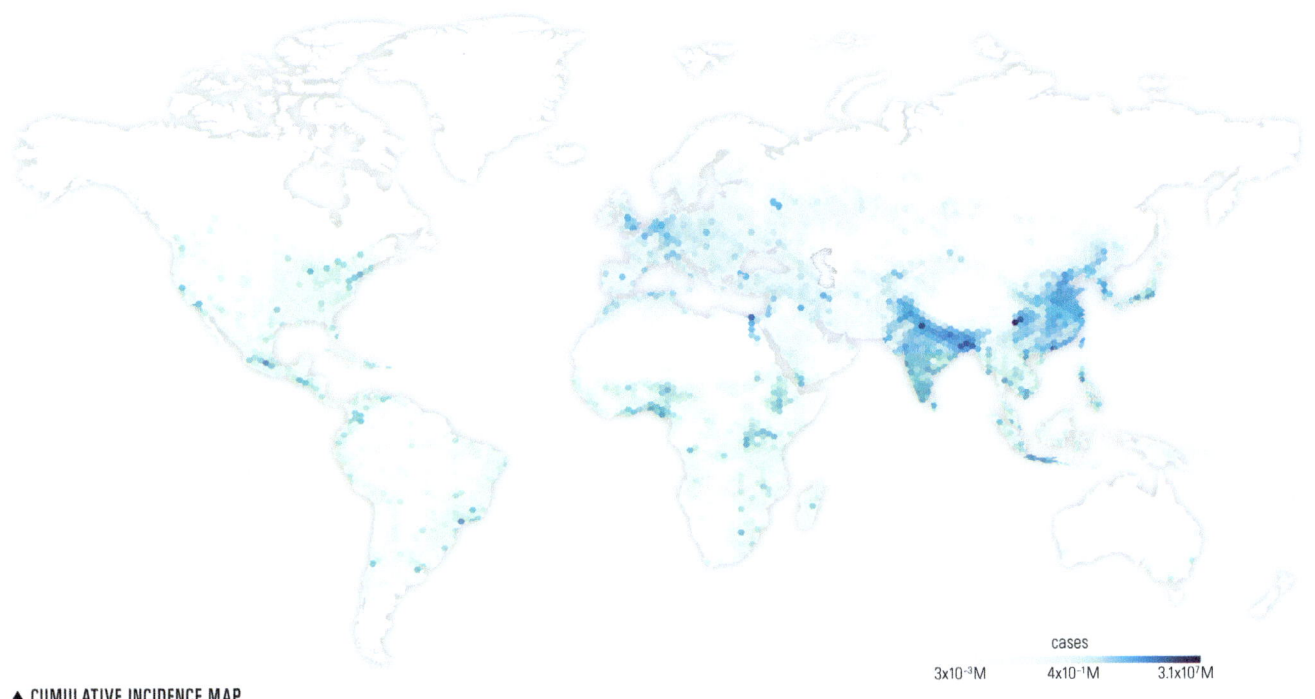

cases

3×10^{-3}M 4×10^{-1}M 3.1×10^{7}M

▲ CUMULATIVE INCIDENCE MAP

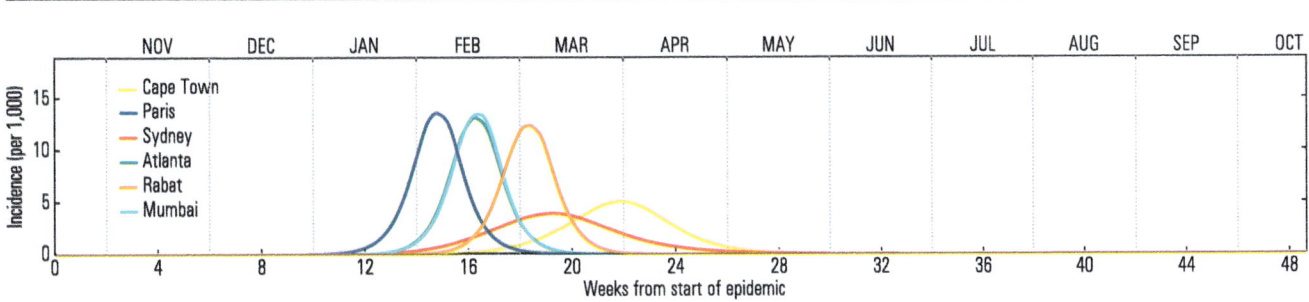

▲ URBAN LEVEL INCIDENCE

▼ RISK QUAD CHART

▼ ATTACK RATE COMPOSITION (ONE YEAR)

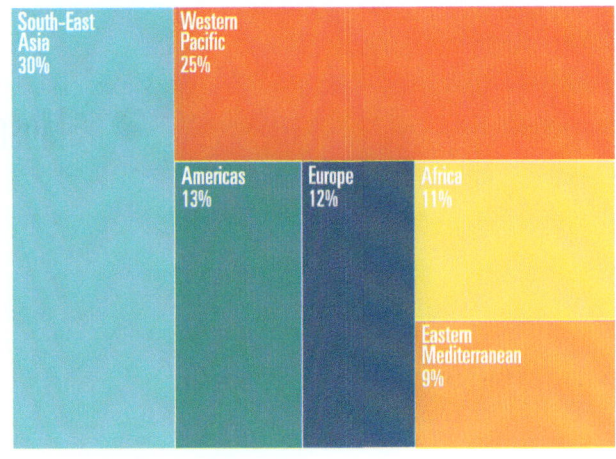

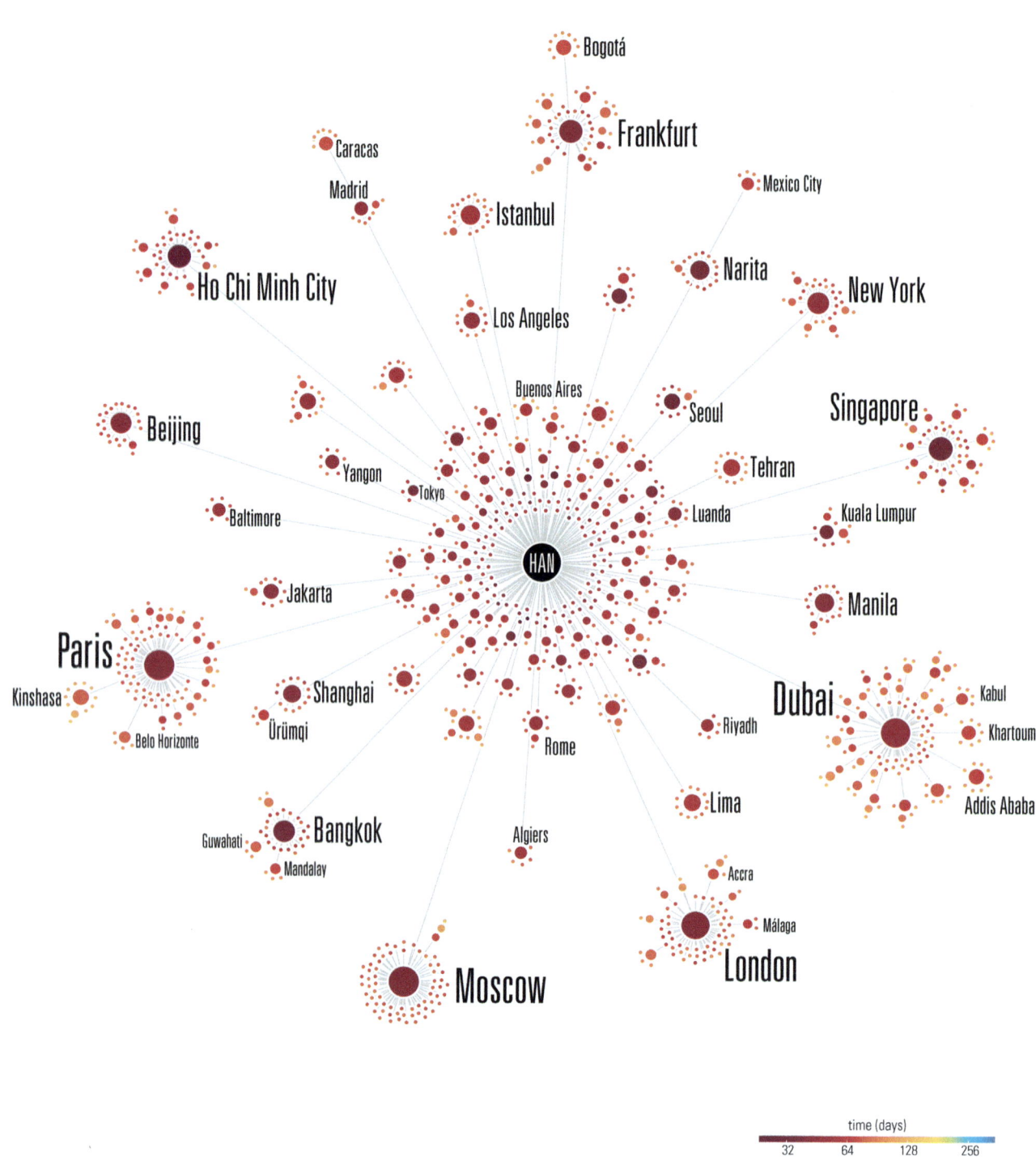

HANOI, VIETNAM

Bogotá

Frankfurt

Caracas

Madrid

Istanbul

Mexico City

Ho Chi Minh City

Narita

New York

Los Angeles

Beijing

Buenos Aires

Seoul

Singapore

Yangon

Tehran

Tokyo

Kuala Lumpur

Baltimore

Luanda

Jakarta

HAN

Manila

Paris

Shanghai

Dubai

Kabul

Kinshasa

Ürümqi

Riyadh

Khartoum

Belo Horizonte

Rome

Addis Ababa

Bangkok

Lima

Guwahati

Algiers

Accra

Mandalay

Málaga

Moscow

London

time (days)

32   64   128   256

▲ INFECTION TREE GRAPH

WHO REGIONS

- Americas
- Africa
- Europe
- E. Mediterranean
- South-East Asia
- Western Pacific

INITIAL CONDITIONS

 OCT
 R₀ 2.0

80% AIR TRAFFIC REDUCTION

▲ CUMULATIVE INCIDENCE AND EPIDEMIC PEAKS

▼ ATTACK RATE COMPOSITION (ONE YEAR)

South-East Asia 29% | Western Pacific 25% | Americas 14% | Europe 12% | Africa 11% | Eastern Mediterranean 9%

▼ EPIDEMIC ACTIVITY

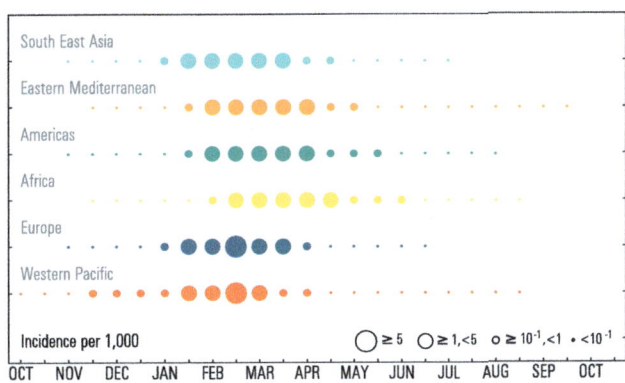

Incidence per 1,000   ○ ≥ 5   ○ ≥ 1,<5   ○ ≥ 10⁻¹,<1   • <10⁻¹

VACCINATIONS

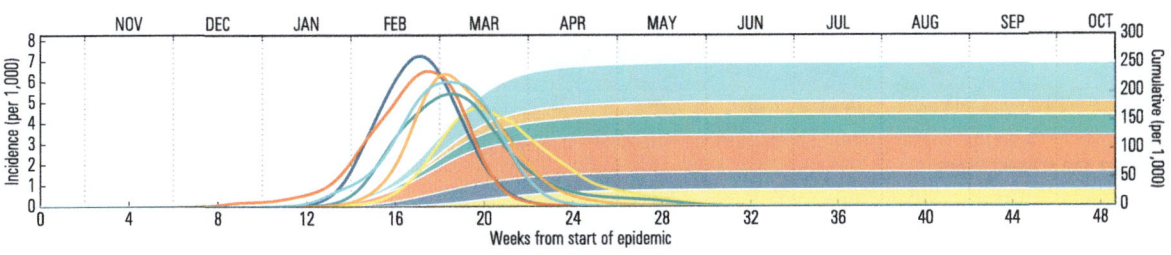

▲ CUMULATIVE INCIDENCE AND EPIDEMIC PEAKS

▼ ATTACK RATE COMPOSITION (ONE YEAR)

South-East Asia 30% | Western Pacific 25% | Americas 13% | Europe 12% | Africa 11% | Eastern Mediterranean 9%

▼ EPIDEMIC ACTIVITY

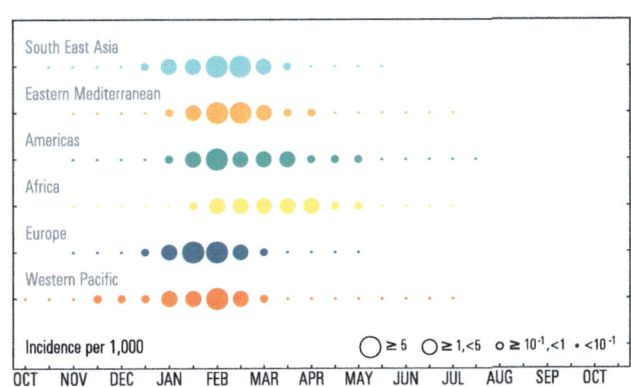

Incidence per 1,000   ○ ≥ 5   ○ ≥ 1,<5   ○ ≥ 10⁻¹,<1   • <10⁻¹

HANOI, VIETNAM

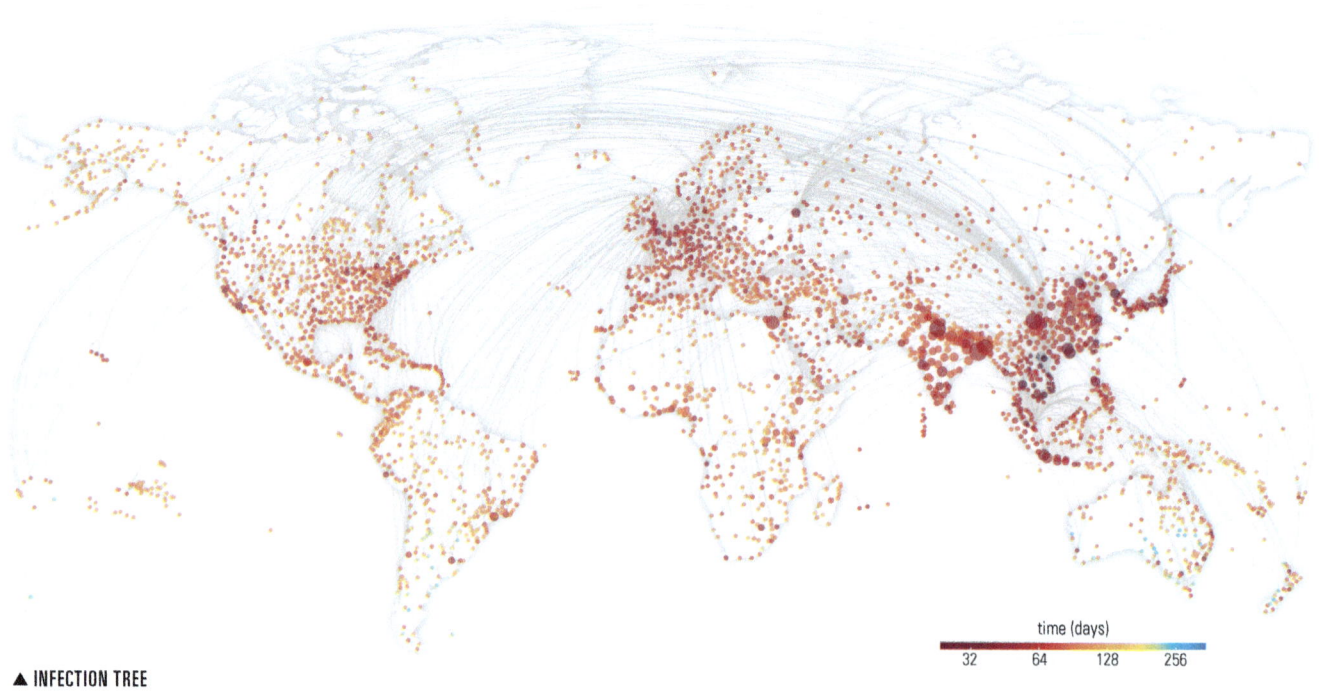

▲ INFECTION TREE

time (days)

32   64   128   256

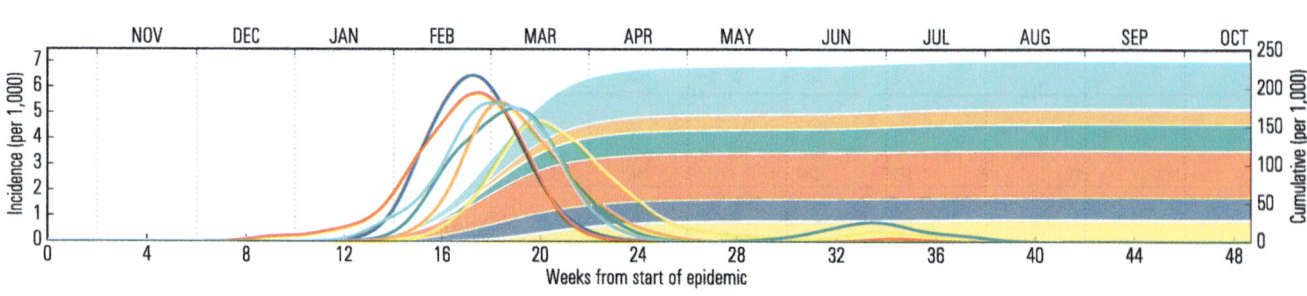

NOV   DEC   JAN   FEB   MAR   APR   MAY   JUN   JUL   AUG   SEP   OCT

Incidence (per 1,000)

Cumulative (per 1,000)

Weeks from start of epidemic

▲ CUMULATIVE INCIDENCE AND EPIDEMIC PEAKS

▼ AFFECTED COUNTRIES

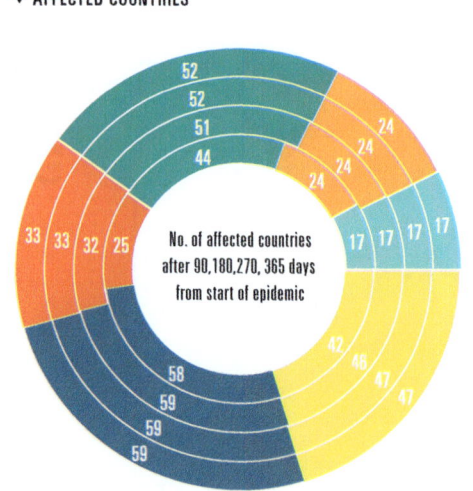

No. of affected countries after 90,180,270, 365 days from start of epidemic

▼ EPIDEMIC ACTIVITY

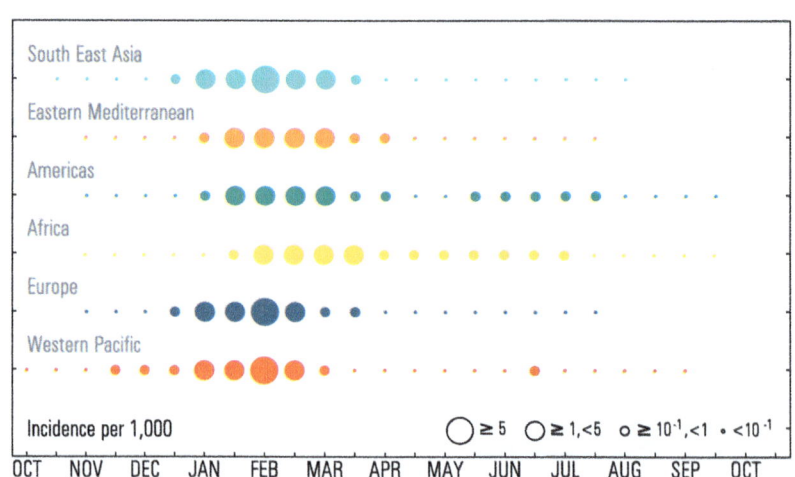

South East Asia

Eastern Mediterranean

Americas

Africa

Europe

Western Pacific

Incidence per 1,000

≥ 5   ≥ 1,<5   ≥ 10⁻¹,<1   < 10⁻¹

OCT   NOV   DEC   JAN   FEB   MAR   APR   MAY   JUN   JUL   AUG   SEP   OCT

- **Americas**
- **Africa**
- **Europe**
- **E. Mediterranean**
- **South-East Asia**
- **Western Pacific**

 OCT  R₀ 2.0

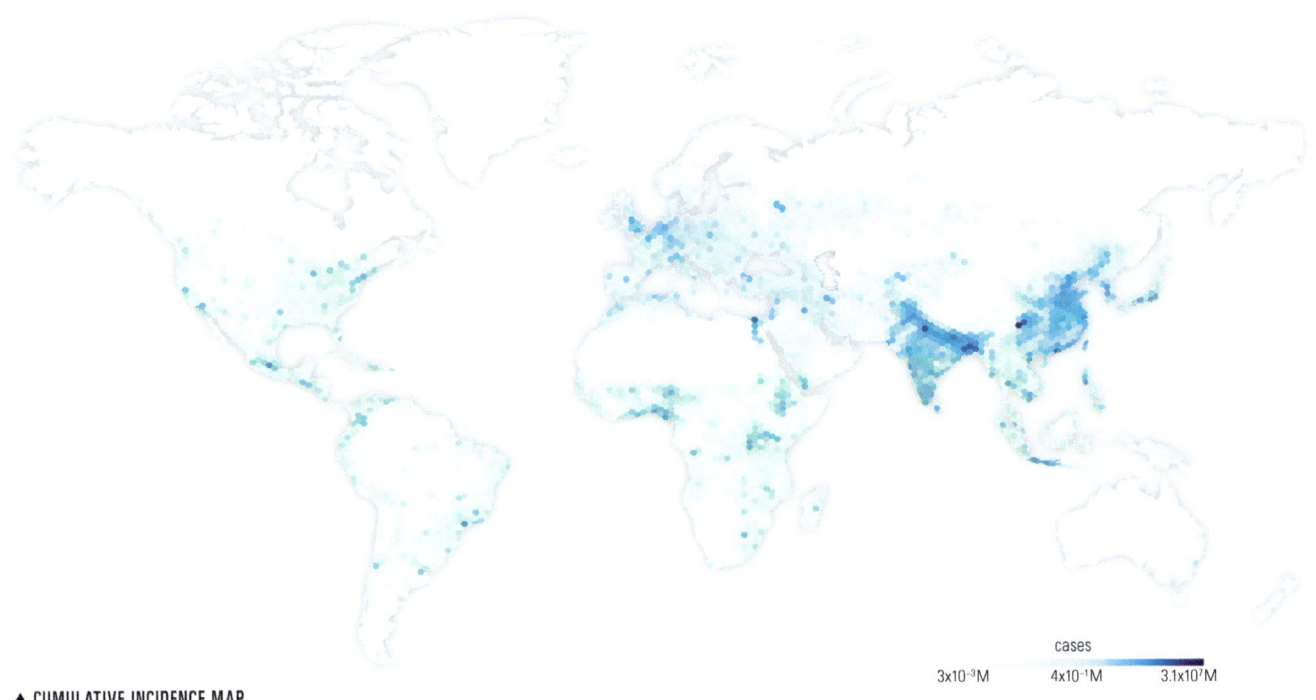

cases

3x10⁻³M   4x10⁻¹M   3.1x10⁷M

▲ CUMULATIVE INCIDENCE MAP

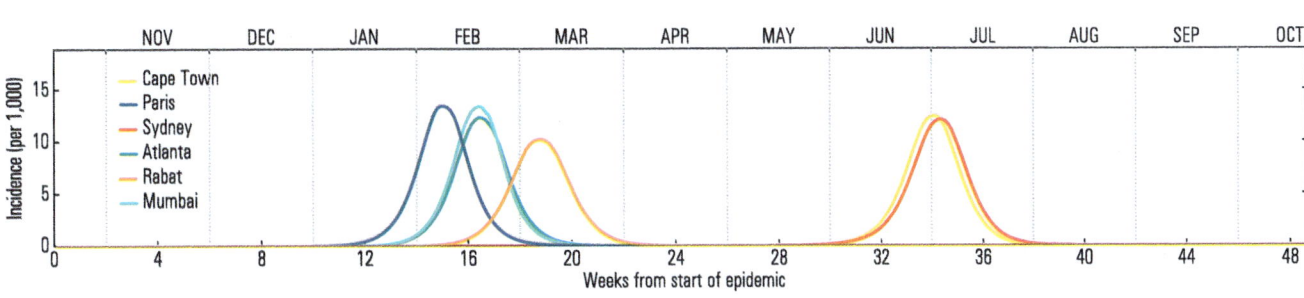

Legend:
- Cape Town
- Paris
- Sydney
- Atlanta
- Rabat
- Mumbai

Incidence (per 1,000)

Weeks from start of epidemic

▲ URBAN LEVEL INCIDENCE

▼ RISK QUAD CHART

LOW RISK   MEDIUM RISK

vaccine

Argentina

Brazil, Italy, China, USA, UK, Mexico, Nigeria, India, Indonesia

MEDIUM RISK   HIGH RISK

Peak time (days)

Health care stress index

▼ ATTACK RATE COMPOSITION (ONE YEAR)

- South-East Asia 29%
- Western Pacific 25%
- Americas 14%
- Africa 12%
- Europe 12%
- Eastern Mediterranean 8%

PANDEMIC CHARTS: PANDEMIC INFLUENZA   **139**

HANOI, VIETNAM

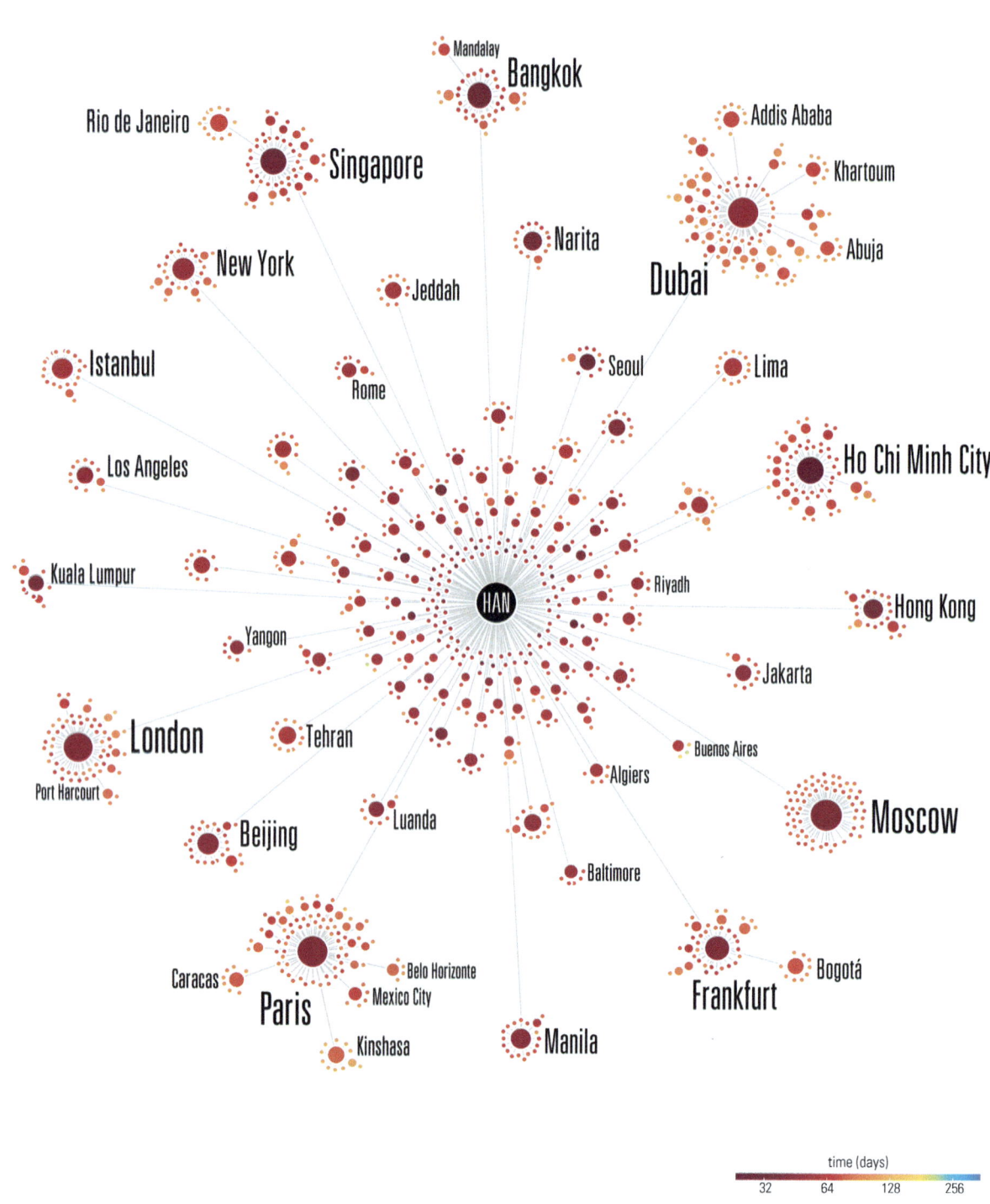

Mandalay

Bangkok

Rio de Janeiro

Singapore

Addis Ababa

Khartoum

New York

Narita

Jeddah

Dubai

Abuja

Istanbul

Rome

Seoul

Lima

Los Angeles

Ho Chi Minh City

Kuala Lumpur

HAN

Riyadh

Hong Kong

Yangon

Jakarta

London

Tehran

Buenos Aires

Port Harcourt

Algiers

Moscow

Beijing

Luanda

Baltimore

Caracas

Belo Horizonte

Bogotá

Mexico City

Frankfurt

Paris

Manila

Kinshasa

time (days)

32   64   128   256

▲ INFECTION TREE GRAPH

■ Americas ■ E. Mediterranean
■ Africa ■ South-East Asia
■ Europe ■ Western Pacific

 OCT  2.0

 **80% AIR TRAFFIC REDUCTION**

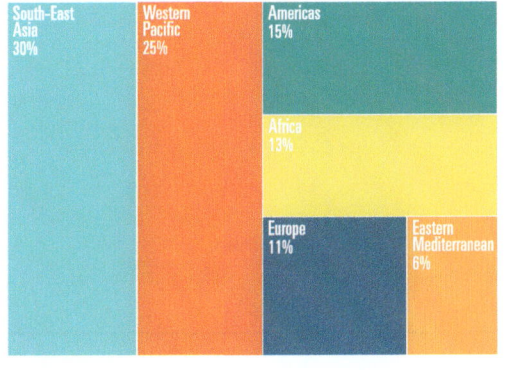

▲ CUMULATIVE INCIDENCE AND EPIDEMIC PEAKS

▼ ATTACK RATE COMPOSITION (ONE YEAR)

South-East Asia 30%
Western Pacific 25%
Americas 15%
Africa 13%
Europe 11%
Eastern Mediterranean 6%

▼ EPIDEMIC ACTIVITY

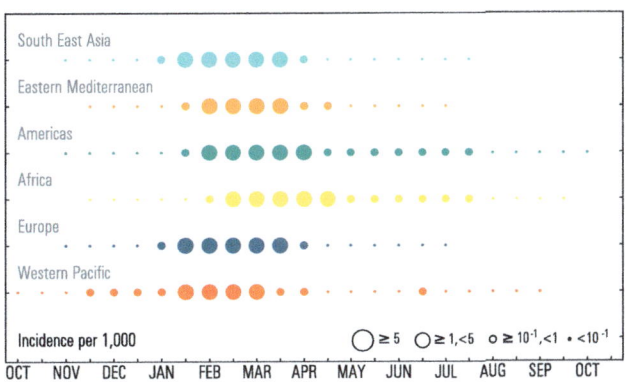

South East Asia
Eastern Mediterranean
Americas
Africa
Europe
Western Pacific

Incidence per 1,000   ○ ≥ 5 ○ ≥ 1,<5 ○ ≥ 10⁻¹,<1 • <10⁻¹

OCT NOV DEC JAN FEB MAR APR MAY JUN JUL AUG SEP OCT

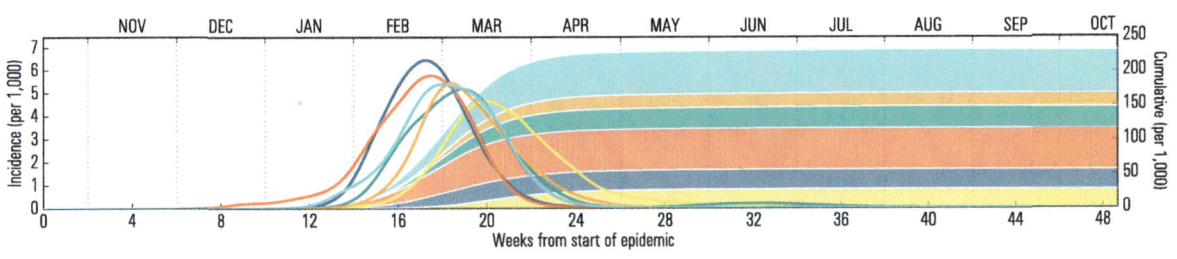

VACCINATIONS

▲ CUMULATIVE INCIDENCE AND EPIDEMIC PEAKS

▼ ATTACK RATE COMPOSITION (ONE YEAR)

South-East Asia 29%
Western Pacific 26%
Americas 13%
Europe 12%
Africa 12%
Eastern Mediterranean 8%

▼ EPIDEMIC ACTIVITY

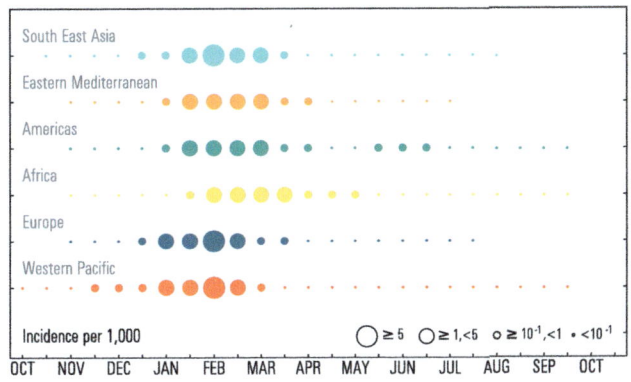

South East Asia
Eastern Mediterranean
Americas
Africa
Europe
Western Pacific

Incidence per 1,000   ○ ≥ 5 ○ ≥ 1,<5 ○ ≥ 10⁻¹,<1 • <10⁻¹

OCT NOV DEC JAN FEB MAR APR MAY JUN JUL AUG SEP OCT

JOHANNESBURG, SOUTH AFRICA

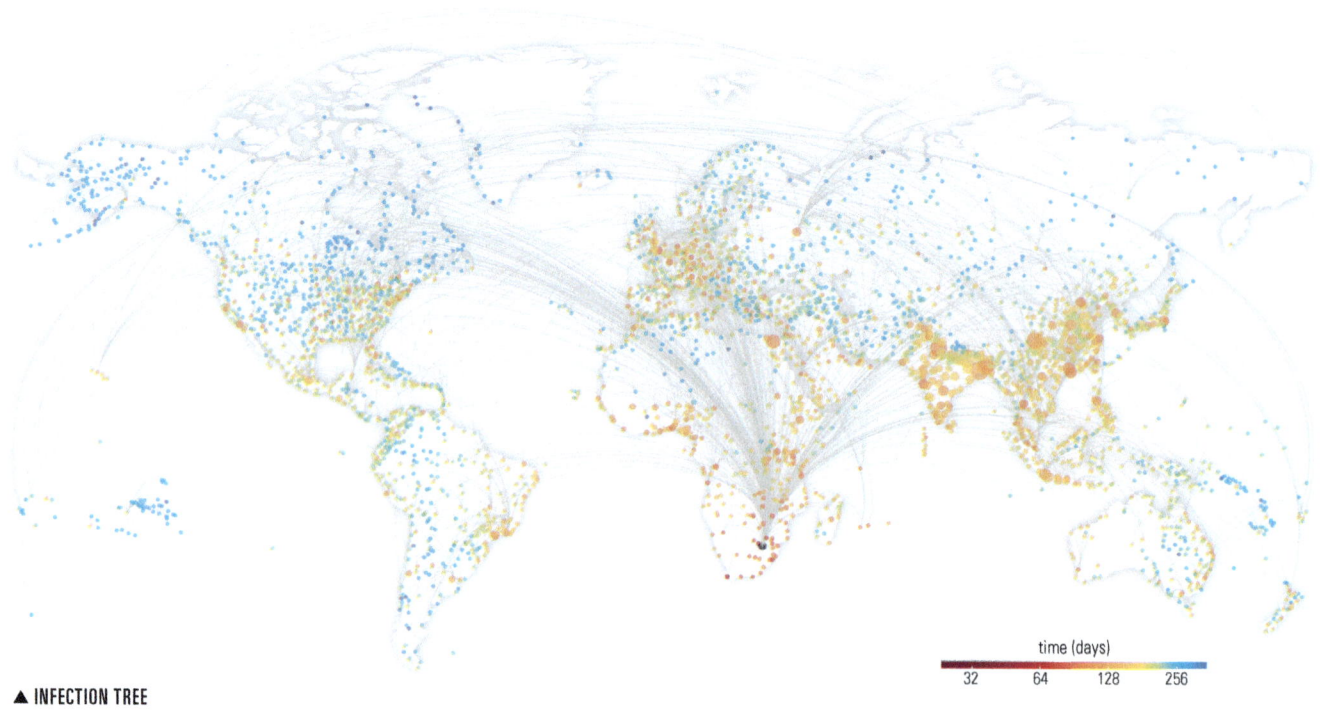

time (days)

| 32 | 64 | 128 | 256 |

▲ INFECTION TREE

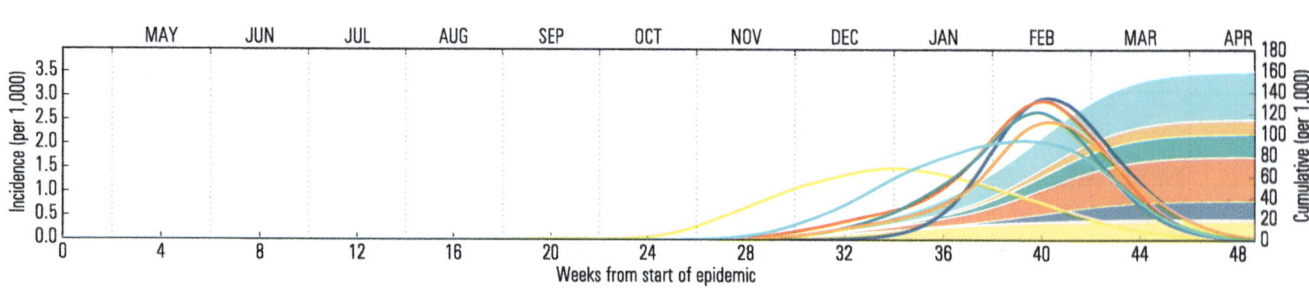

▲ CUMULATIVE INCIDENCE AND EPIDEMIC PEAKS

▼ AFFECTED COUNTRIES

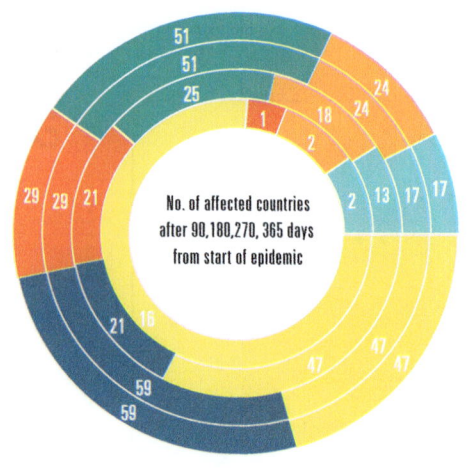

No. of affected countries after 90,180,270, 365 days from start of epidemic

▼ EPIDEMIC ACTIVITY

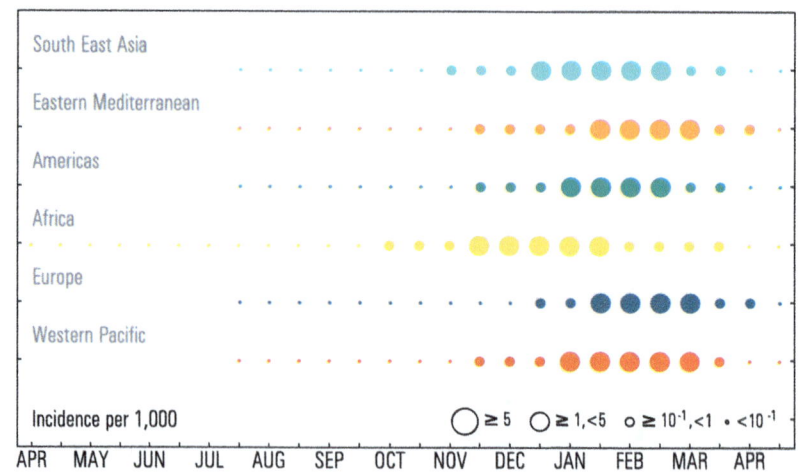

South East Asia

Eastern Mediterranean

Americas

Africa

Europe

Western Pacific

Incidence per 1,000   ◯ ≥ 5   ◯ ≥ 1,<5   ○ ≥ 10⁻¹,<1   · <10⁻¹

APR MAY JUN JUL AUG SEP OCT NOV DEC JAN FEB MAR APR

WHO REGIONS

■ Americas	■ E. Mediterranean
■ Africa	■ South-East Asia
■ Europe	■ Western Pacific

INITIAL CONDITIONS
 APR  R₀ 1.5

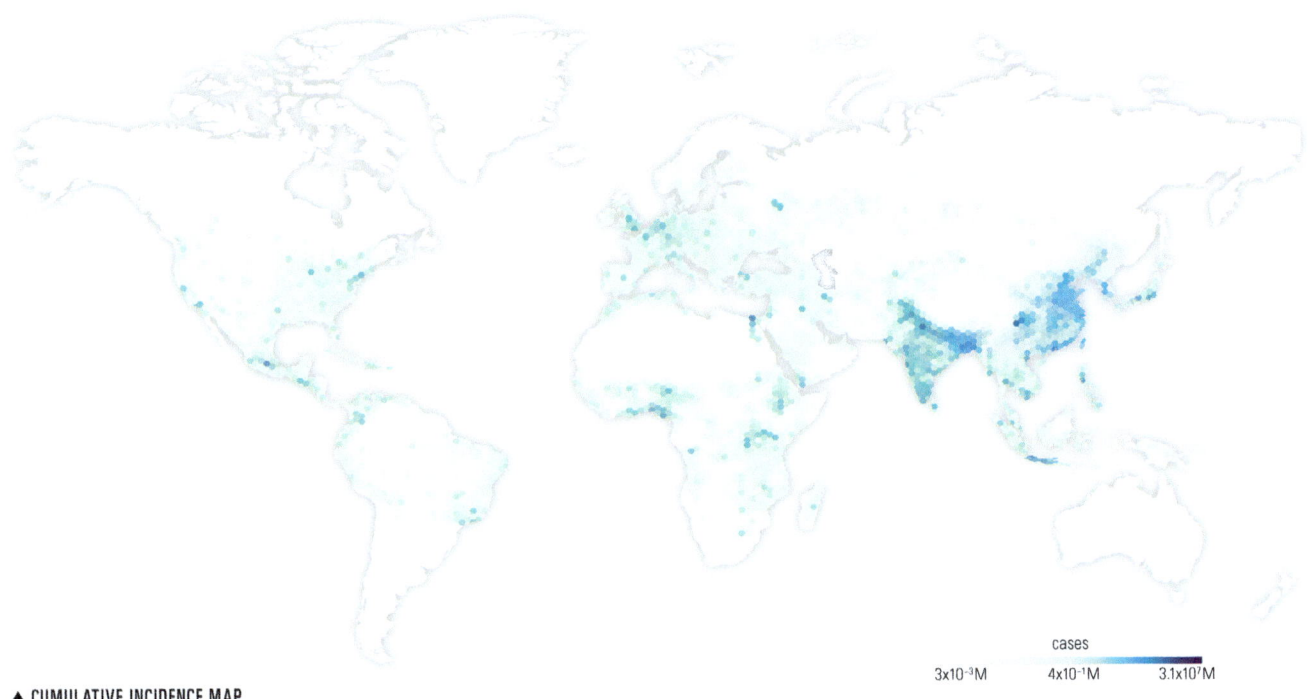

▲ CUMULATIVE INCIDENCE MAP

cases
3x10⁻³M   4x10⁻¹M   3.1x10⁷M

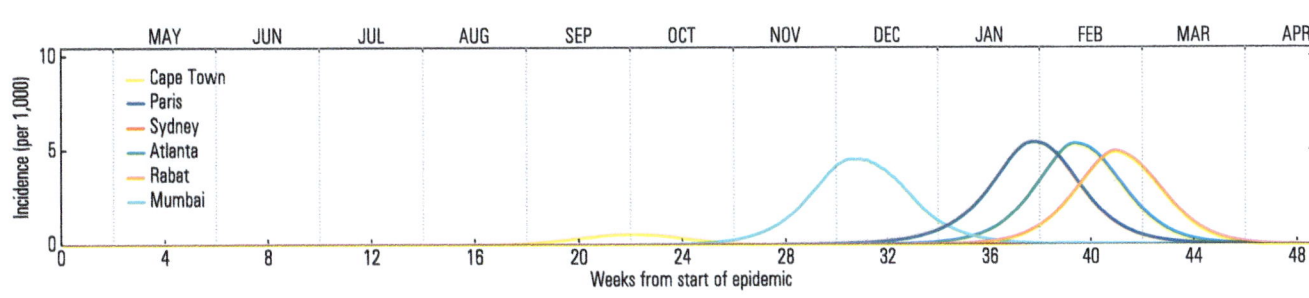

▲ URBAN LEVEL INCIDENCE

▼ RISK QUAD CHART

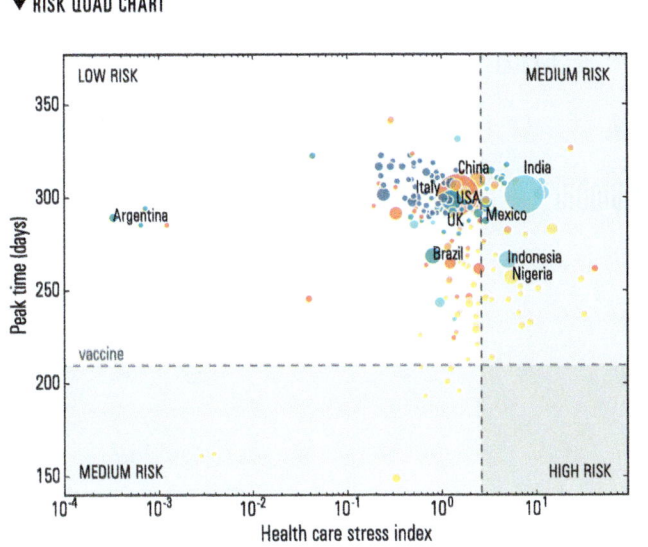

▼ ATTACK RATE COMPOSITION (ONE YEAR)

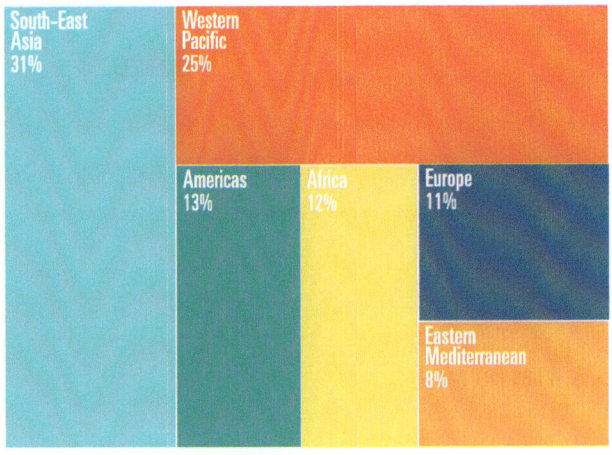

JOHANNESBURG, SOUTH AFRICA

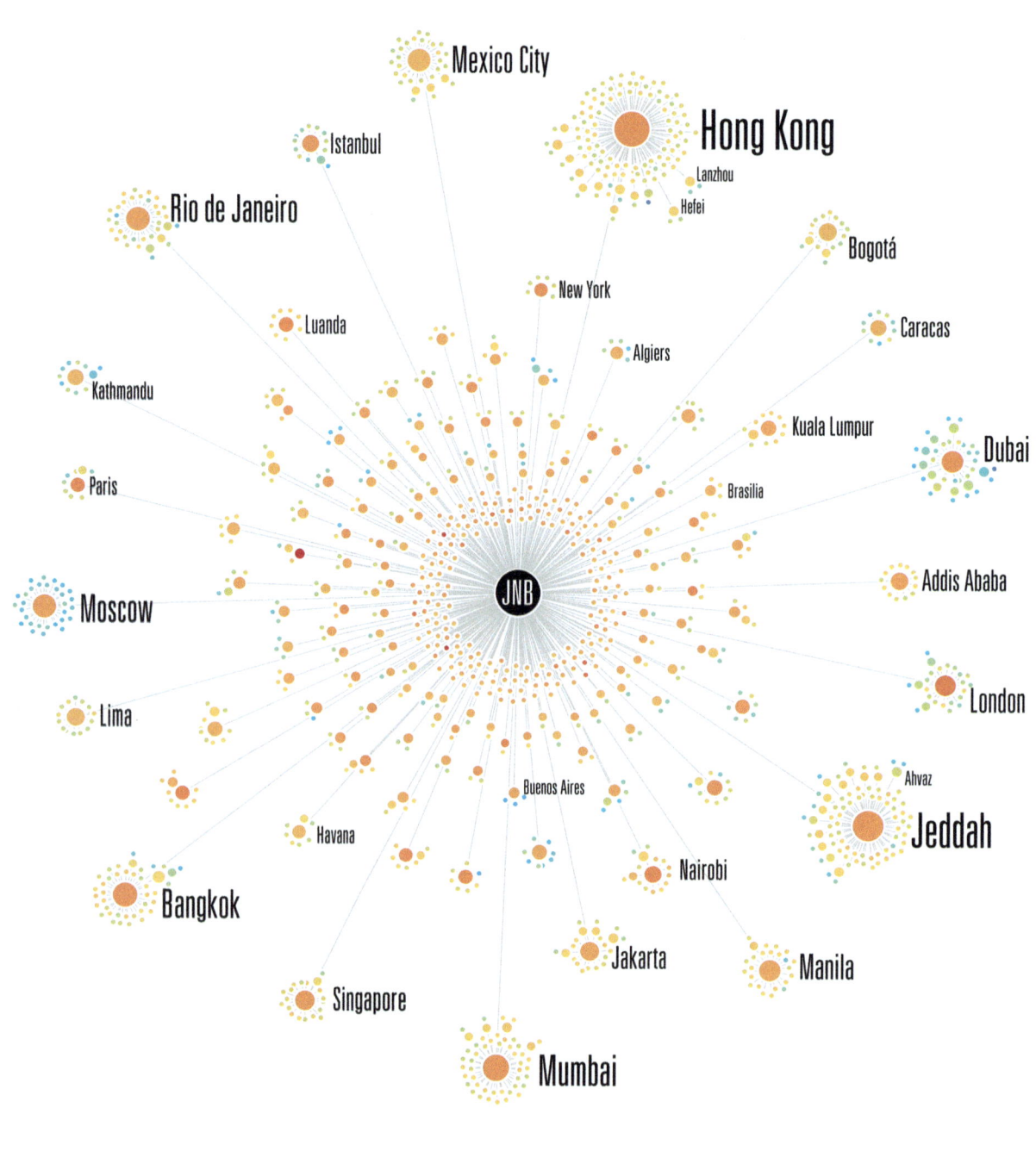

Mexico City

Istanbul

Rio de Janeiro

Hong Kong

Lanzhou

Hefei

Bogotá

Luanda

New York

Caracas

Algiers

Kathmandu

Kuala Lumpur

Dubai

Paris

Brasilia

Addis Ababa

Moscow

JNB

London

Lima

Ahvaz

Buenos Aires

Jeddah

Havana

Nairobi

Bangkok

Jakarta

Manila

Singapore

Mumbai

time (days)

32   64   128   256

▲ INFECTION TREE GRAPH

■ Americas ■ E. Mediterranean
■ Africa ■ South-East Asia
■ Europe ■ Western Pacific

INITIAL CONDITIONS

 APR
 R₀ 1.5

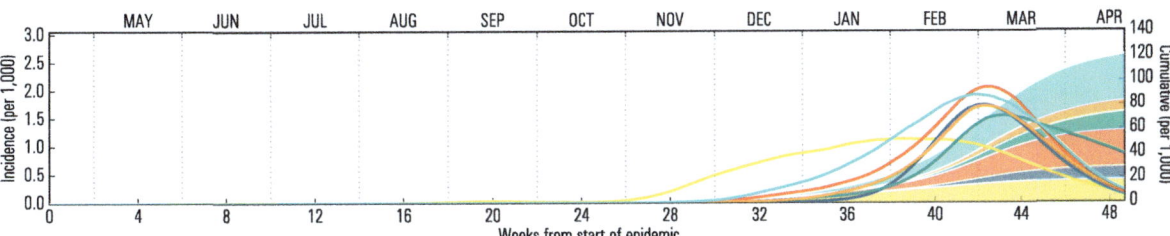

▲ CUMULATIVE INCIDENCE AND EPIDEMIC PEAKS

▼ ATTACK RATE COMPOSITION (ONE YEAR)

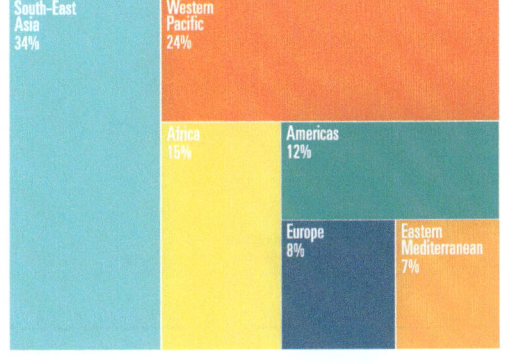

▼ EPIDEMIC ACTIVITY

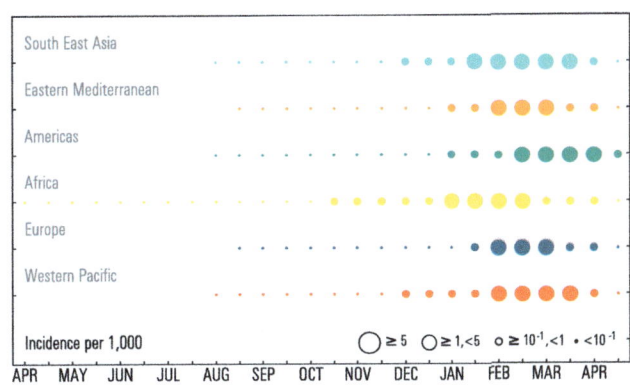

80% AIR TRAFFIC REDUCTION

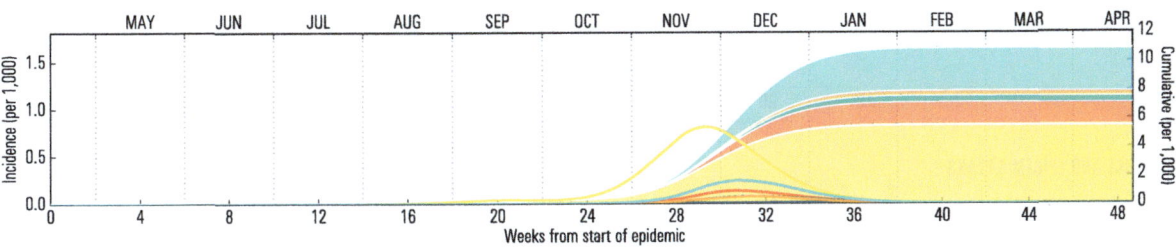

▲ CUMULATIVE INCIDENCE AND EPIDEMIC PEAKS

▼ ATTACK RATE COMPOSITION (ONE YEAR)

▼ EPIDEMIC ACTIVITY

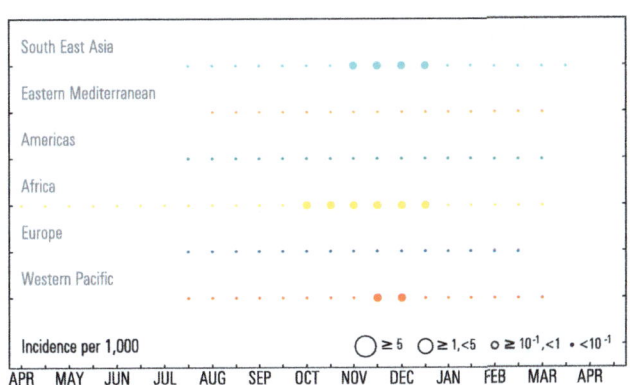

VACCINATIONS

JOHANNESBURG, SOUTH AFRICA

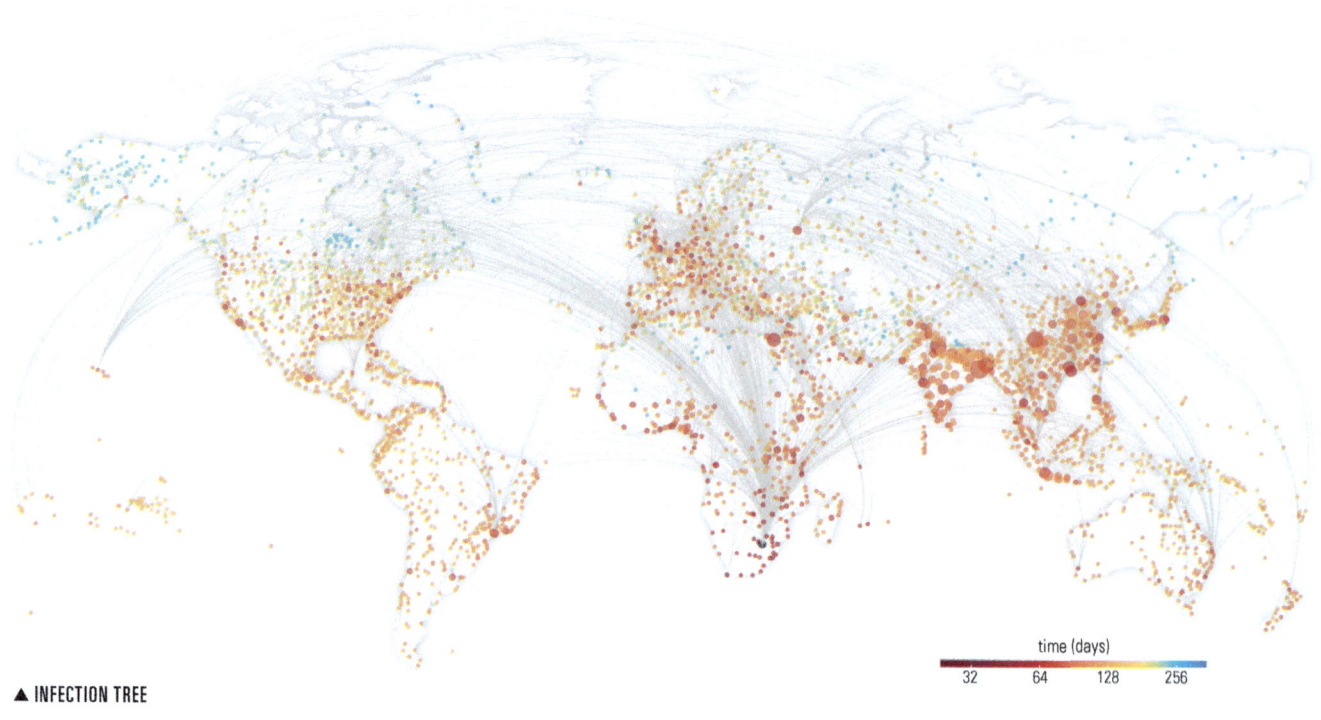

time (days)

32   64   128   256

▲ INFECTION TREE

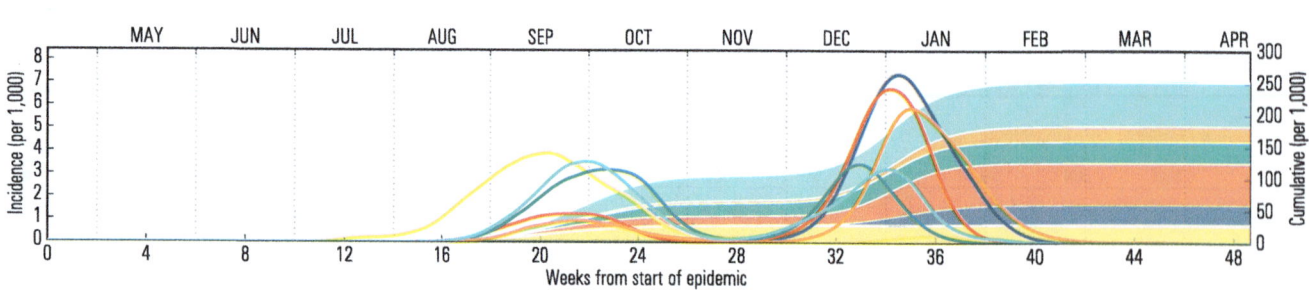

▲ CUMULATIVE INCIDENCE AND EPIDEMIC PEAKS

▼ AFFECTED COUNTRIES

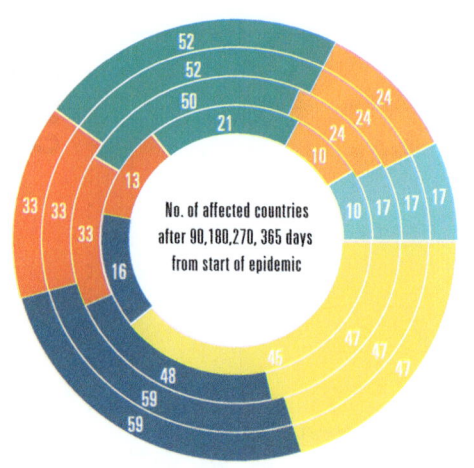

No. of affected countries
after 90,180,270, 365 days
from start of epidemic

▼ EPIDEMIC ACTIVITY

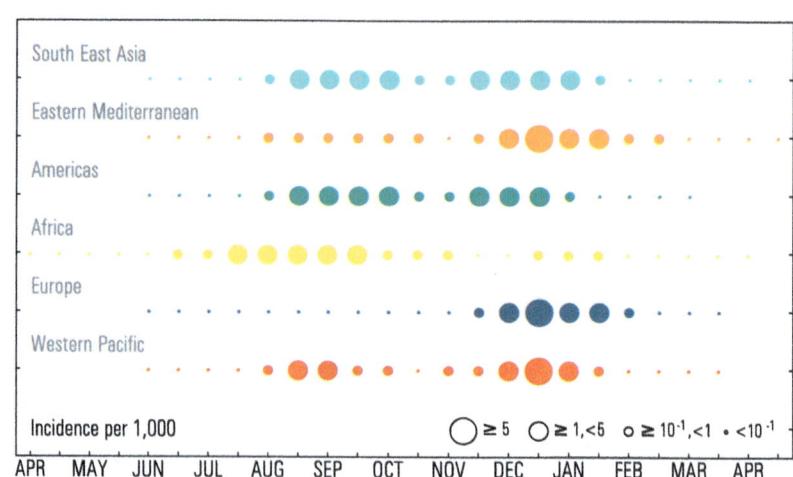

WHO REGIONS
- Americas
- Africa
- Europe
- E. Mediterranean
- South-East Asia
- Western Pacific

INITIAL CONDITIONS

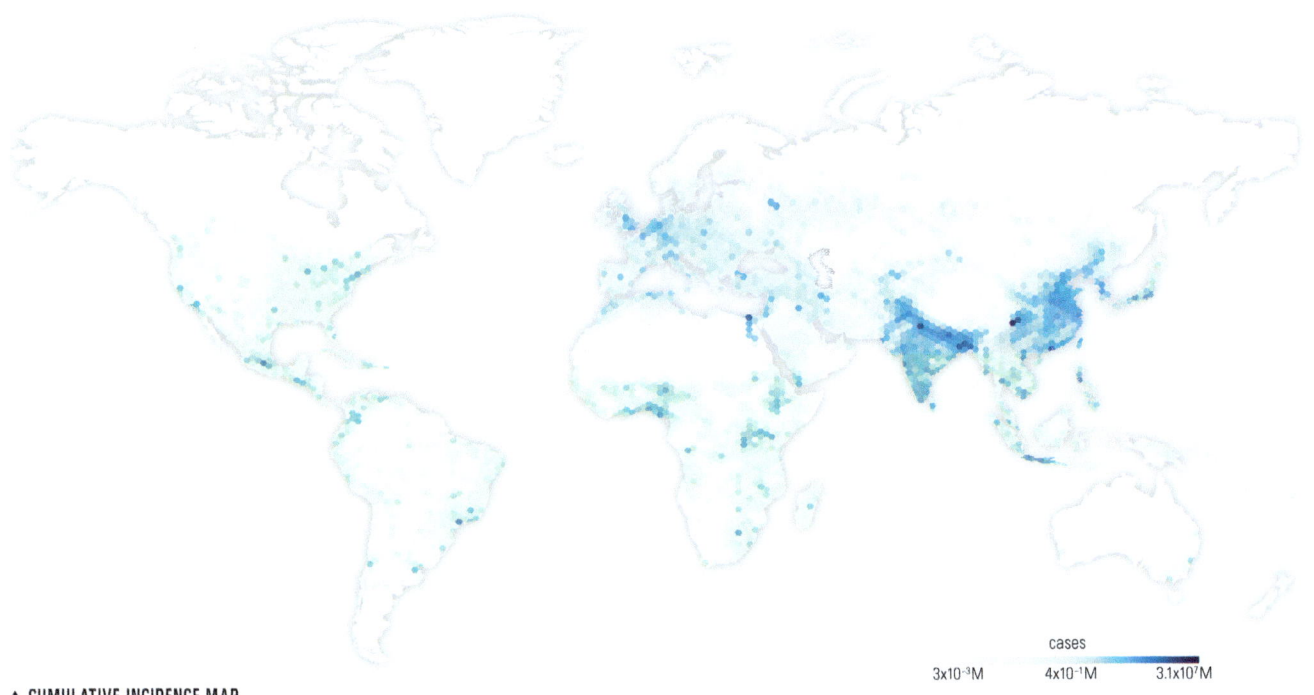

cases

3x10⁻³M   4x10⁻¹M   3.1x10⁷M

▲ CUMULATIVE INCIDENCE MAP

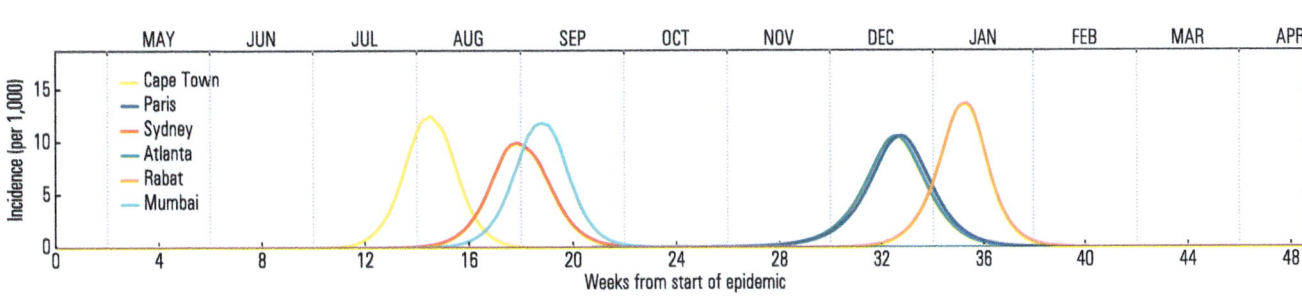

▲ URBAN LEVEL INCIDENCE

▼ RISK QUAD CHART

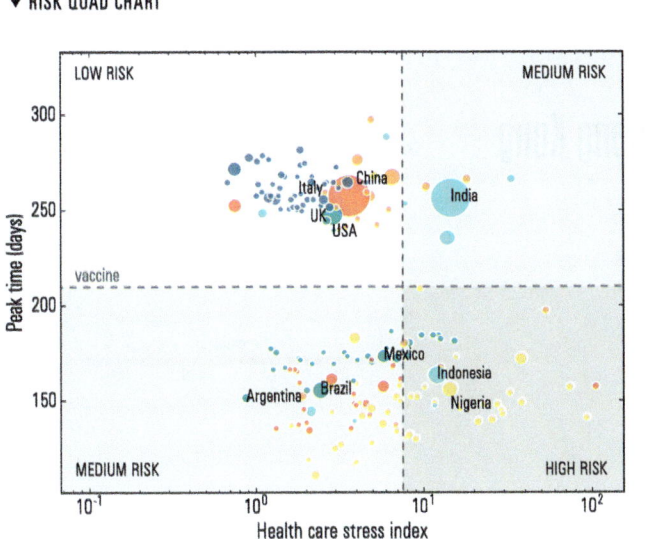

▼ ATTACK RATE COMPOSITION (ONE YEAR)

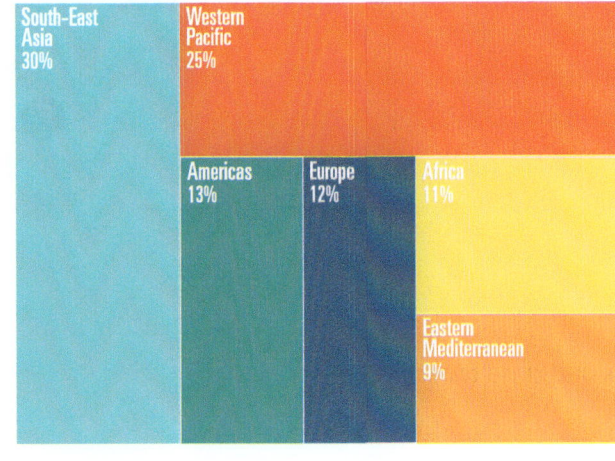

JOHANNESBURG, SOUTH AFRICA

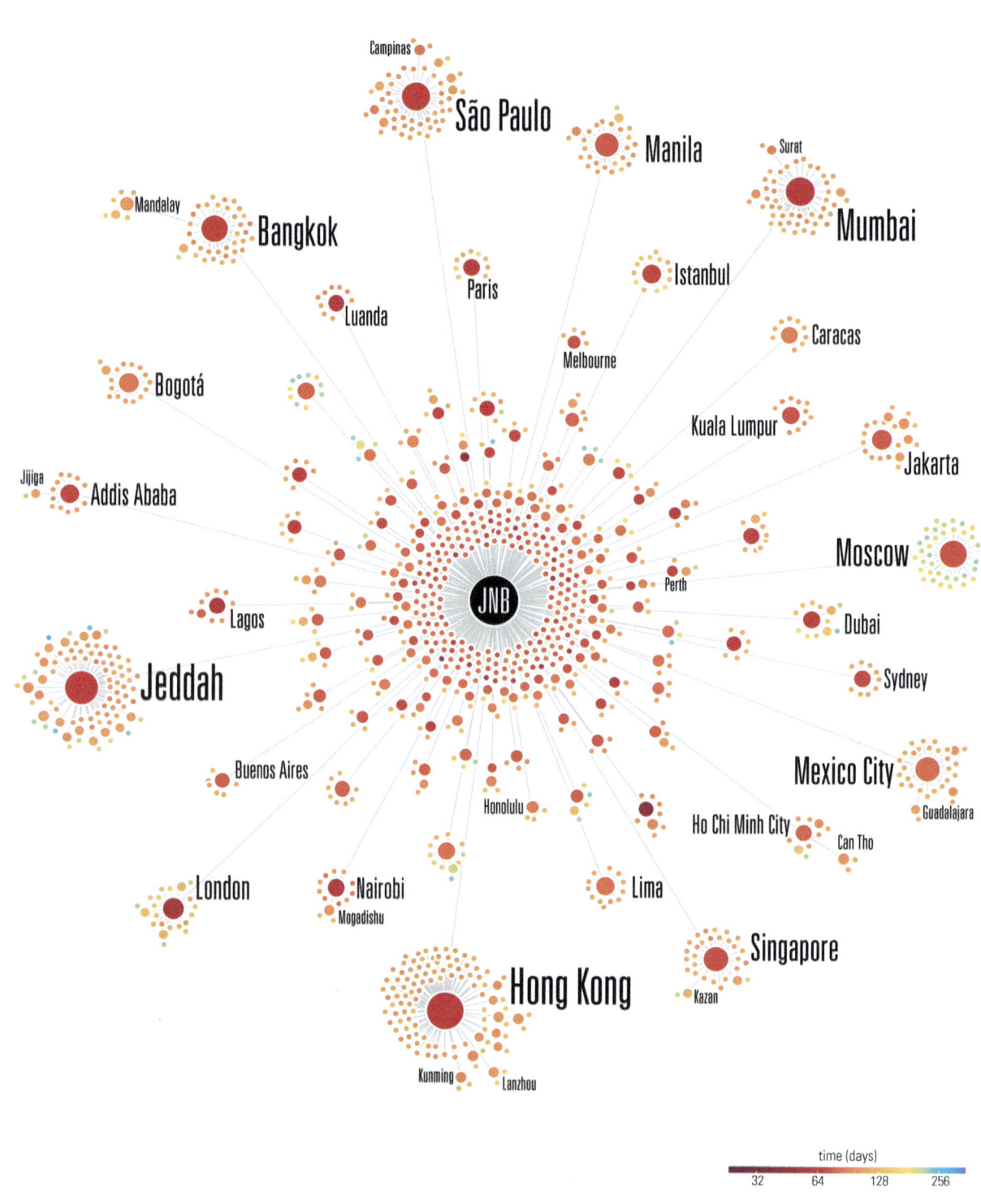

Campinas

São Paulo

Manila

Surat

Mumbai

Mandalay

Bangkok

Paris

Luanda

Istanbul

Caracas

Melbourne

Bogotá

Kuala Lumpur

Jakarta

Jijiga

Addis Ababa

Moscow

JNB

Perth

Dubai

Lagos

Sydney

Jeddah

Buenos Aires

Mexico City

Honolulu

Ho Chi Minh City

Guadalajara

Can Tho

London

Nairobi

Lima

Mogadishu

Singapore

Kazan

Hong Kong

Kunming

Lanzhou

time (days)

32   64   128   256

▲ INFECTION TREE GRAPH

WHO REGIONS

■ Americas
■ Africa
■ Europe
■ E. Mediterranean
■ South-East Asia
■ Western Pacific

INITIAL CONDITIONS

80% AIR TRAFFIC REDUCTION

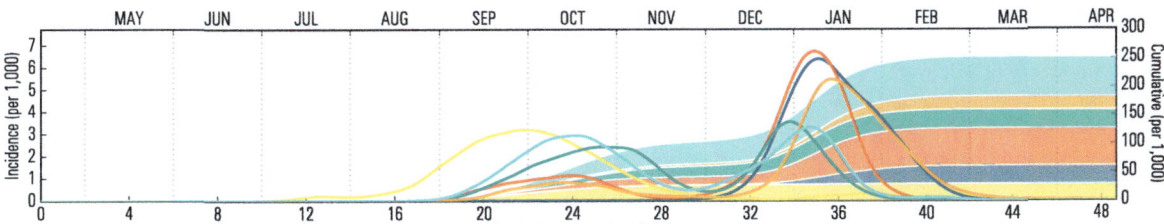

▲ CUMULATIVE INCIDENCE AND EPIDEMIC PEAKS

▼ ATTACK RATE COMPOSITION (ONE YEAR)

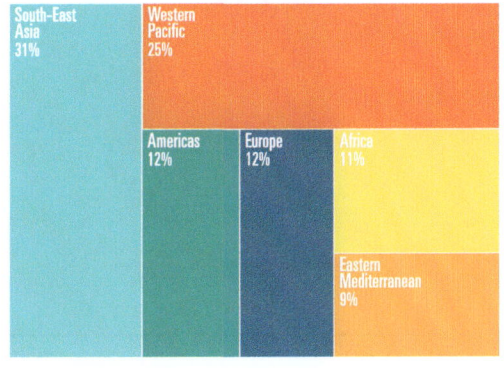

▼ EPIDEMIC ACTIVITY

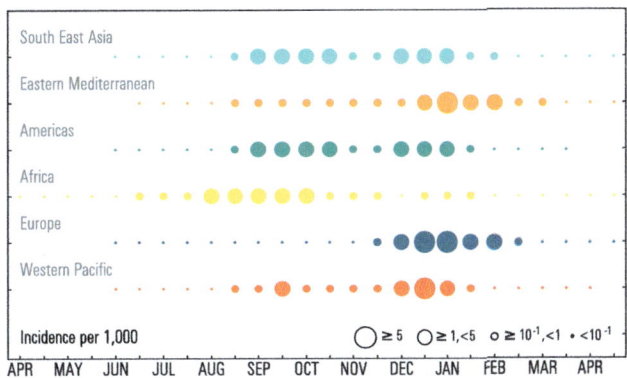

VACCINATIONS

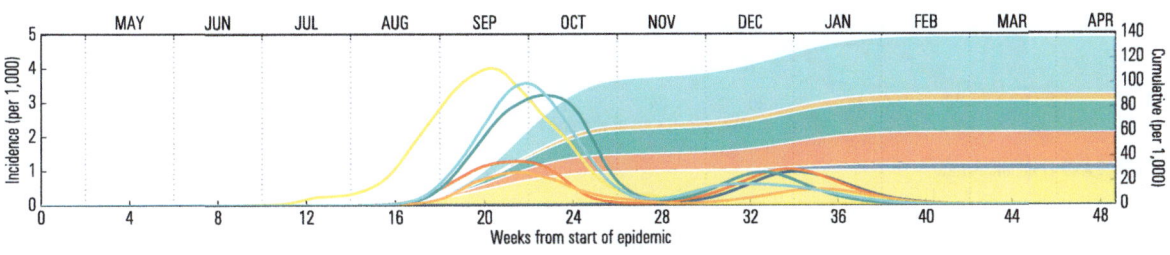

▲ CUMULATIVE INCIDENCE AND EPIDEMIC PEAKS

▼ ATTACK RATE COMPOSITION (ONE YEAR)

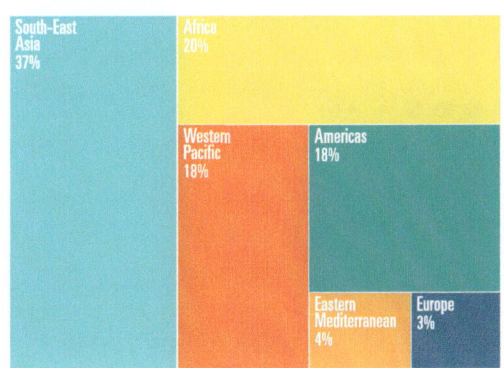

▼ EPIDEMIC ACTIVITY

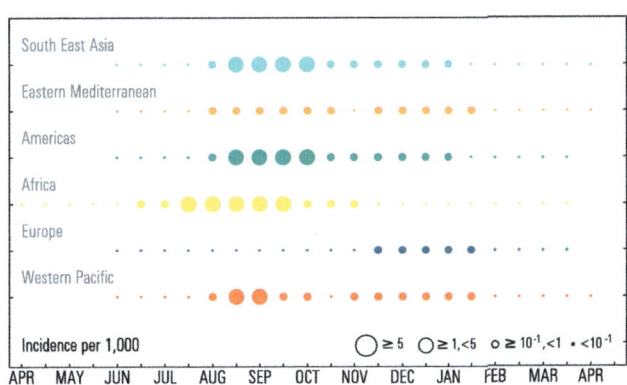

MELBOURNE, AUSTRALIA

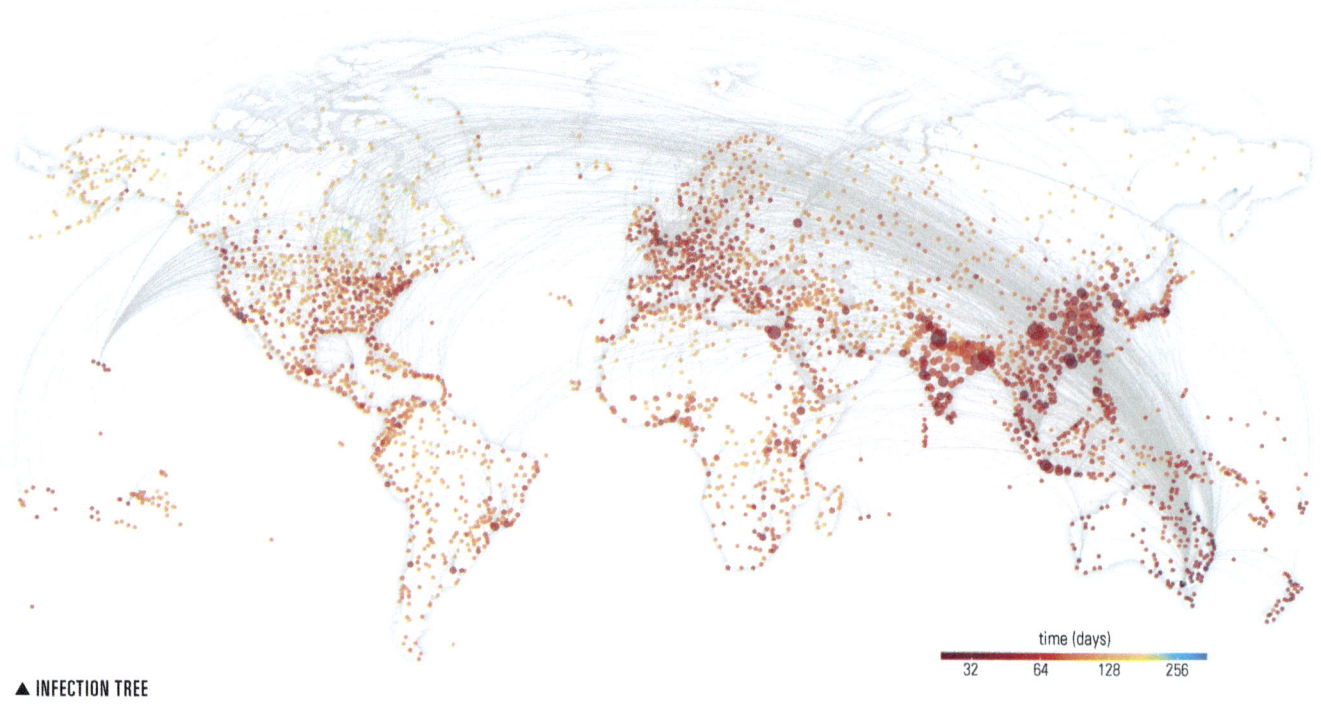

▲ INFECTION TREE

time (days)
32 64 128 256

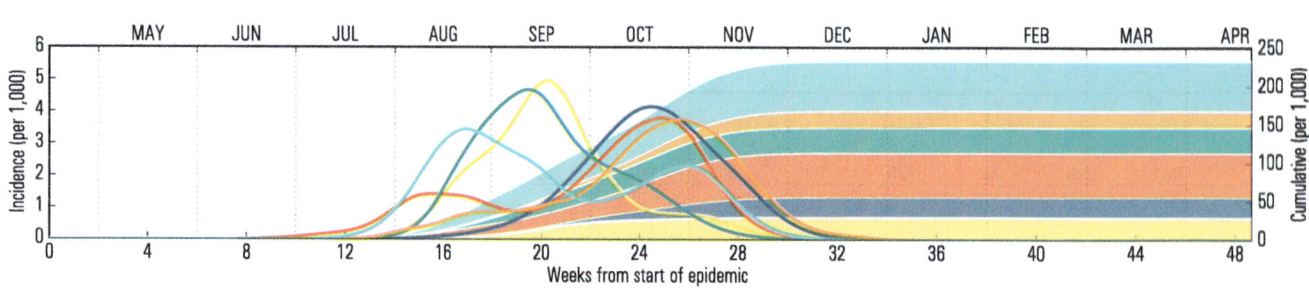

▲ CUMULATIVE INCIDENCE AND EPIDEMIC PEAKS

▼ AFFECTED COUNTRIES

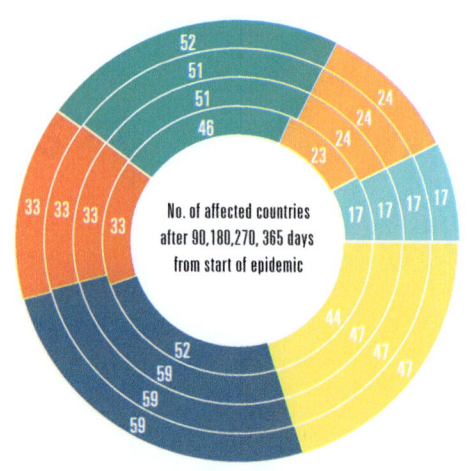

No. of affected countries after 90,180,270, 365 days from start of epidemic

▼ EPIDEMIC ACTIVITY

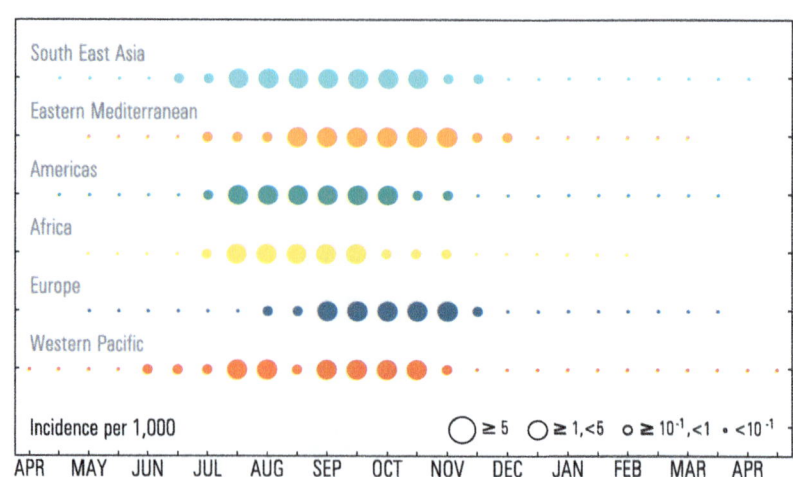

■ Americas   ■ E. Mediterranean
■ Africa   ■ South-East Asia
■ Europe   ■ Western Pacific

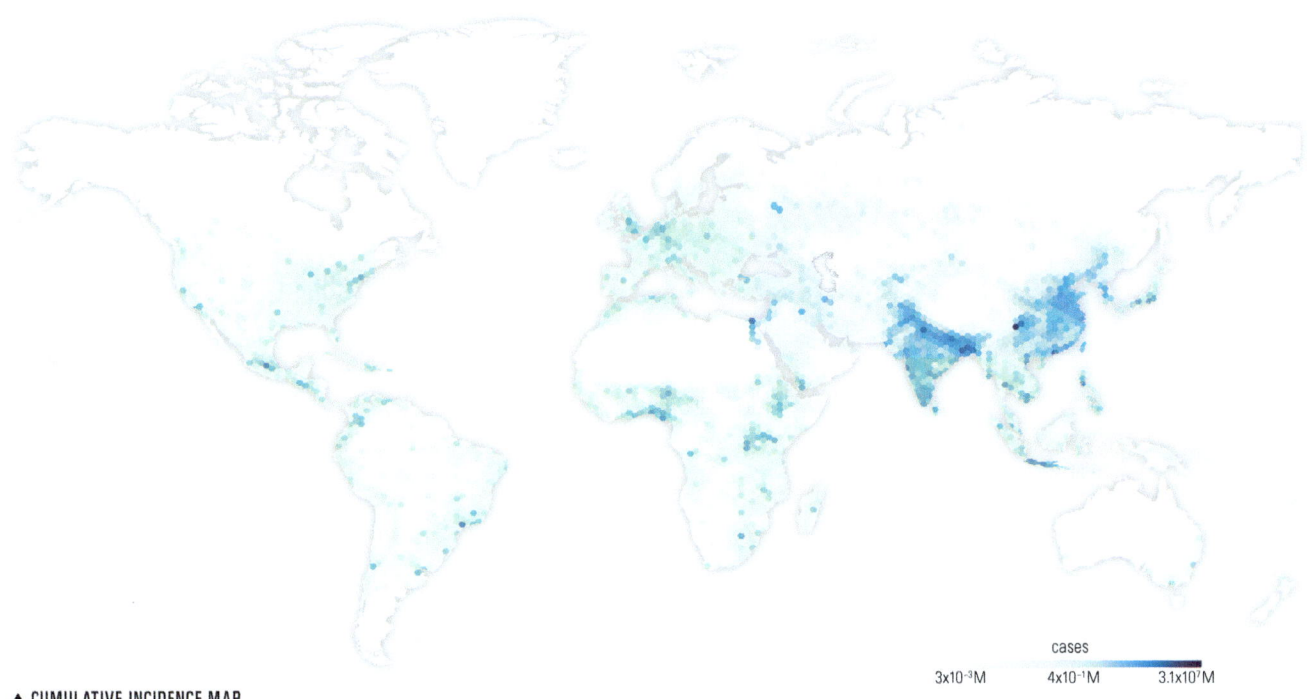

cases

3x10⁻³M   4x10⁻¹M   3.1x10⁷M

▲ CUMULATIVE INCIDENCE MAP

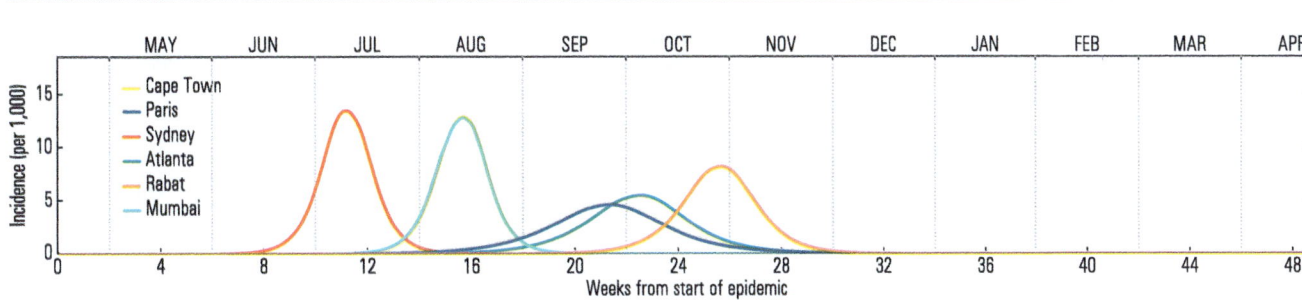

▲ URBAN LEVEL INCIDENCE

▼ RISK QUAD CHART

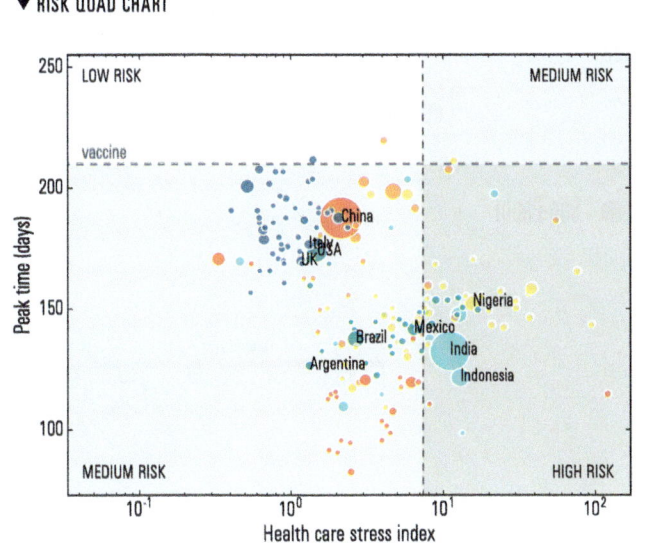

▼ ATTACK RATE COMPOSITION (ONE YEAR)

MELBOURNE, AUSTRALIA

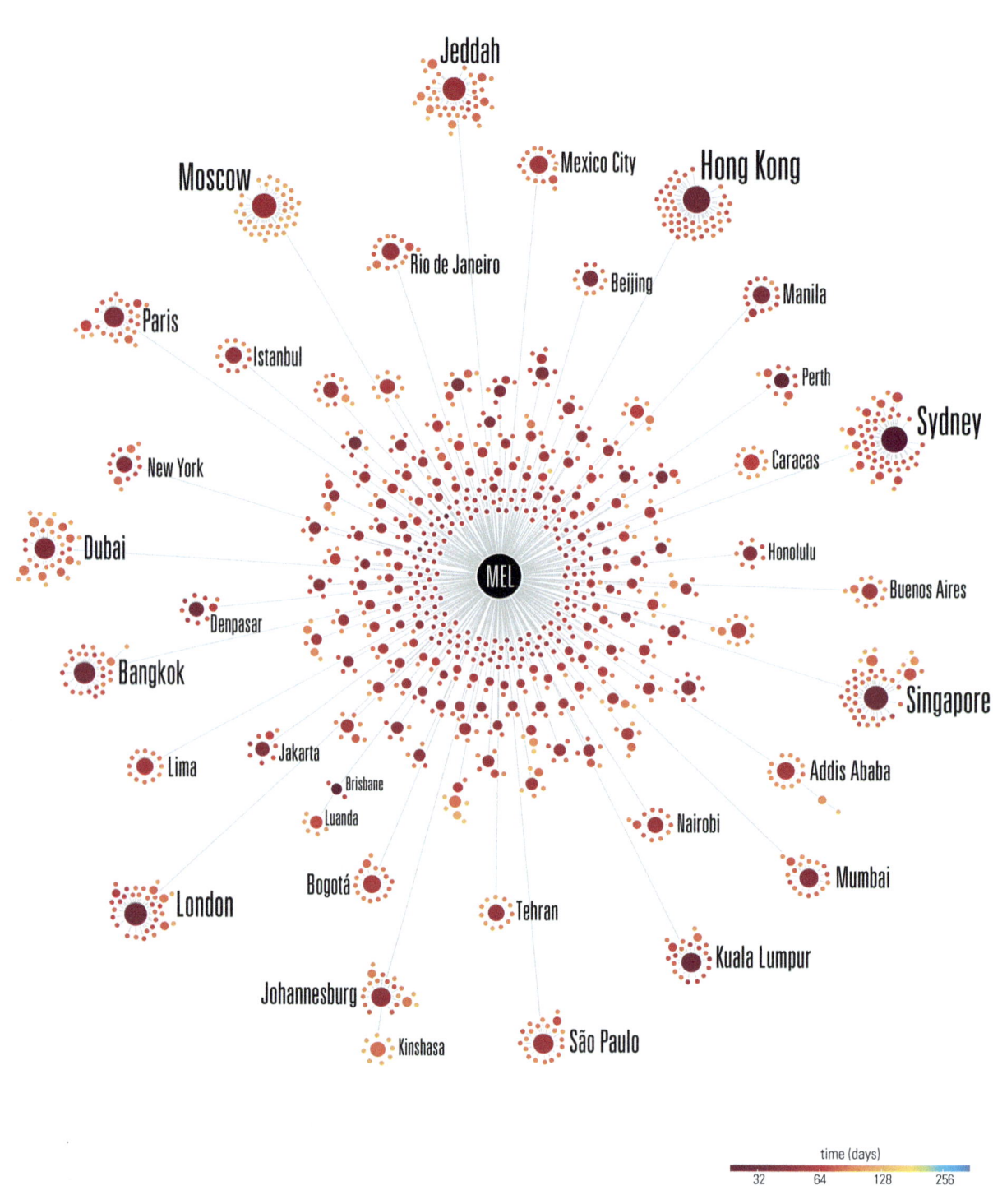

Jeddah

Mexico City
Hong Kong

Moscow

Rio de Janeiro
Beijing

Paris

Istanbul
Manila

Perth

New York
Sydney

Caracas

Dubai

Honolulu
Denpasar
Buenos Aires

Bangkok
Singapore

Jakarta
Addis Ababa
Lima

Brisbane
Luanda
Nairobi

Bogotá
Mumbai
London
Tehran

Kuala Lumpur

Johannesburg

Kinshasa
São Paulo

MEL

time (days)

32   64   128   256

▲ INFECTION TREE GRAPH

WHO REGIONS

- Americas
- Africa
- Europe
- E. Mediterranean
- South-East Asia
- Western Pacific

INITIAL CONDITIONS

80% AIR TRAFFIC REDUCTION

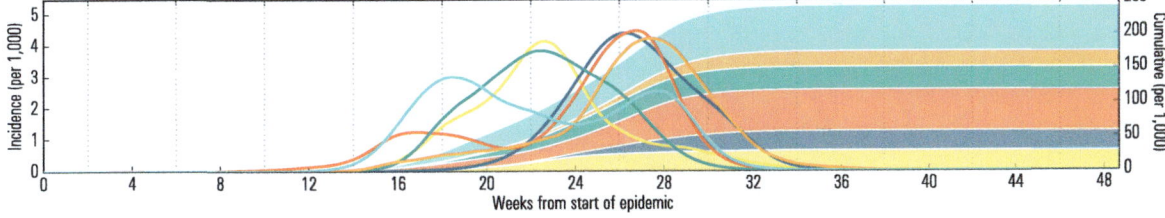

▲ CUMULATIVE INCIDENCE AND EPIDEMIC PEAKS

▼ ATTACK RATE COMPOSITION (ONE YEAR)

▼ EPIDEMIC ACTIVITY

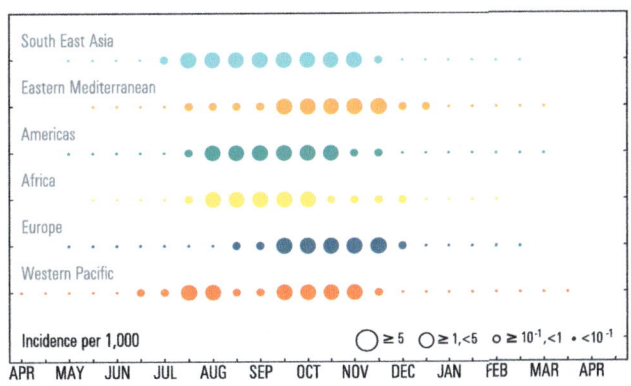

VACCINATIONS

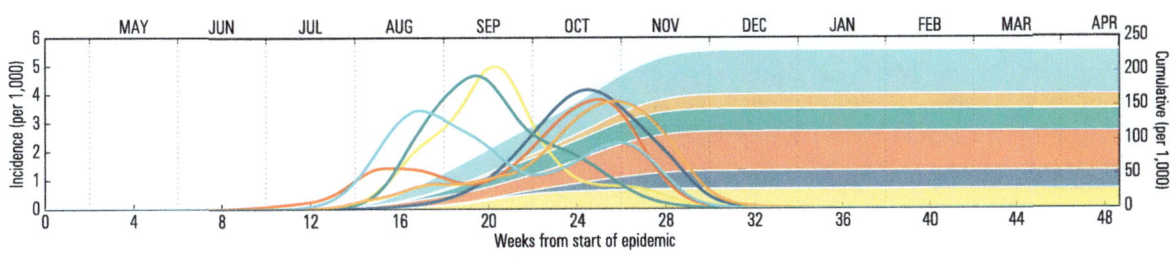

▲ CUMULATIVE INCIDENCE AND EPIDEMIC PEAKS

▼ ATTACK RATE COMPOSITION (ONE YEAR)

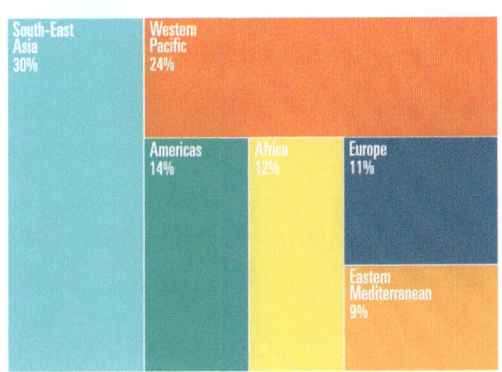

▼ EPIDEMIC ACTIVITY

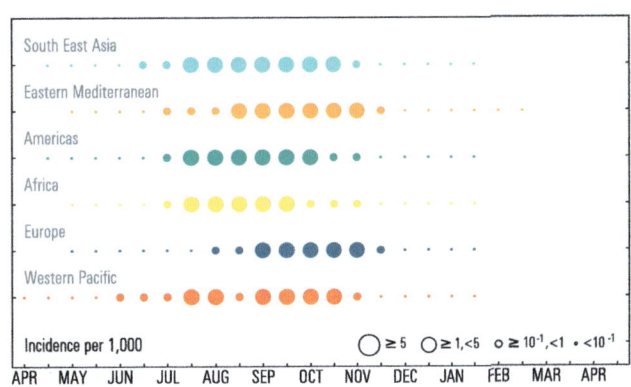

MELBOURNE, AUSTRALIA

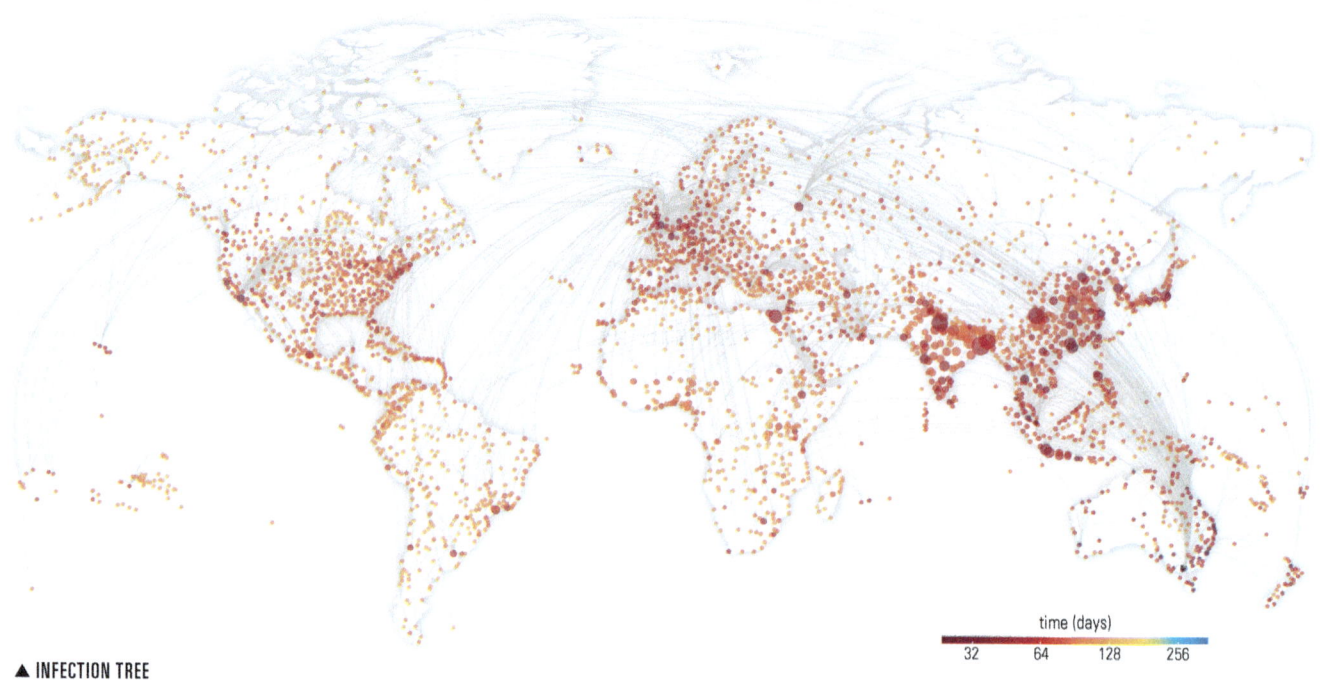

▲ INFECTION TREE

time (days)
32   64   128   256

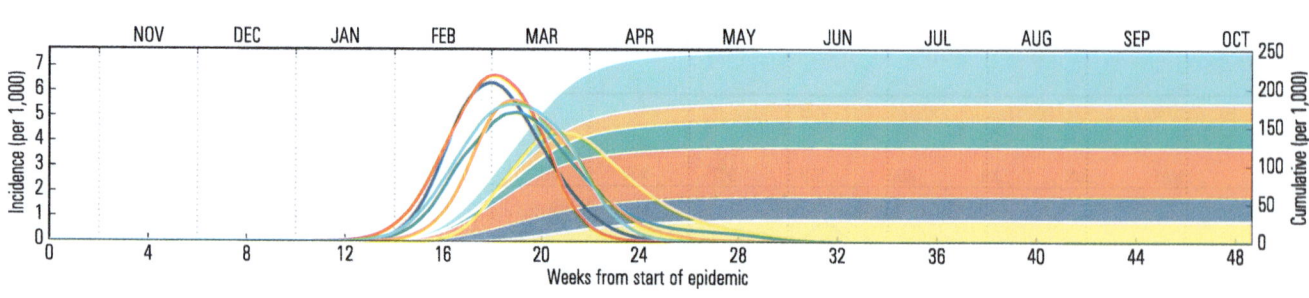

NOV   DEC   JAN   FEB   MAR   APR   MAY   JUN   JUL   AUG   SEP   OCT

Incidence (per 1,000)

Cumulative (per 1,000)

Weeks from start of epidemic

▲ CUMULATIVE INCIDENCE AND EPIDEMIC PEAKS

▼ AFFECTED COUNTRIES

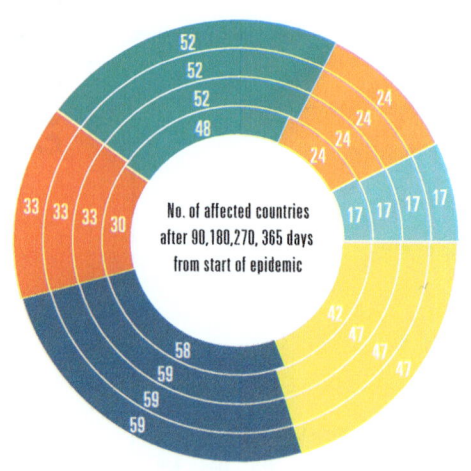

52
52
52
48

24
24
24
24

17 17 17 17

33 33 33 30

42
47
47
47

58
59
59
59

No. of affected countries
after 90,180,270, 365 days
from start of epidemic

▼ EPIDEMIC ACTIVITY

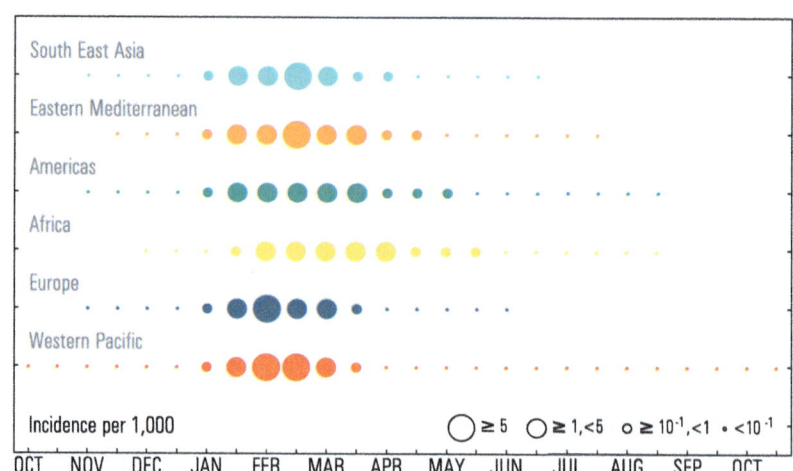

South East Asia

Eastern Mediterranean

Americas

Africa

Europe

Western Pacific

Incidence per 1,000

≥ 5 $\geq 1, <5$ $\geq 10^{-1}, <1$ $< 10^{-1}$

OCT   NOV   DEC   JAN   FEB   MAR   APR   MAY   JUN   JUL   AUG   SEP   OCT

■ Americas   ■ E. Mediterranean
■ Africa   ■ South-East Asia
■ Europe   ■ Western Pacific

INITIAL CONDITIONS

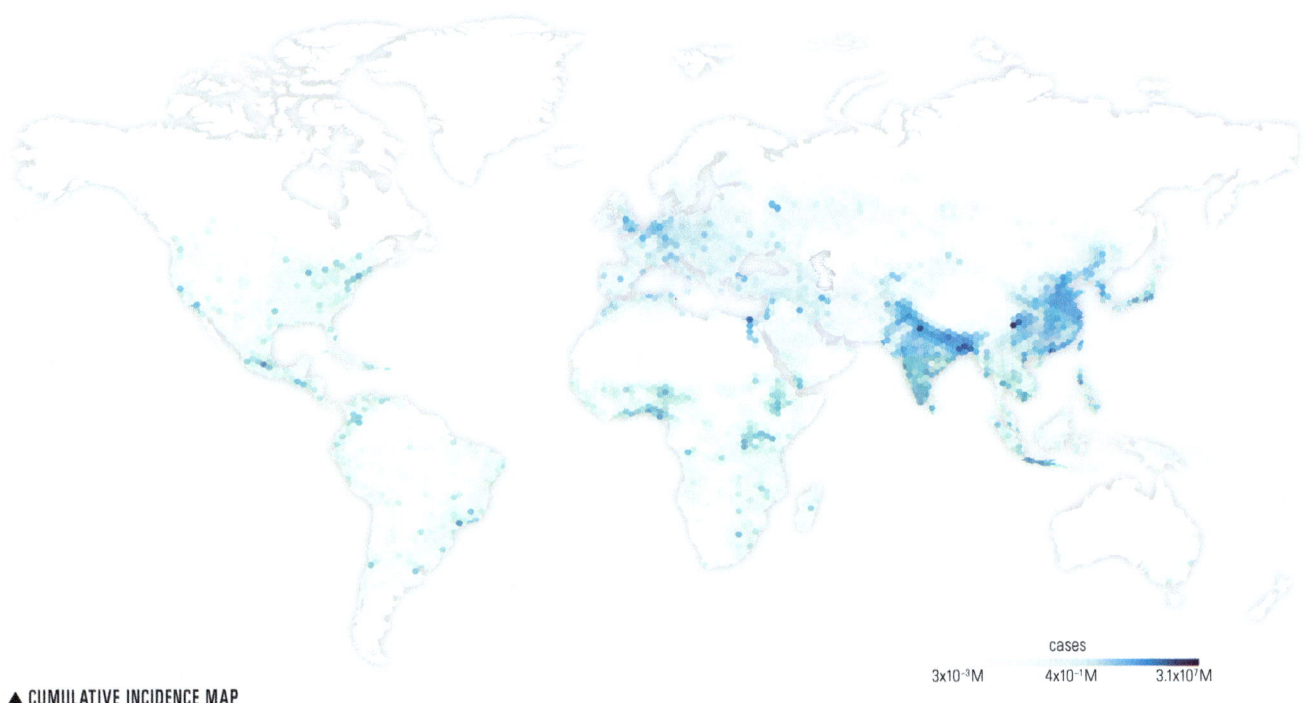

cases

3x10⁻³M   4x10⁻¹M   3.1x10⁷M

▲ CUMULATIVE INCIDENCE MAP

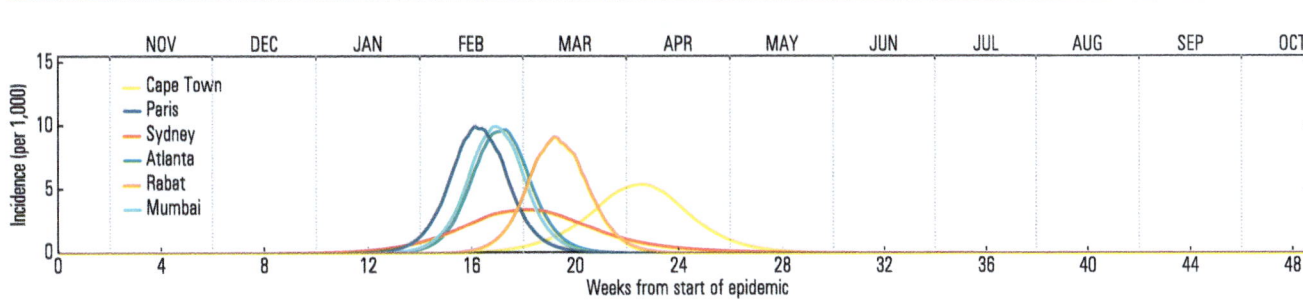

Incidence (per 1,000)

— Cape Town
— Paris
— Sydney
— Atlanta
— Rabat
— Mumbai

Weeks from start of epidemic

▲ URBAN LEVEL INCIDENCE

▼ RISK QUAD CHART

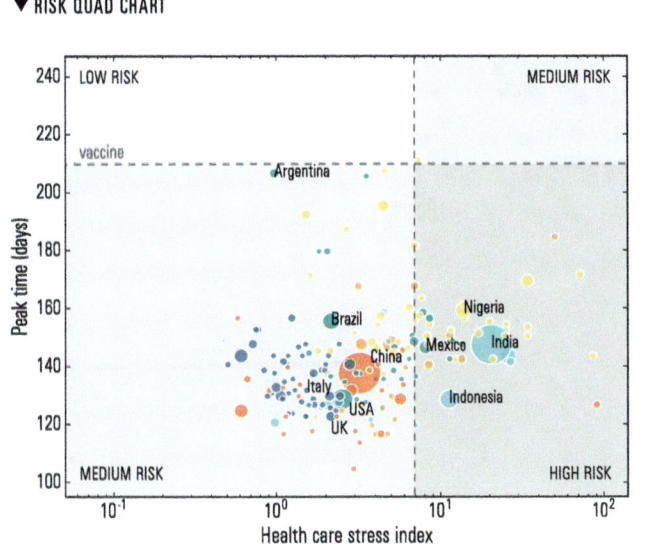

LOW RISK   MEDIUM RISK

vaccine
Argentina

Brazil
Nigeria
Mexico   India
China
Italy
Indonesia
USA
UK

MEDIUM RISK   HIGH RISK

Peak time (days)
Health care stress index

▼ ATTACK RATE COMPOSITION (ONE YEAR)

South-East Asia 30%
Western Pacific 25%
Americas 13%
Europe 12%
Africa 11%
Eastern Mediterranean 9%

MELBOURNE, AUSTRALIA

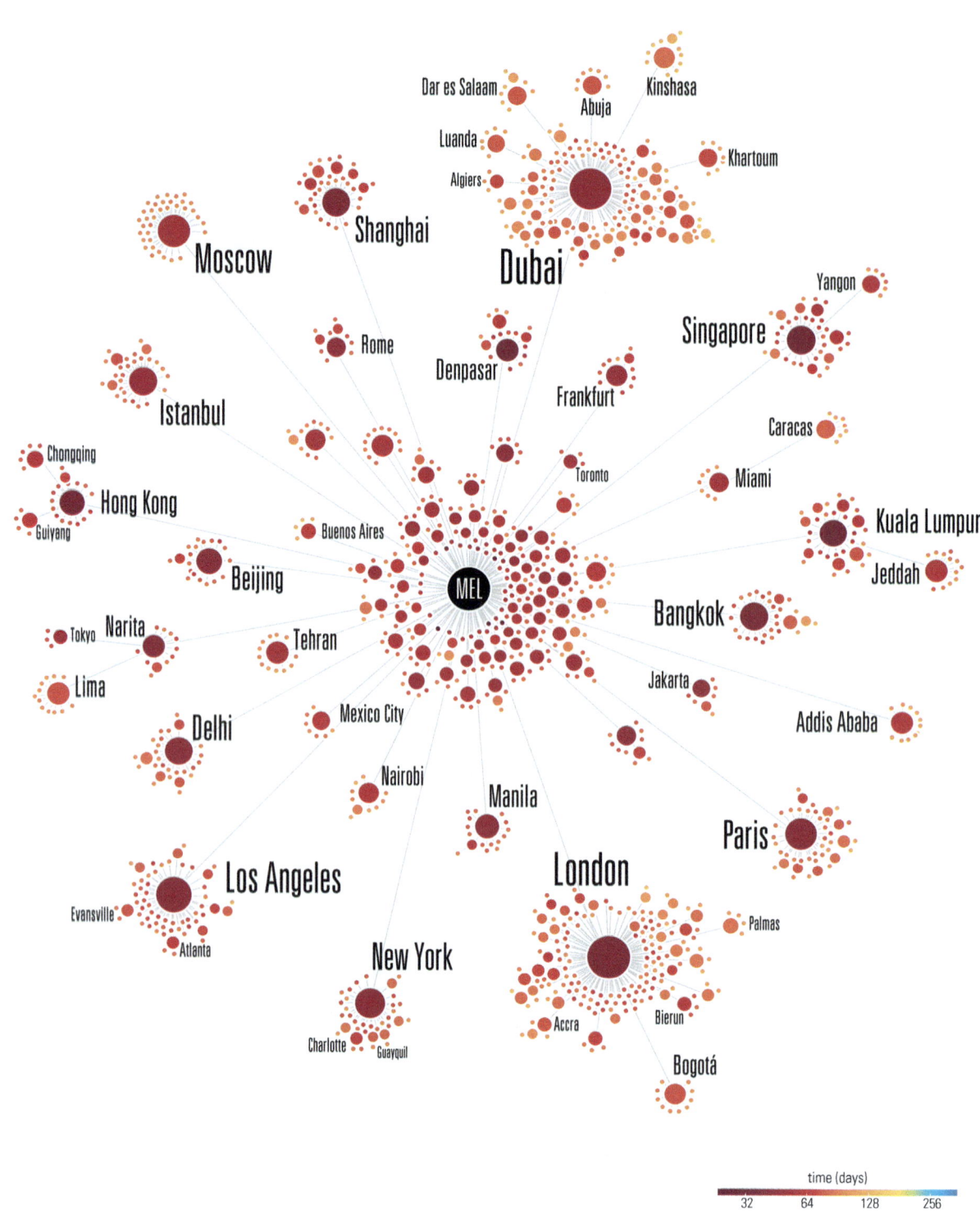

Dar es Salaam
Kinshasa
Abuja
Luanda
Khartoum
Algiers

Shanghai

Dubai

Moscow

Yangon

Singapore

Rome

Denpasar

Frankfurt

Istanbul

Caracas

Chongqing

Miami

Hong Kong

Kuala Lumpur

Guiyang

Toronto

Jeddah

Beijing

Buenos Aires

MEL

Bangkok

Tokyo Narita

Tehran

Lima

Jakarta

Addis Ababa

Delhi

Mexico City

Nairobi

Manila

Paris

Los Angeles

London

Evansville

Palmas

Atlanta

New York

Accra

Bierun

Charlotte Guayaquil

Bogotá

time (days)
32   64   128   256

▲ INFECTION TREE GRAPH

- Americas
- Africa
- Europe
- E. Mediterranean
- South-East Asia
- Western Pacific

INITIAL CONDITIONS

 OCT
 R₀ 2.0

 80% AIR TRAFFIC REDUCTION

NOV DEC JAN FEB MAR APR MAY JUN JUL AUG SEP OCT

Incidence (per 1,000)
Weeks from start of epidemic
Cumulative (per 1,000)

▲ CUMULATIVE INCIDENCE AND EPIDEMIC PEAKS

▼ ATTACK RATE COMPOSITION (ONE YEAR)

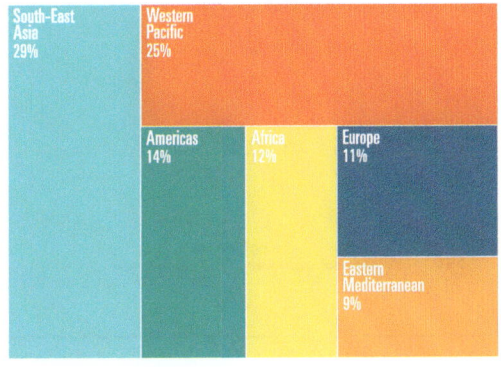

South-East Asia 29%
Western Pacific 25%
Americas 14%
Africa 12%
Europe 11%
Eastern Mediterranean 9%

▼ EPIDEMIC ACTIVITY

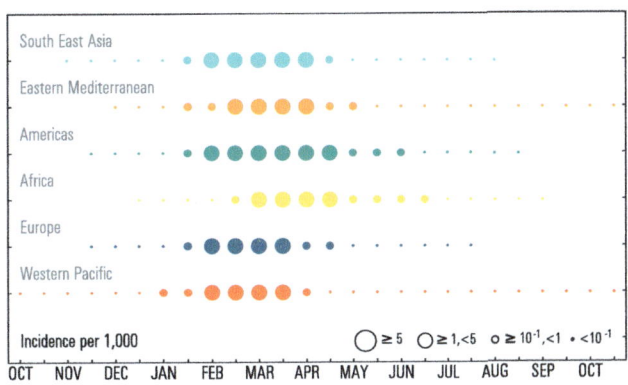

South East Asia
Eastern Mediterranean
Americas
Africa
Europe
Western Pacific

Incidence per 1,000   ≥ 5   ≥ 1,<5   ≥ 10⁻¹,<1   <10⁻¹

OCT NOV DEC JAN FEB MAR APR MAY JUN JUL AUG SEP OCT

VACCINATIONS

NOV DEC JAN FEB MAR APR MAY JUN JUL AUG SEP OCT

Incidence (per 1,000)
Weeks from start of epidemic
Cumulative (per 1,000)

▲ CUMULATIVE INCIDENCE AND EPIDEMIC PEAKS

▼ ATTACK RATE COMPOSITION (ONE YEAR)

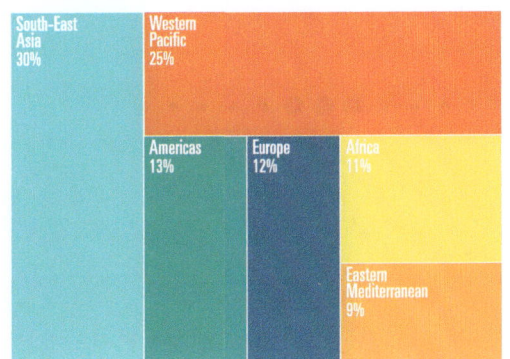

South-East Asia 30%
Western Pacific 25%
Americas 13%
Europe 12%
Africa 11%
Eastern Mediterranean 9%

▼ EPIDEMIC ACTIVITY

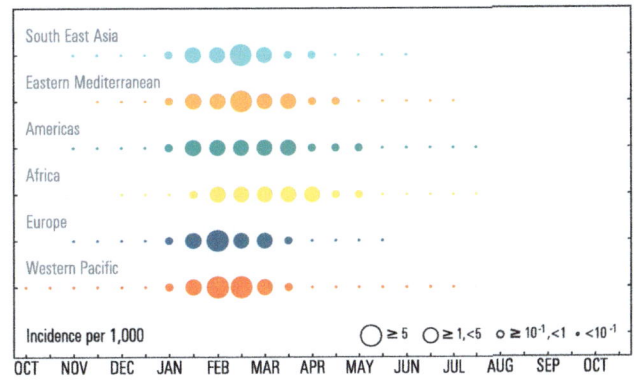

South East Asia
Eastern Mediterranean
Americas
Africa
Europe
Western Pacific

Incidence per 1,000   ≥ 5   ≥ 1,<5   ≥ 10⁻¹,<1   <10⁻¹

OCT NOV DEC JAN FEB MAR APR MAY JUN JUL AUG SEP OCT

NEW YORK CITY, USA

time (days)

32   64   128   256

▲ INFECTION TREE

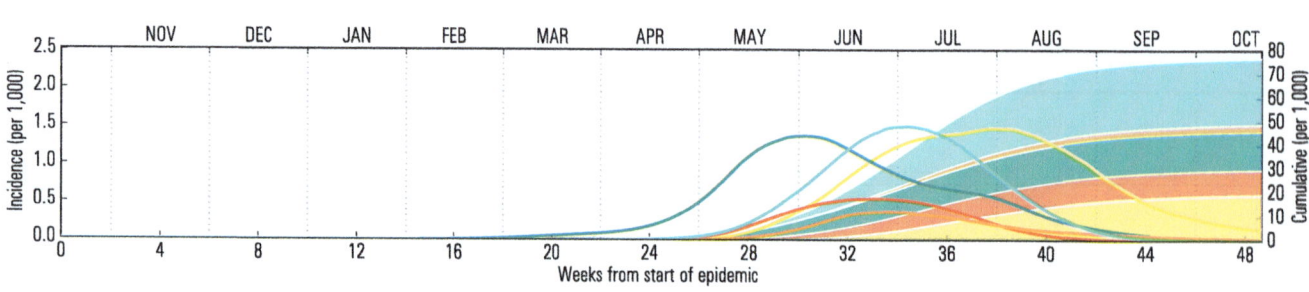

▲ CUMULATIVE INCIDENCE AND EPIDEMIC PEAKS

▼ AFFECTED COUNTRIES

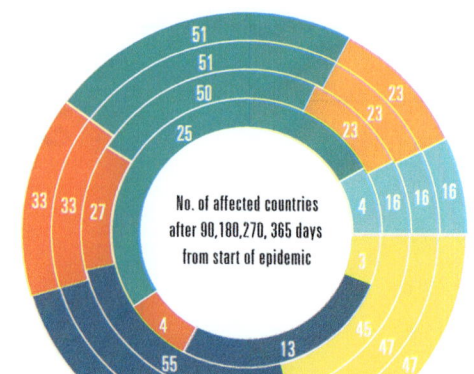

No. of affected countries
after 90,180,270,365 days
from start of epidemic

▼ EPIDEMIC ACTIVITY

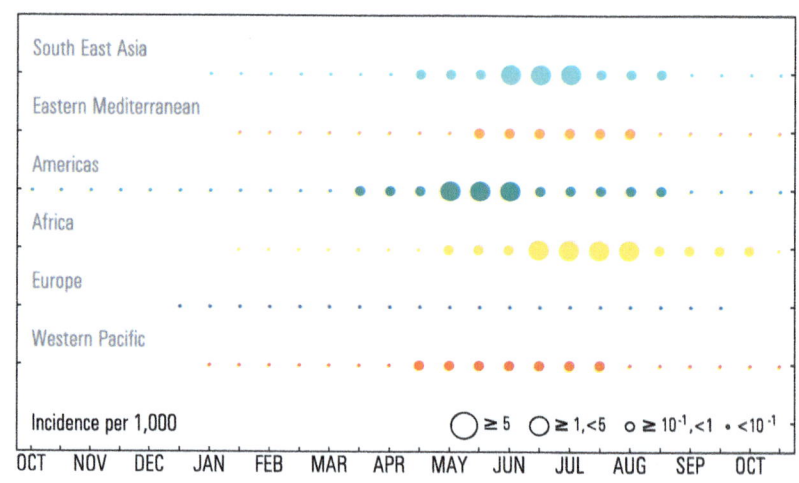

South East Asia

Eastern Mediterranean

Americas

Africa

Europe

Western Pacific

Incidence per 1,000

○ ≥ 5   ○ ≥ 1,<5   ○ ≥ 10⁻¹,<1   · < 10⁻¹

OCT NOV DEC JAN FEB MAR APR MAY JUN JUL AUG SEP OCT

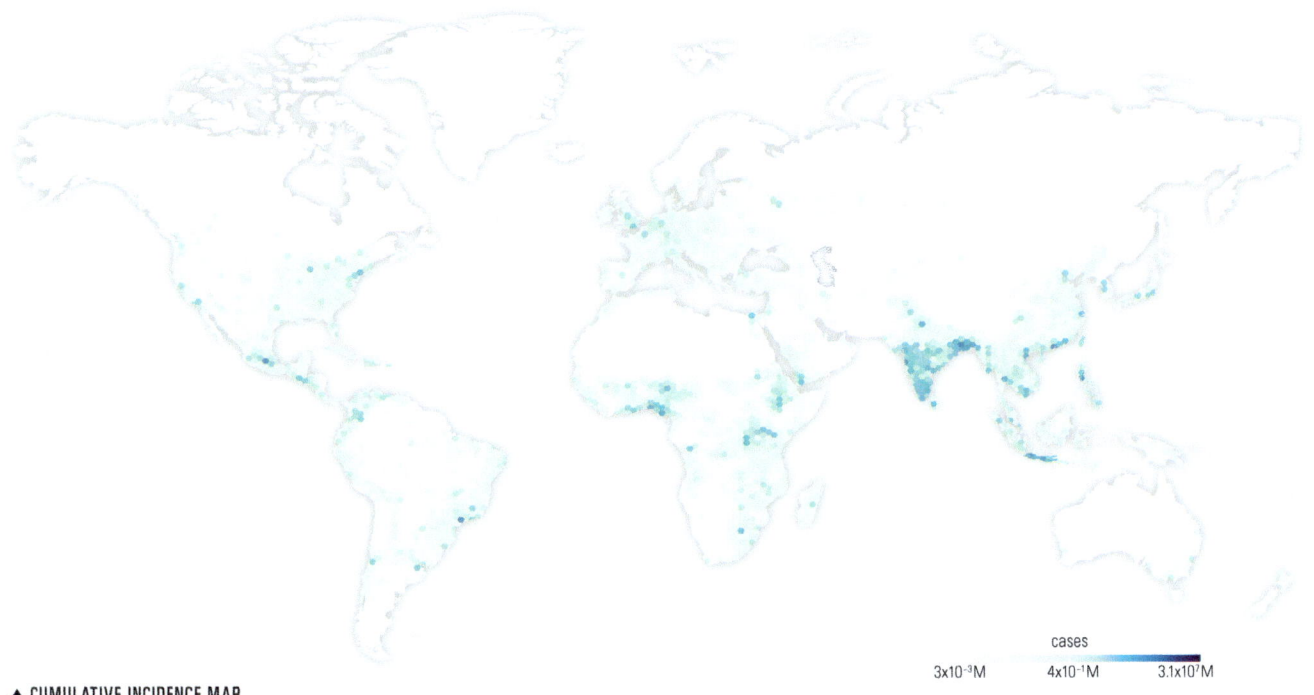

cases
3×10^{-3}M 4×10^{-1}M 3.1×10^{7}M

▲ CUMULATIVE INCIDENCE MAP

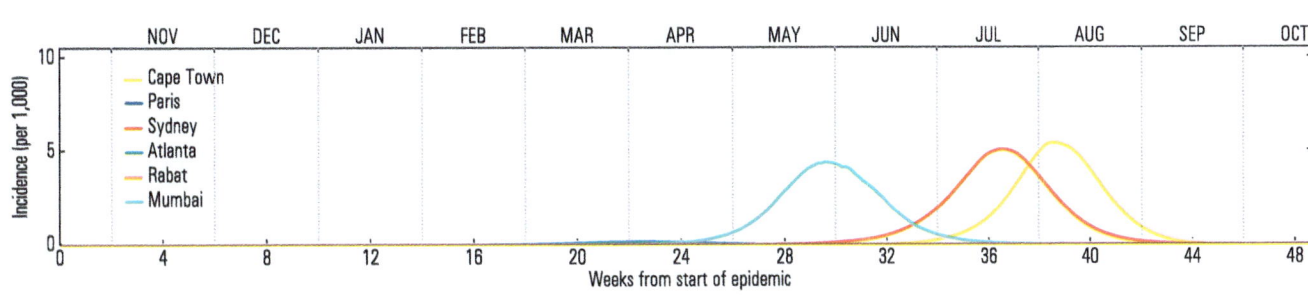

▲ URBAN LEVEL INCIDENCE

▼ RISK QUAD CHART

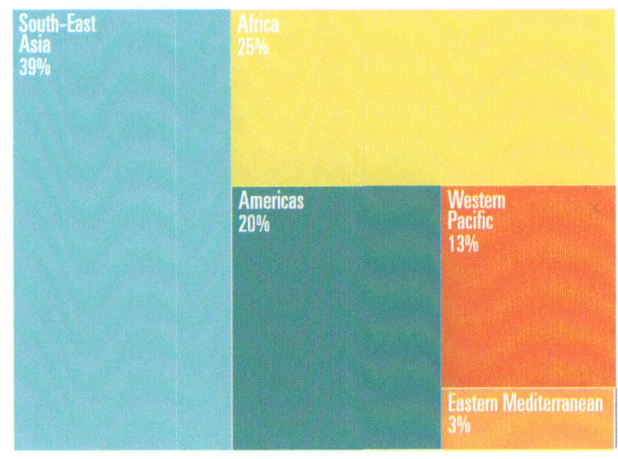

▼ ATTACK RATE COMPOSITION (ONE YEAR)

NEW YORK CITY, USA

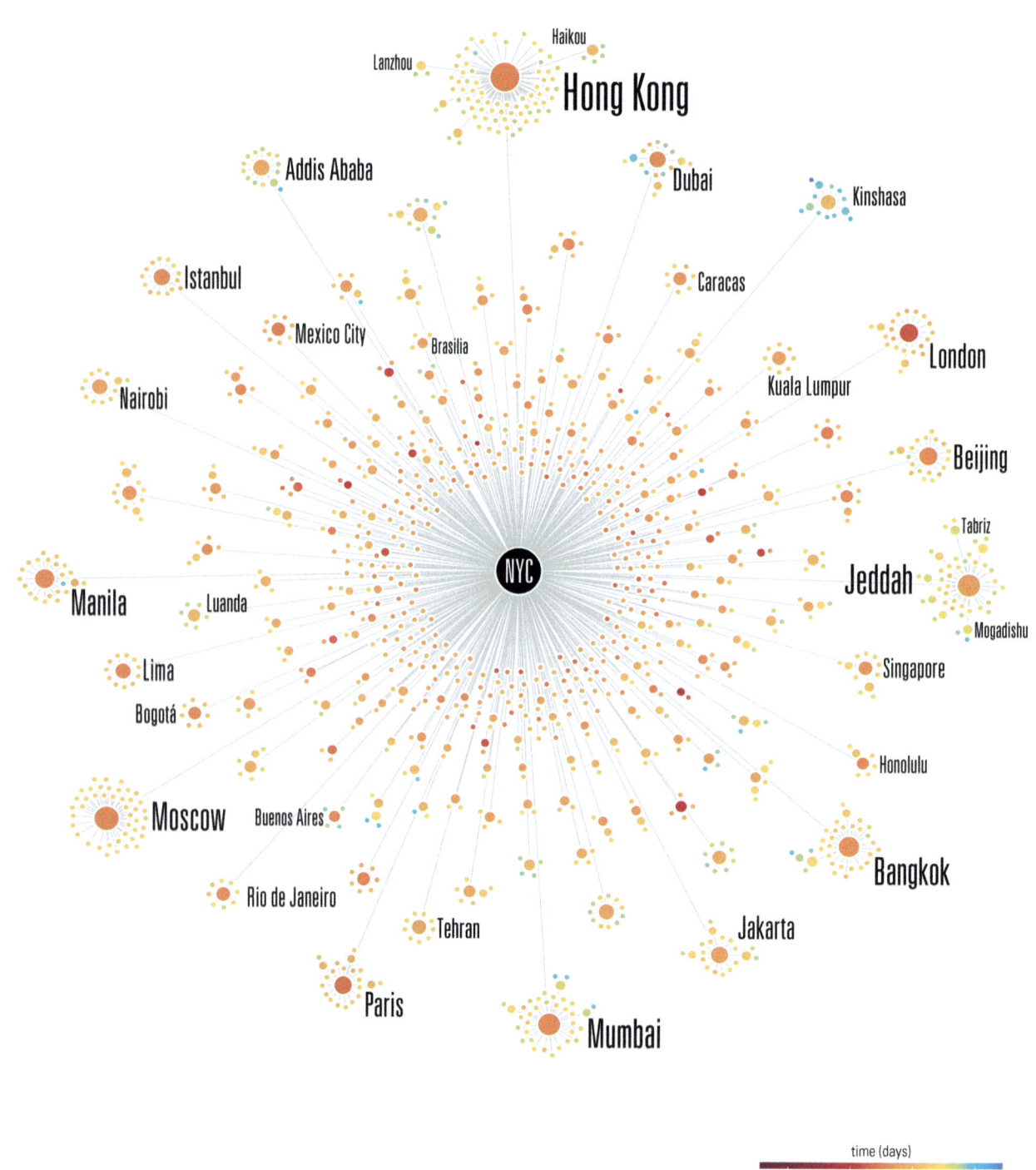

Haikou

Lanzhou

Hong Kong

Addis Ababa

Dubai

Kinshasa

Istanbul

Caracas

Mexico City

Brasilia

London

Nairobi

Kuala Lumpur

Beijing

Tabriz

Manila

Luanda

Jeddah

Mogadishu

Lima

Singapore

Bogotá

Honolulu

Moscow

Buenos Aires

Bangkok

Rio de Janeiro

Tehran

Jakarta

Paris

Mumbai

NYC

time (days)

32   64   128   256

▲ INFECTION TREE GRAPH

■ Americas ■ E. Mediterranean
■ Africa ■ South-East Asia
■ Europe ■ Western Pacific

 OCT  R₀ 1.5

80% AIR TRAFFIC REDUCTION

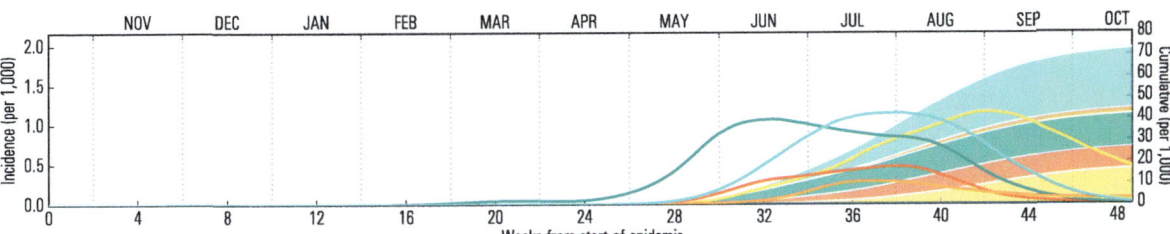

▲ CUMULATIVE INCIDENCE AND EPIDEMIC PEAKS

▼ ATTACK RATE COMPOSITION (ONE YEAR)

▼ EPIDEMIC ACTIVITY

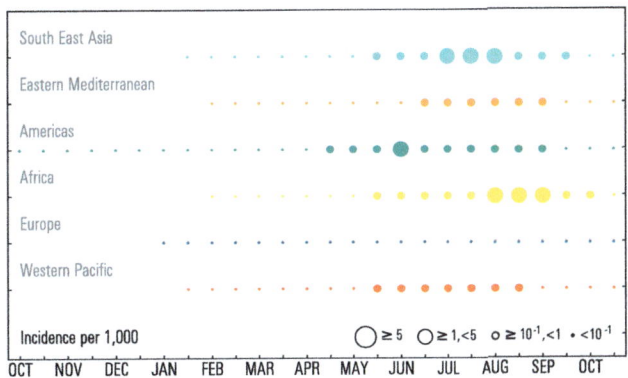

VACCINATIONS

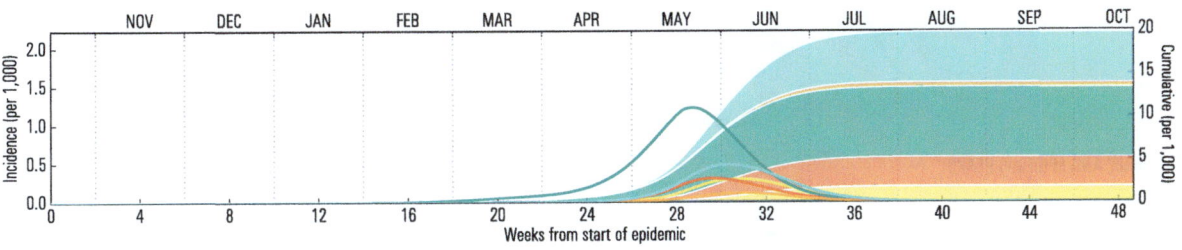

▲ CUMULATIVE INCIDENCE AND EPIDEMIC PEAKS

▼ ATTACK RATE COMPOSITION (ONE YEAR)

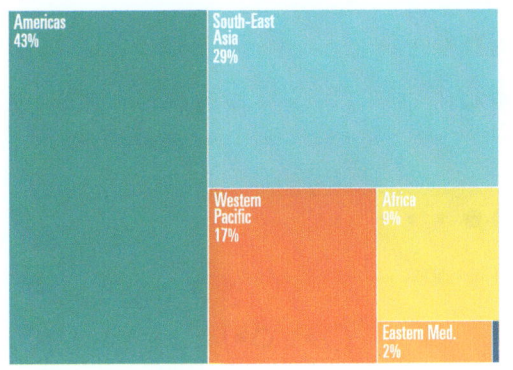

▼ EPIDEMIC ACTIVITY

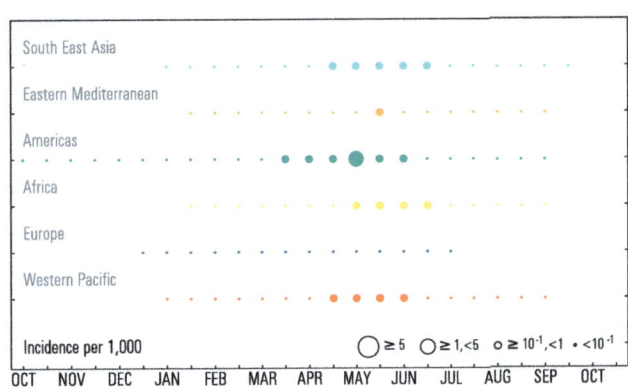

NEW YORK CITY, USA

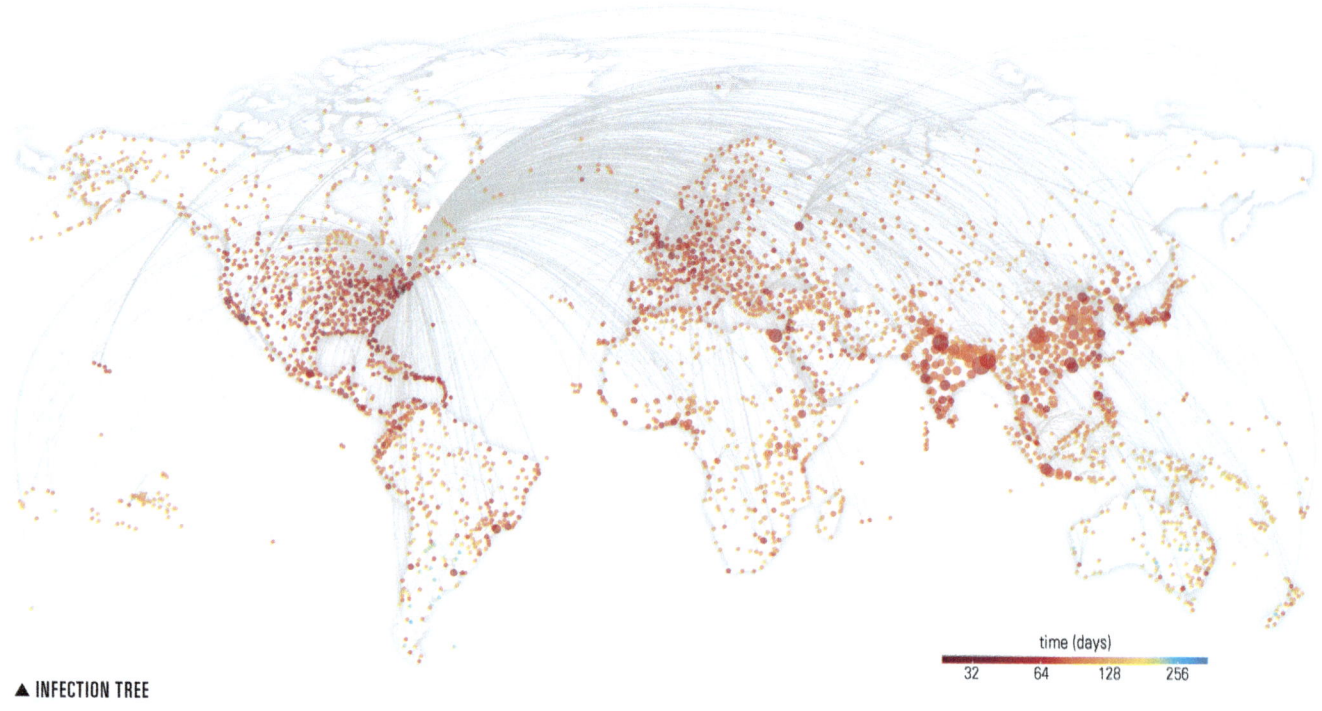

▲ INFECTION TREE

time (days)

| 32 | 64 | 128 | 256 |

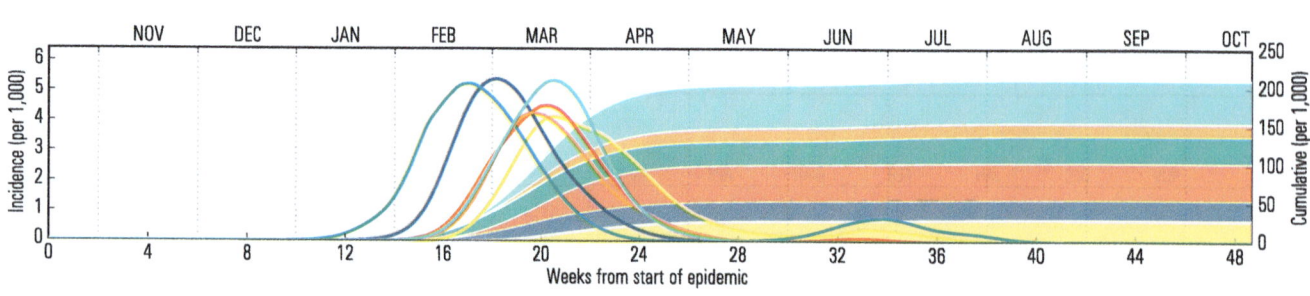

▲ CUMULATIVE INCIDENCE AND EPIDEMIC PEAKS

▼ AFFECTED COUNTRIES

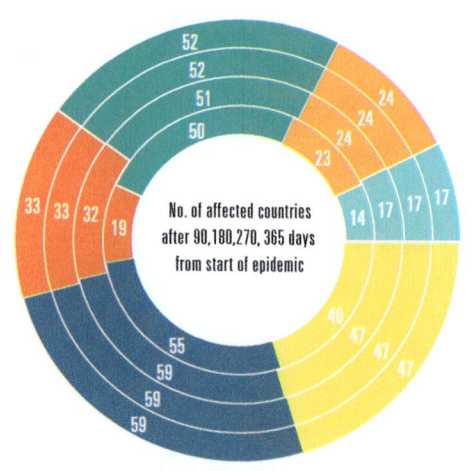

No. of affected countries
after 90,180,270, 365 days
from start of epidemic

▼ EPIDEMIC ACTIVITY

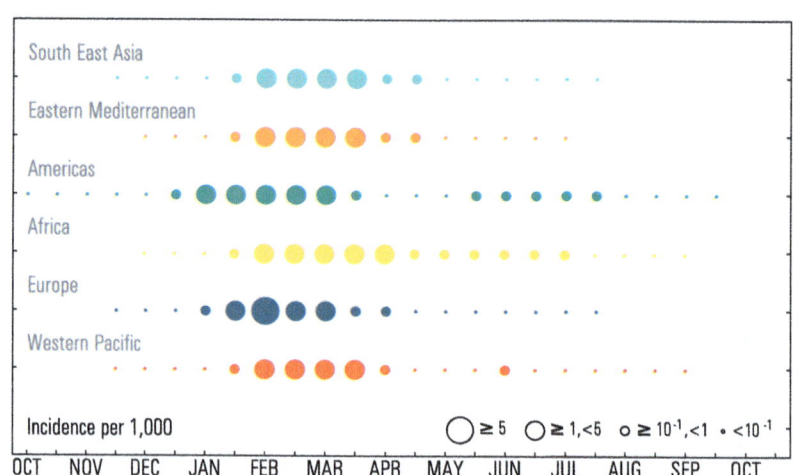

Incidence per 1,000

≥ 5   ≥ 1,<5   ≥ 10⁻¹,<1   · <10⁻¹

WHO REGIONS

- Americas
- Africa
- Europe
- E. Mediterranean
- South-East Asia
- Western Pacific

INITIAL CONDITIONS

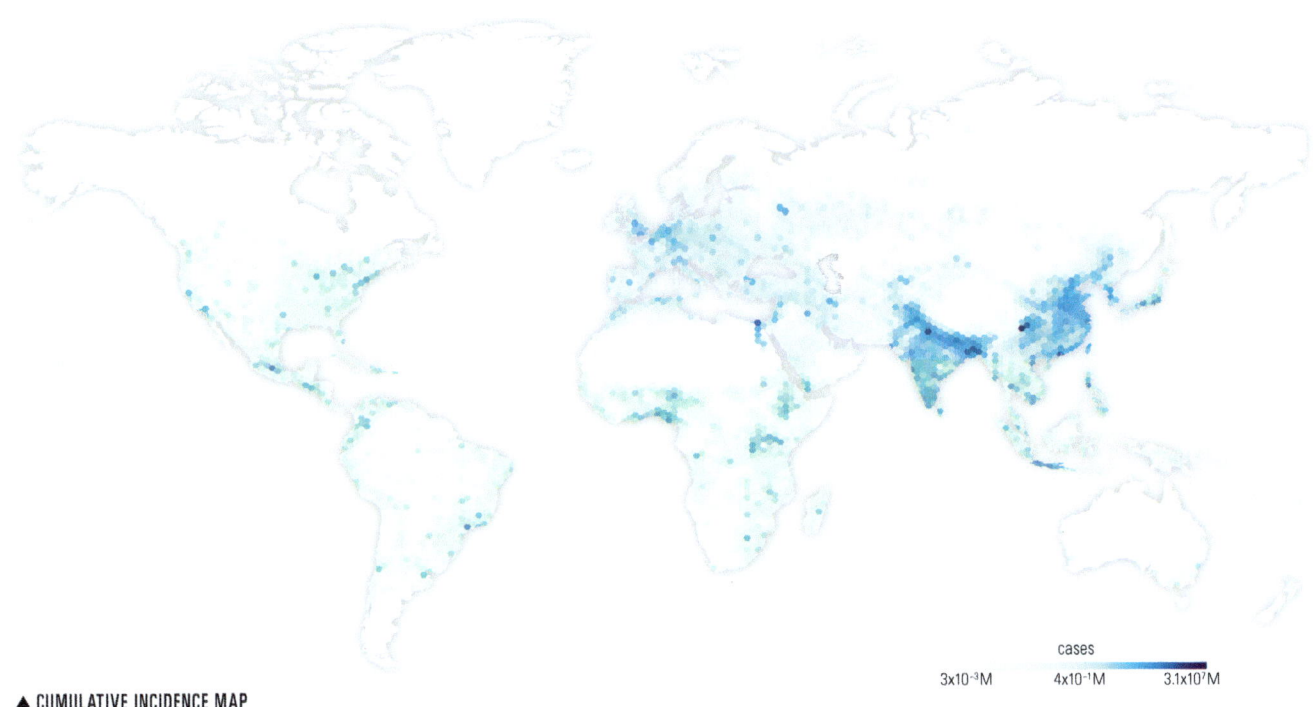

cases

3x10⁻³M   4x10⁻¹M   3.1x10⁷M

▲ CUMULATIVE INCIDENCE MAP

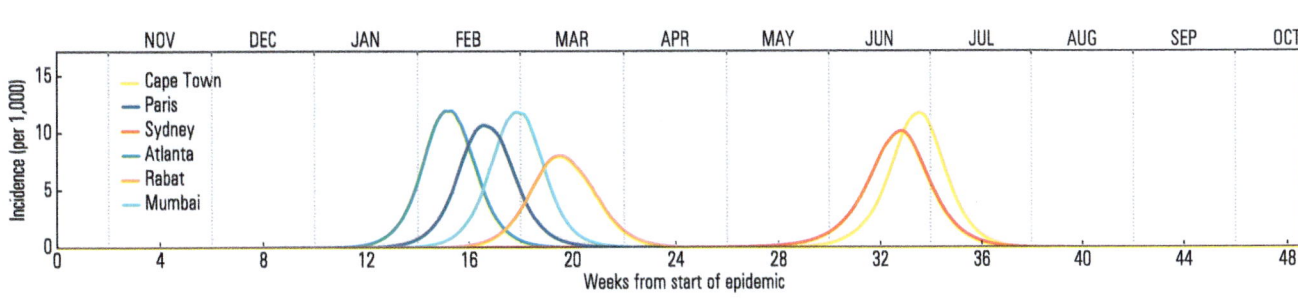

▲ URBAN LEVEL INCIDENCE

▼ RISK QUAD CHART

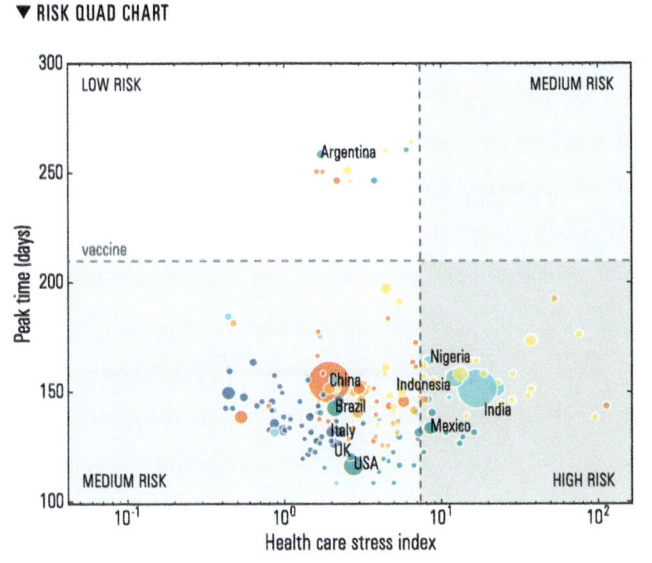

▼ ATTACK RATE COMPOSITION (ONE YEAR)

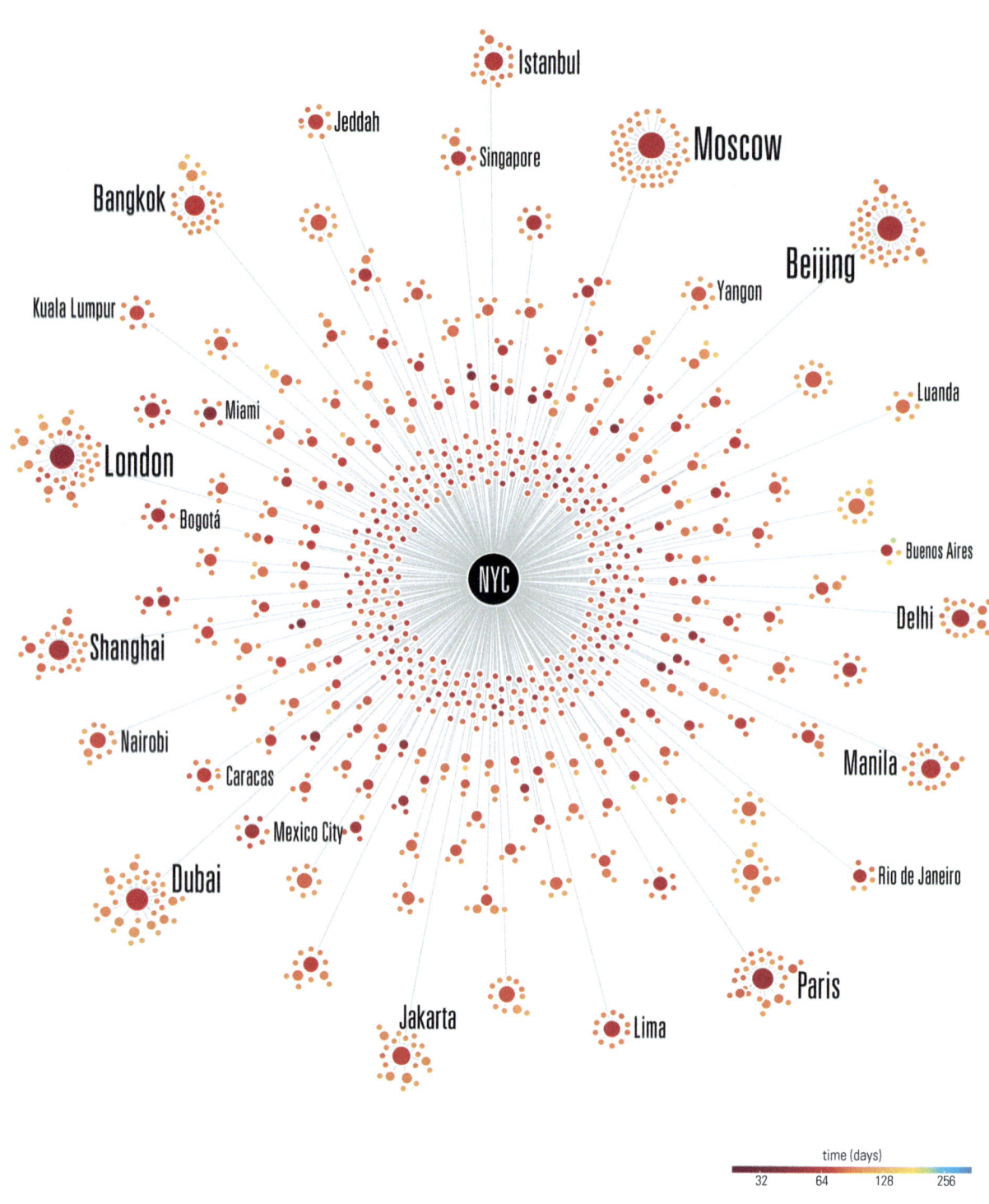

Istanbul

Jeddah

Singapore

Moscow

Bangkok

Beijing

Kuala Lumpur

Yangon

Miami

Luanda

London

Buenos Aires

Bogotá

Delhi

Shanghai

Nairobi

Manila

Caracas

Mexico City

Rio de Janeiro

Dubai

Paris

Jakarta

Lima

NYC

time (days)

32   64   128   256

▲ INFECTION TREE GRAPH

WHO REGIONS

- Americas
- Africa
- Europe
- E. Mediterranean
- South-East Asia
- Western Pacific

INITIAL CONDITIONS

 OCT  R₀ 2.0

 80% AIR TRAFFIC REDUCTION

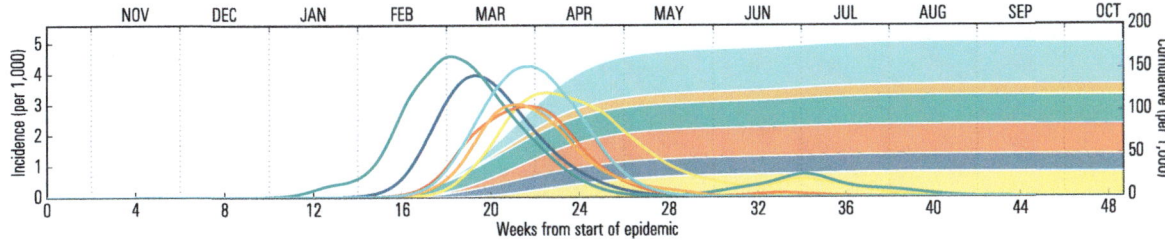

▲ CUMULATIVE INCIDENCE AND EPIDEMIC PEAKS

▼ ATTACK RATE COMPOSITION (ONE YEAR)

▼ EPIDEMIC ACTIVITY

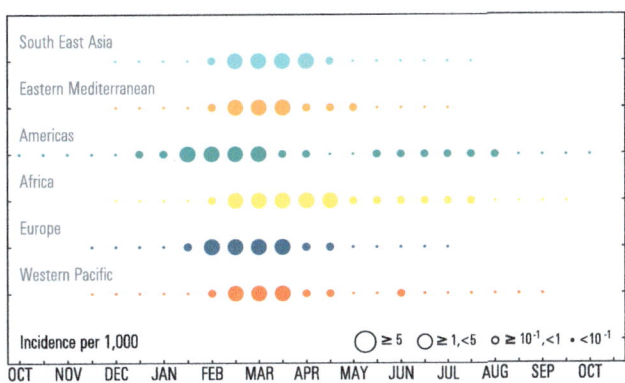

VACCINATIONS

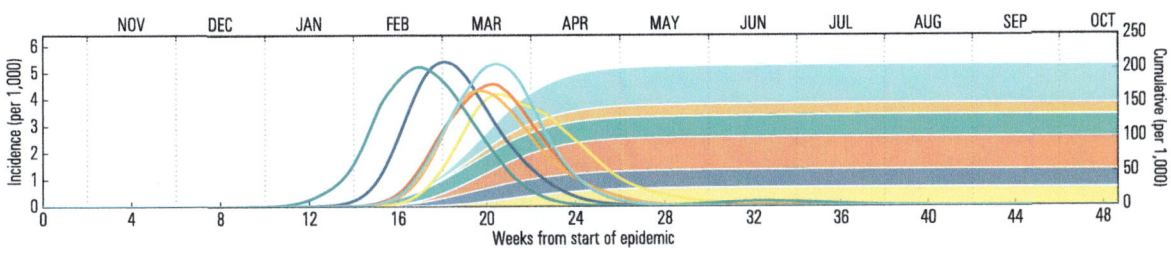

▲ CUMULATIVE INCIDENCE AND EPIDEMIC PEAKS

▼ ATTACK RATE COMPOSITION (ONE YEAR)

▼ EPIDEMIC ACTIVITY

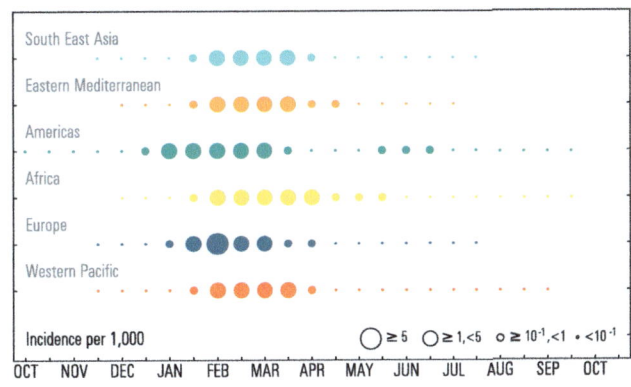

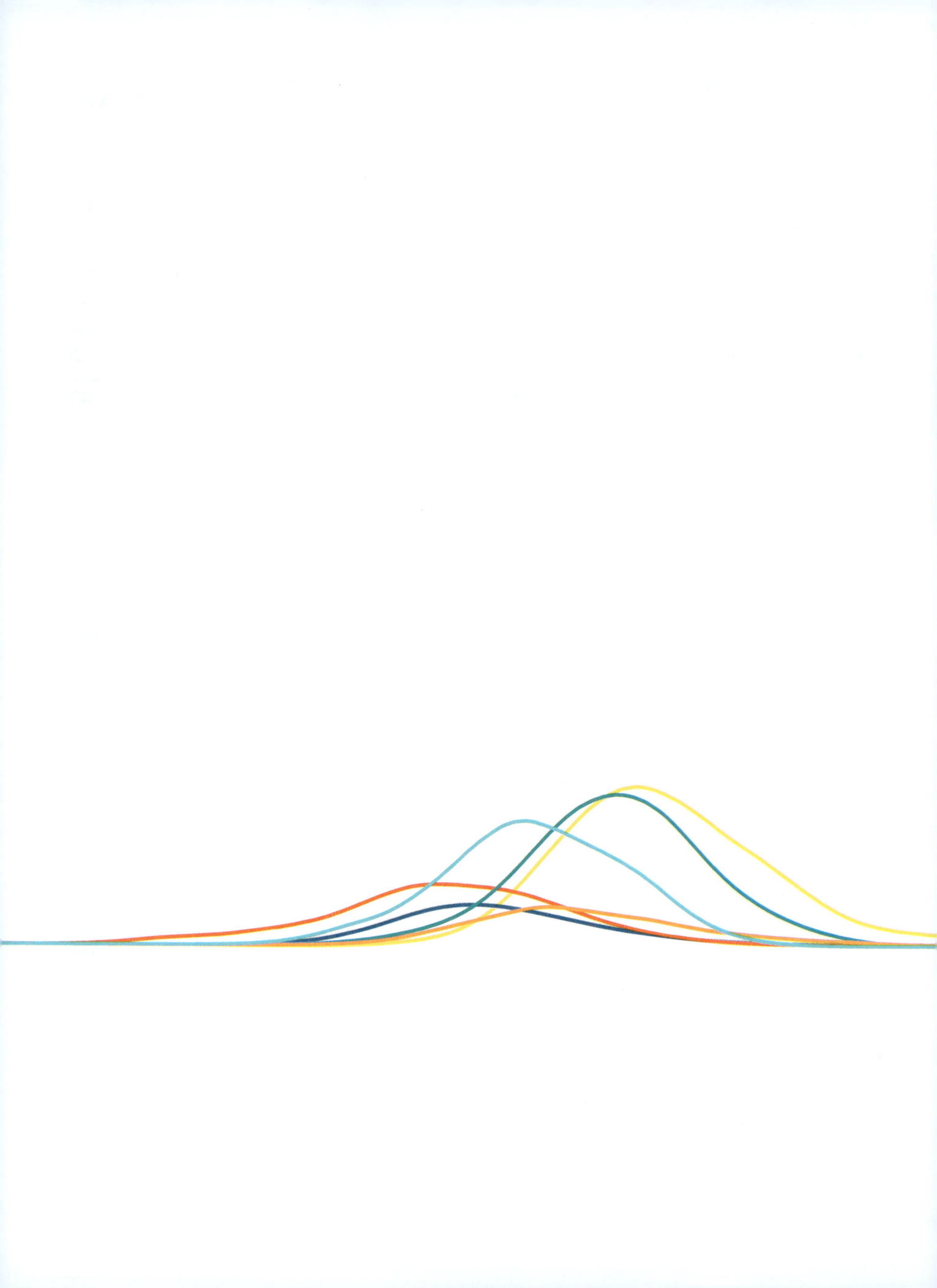

CORONAVIRUS

In the case of coronaviruses, we selected scenarios referring to a case with a transmission rate and natural history of the disease similar to the SARS virus. Thus we assume that the infectiousness of individuals starts only after the onset of clinical symptoms, and we consider the absence of asymptomatic infections.

Although we take into account a relatively high reproduction number $R_0 = 2.7$, the absence of asymptomatic transmission makes all containment measures based on the timely isolation of cases viable and very effective. We therefore consider, for each possible initial condition of the outbreak, two scenarios with a transmissibility reduction, due to prompt case isolation, of 30% and 50% after the first 4 weeks and 50% and 90% after the seventh week, respectively.

Contrary to influenza, we do not consider seasonal variations because seasonal forcing does not appear to have a large impact on SARS. For this reason, we consider only one starting date during the calendar year for each geographical location.

In total we provide six scenarios summarized in the infographic charts. The typical coronavirus chart layout and the "how to read it" guide can be found on the following pages.

STARTING CONDITIONS

- **$R_0 = 2.7$**
- **Starting date:** varies by geographical location

SCENARIOS

Reduction of transmissibility:
- 30% after 4 weeks; 50% after 7 weeks
- 50% after 4 weeks, 90% after 7 weeks

ORIGIN CITIES

Barcelona, Spain

15 million
passengers/year

Guangzhou, China

21 million
passengers/year

Jeddah,
Saudi Arabia

11.3 million
passengers/year

CHART LAYOUT: CORONAVIRUS

Location of pandemic origin

Infection tree

Each link describes an infection path along which the epidemic moved from one population to the other. The color of the nodes refers to the time lag with respect to the initial spreading.

Cumulative incidence and epidemic peaks

The colored regions describe the cumulative incidence at the regional level. The solid lines represent the regional epidemic profiles.

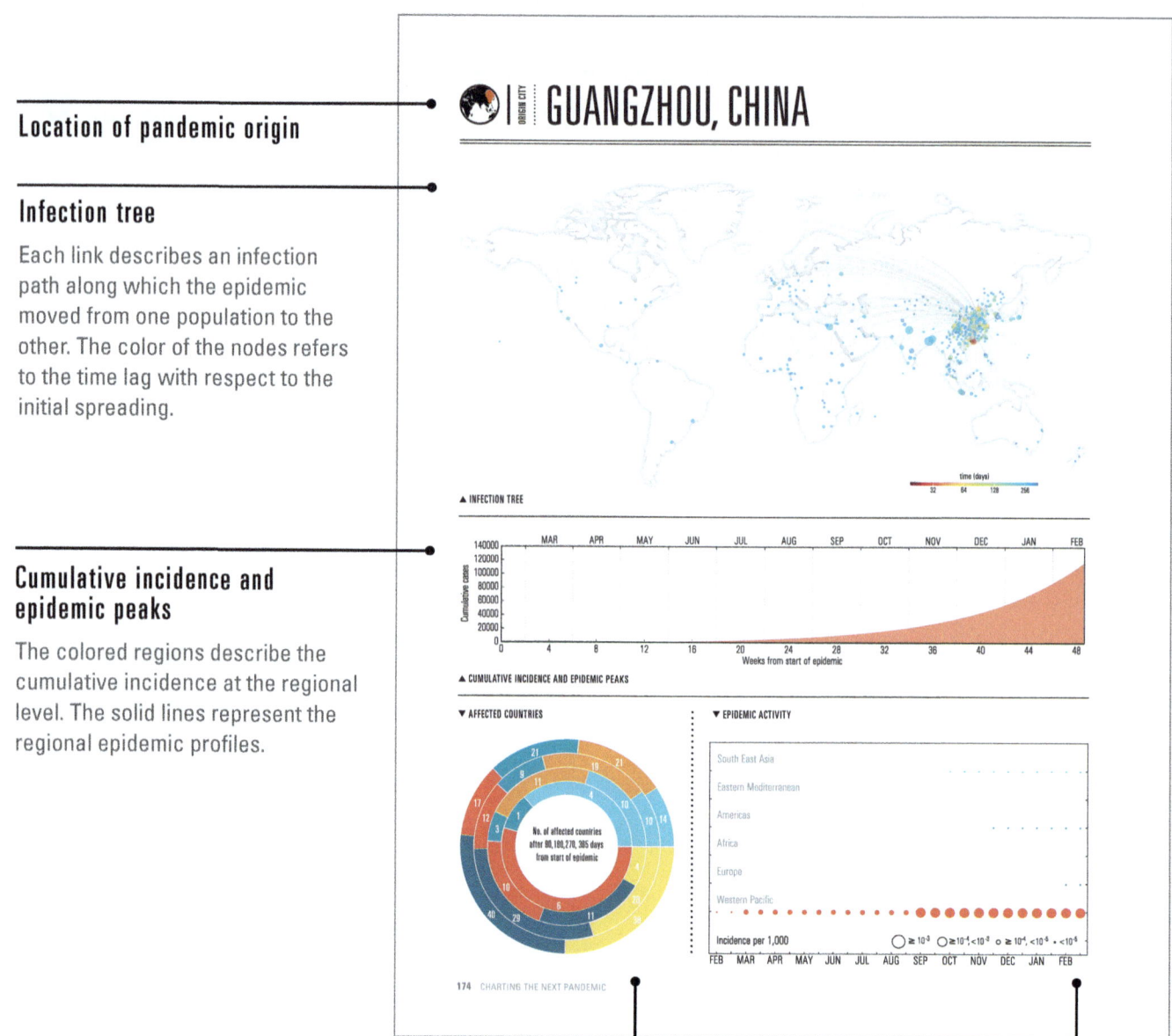

ORIGIN CITY

GUANGZHOU, CHINA

time (days)
32   64   128   256

▲ INFECTION TREE

Cumulative cases

140000
120000
100000
80000
60000
40000
20000
0

MAR   APR   MAY   JUN   JUL   AUG   SEP   OCT   NOV   DEC   JAN   FEB

0   4   8   12   16   20   24   28   32   36   40   44   48
Weeks from start of epidemic

▲ CUMULATIVE INCIDENCE AND EPIDEMIC PEAKS

▼ AFFECTED COUNTRIES

No. of affected countries after 90, 180, 270, 365 days from start of epidemic

▼ EPIDEMIC ACTIVITY

South East Asia
Eastern Mediterranean
Americas
Africa
Europe
Western Pacific

Incidence per 1,000   ○ ≥ 10⁻³   ○ ≥ 10⁻⁴, <10⁻³   ○ ≥ 10⁻⁵, <10⁻⁴   · <10⁻⁵
FEB   MAR   APR   MAY   JUN   JUL   AUG   SEP   OCT   NOV   DEC   JAN   FEB

174   CHARTING THE NEXT PANDEMIC

Affected countries as a function of time

The concentric circles refer to 90, 180, 270, and 365 days from the start of the spreading, respectively. In each circle we report the number of countries affected by the outbreak in each region.

Epidemic activity

Epidemic activity at the regional level. The size of the circles indicates the level of activity, that is proportional to the number of cases.

WHO regions

Interventions

Starting date

R_0 value

Seasonality

Disease type

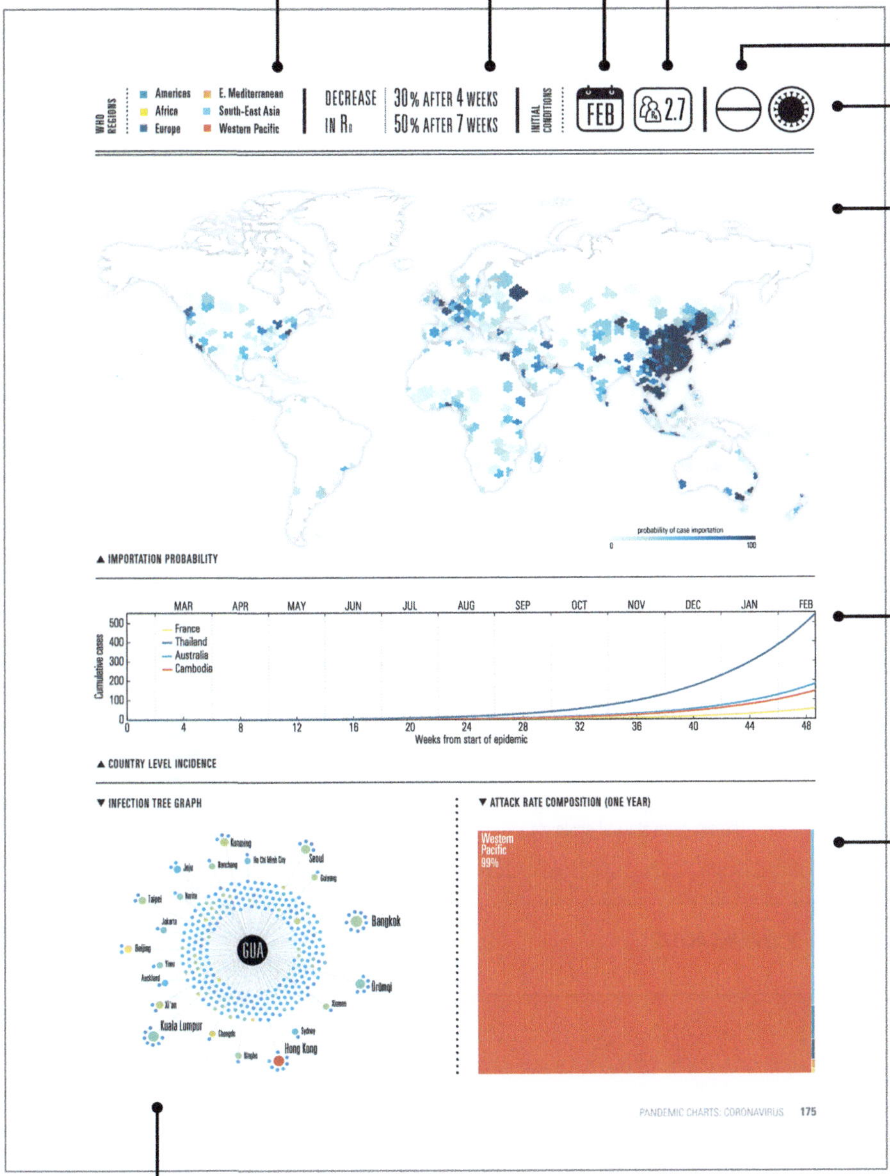

WHO REGIONS		
Americas	E. Mediterranean	
Africa	South-East Asia	
Europe	Western Pacific	

DECREASE IN R_0 | 30% AFTER 4 WEEKS 50% AFTER 7 WEEKS | INITIAL CONDITIONS | FEB | 2.7

Importation probability map

Probability of observing an imported case.

▲ IMPORTATION PROBABILITY

probability of case importation
0   100

MAR APR MAY JUN JUL AUG SEP OCT NOV DEC JAN FEB

Cumulative cases
500
400
300
200
100
0

France
Thailand
Australia
Cambodia

0   4   8   12   16   20   24   28   32   36   40   44   48
Weeks from start of epidemic

▲ COUNTRY LEVEL INCIDENCE

Country level incidence

Daily incidence for selected countries.

▼ INFECTION TREE GRAPH

Kunming
Jixia   Shenzhang   Ho Chi Minh City   Seoul
Taipei   Nanino   Guiyang
Jakarta   Bangkok
Beijing   Tones   GUA
Auckland   Ürümqi
Xi'an   Xiamen
Kuala Lumpur   Chengdu   Sydney
Ningbo   Hong Kong

▼ ATTACK RATE COMPOSITION (ONE YEAR)

Western Pacific 89%

Attack rate composition

We illustrate the geographical impact of the disease after one year considering the share of clinical cases in each WHO region.

Infection tree graph

Each link describes an infection path along which the epidemic has moved from one population to the other. The color refers to the time lag with respect to the start of the epidemic.

BARCELONA, SPAIN

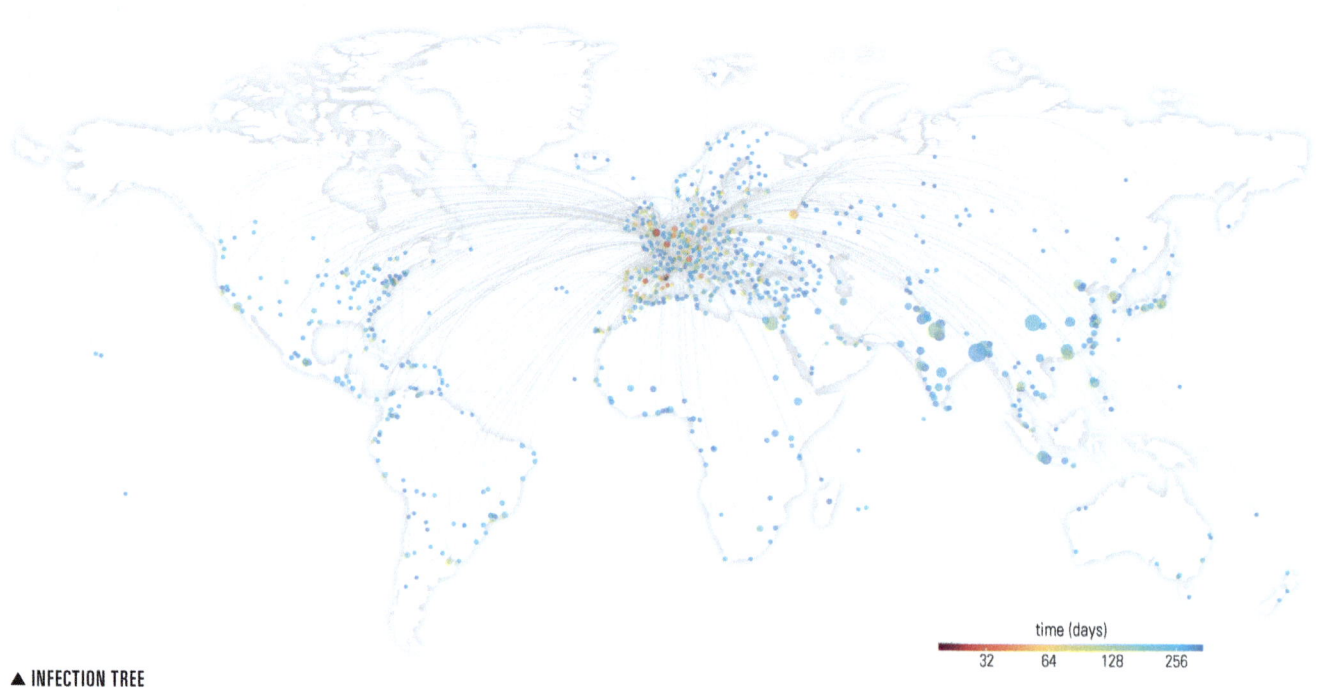

time (days)
32 64 128 256

▲ INFECTION TREE

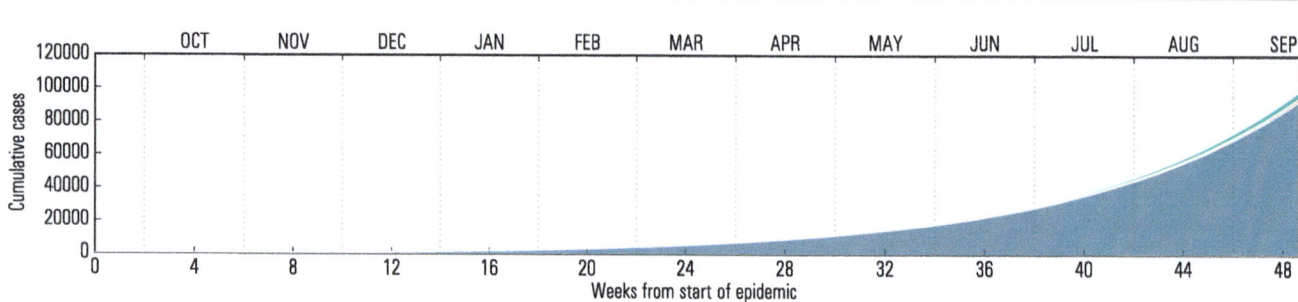

Weeks from start of epidemic

▲ CUMULATIVE INCIDENCE AND EPIDEMIC PEAKS

▼ AFFECTED COUNTRIES

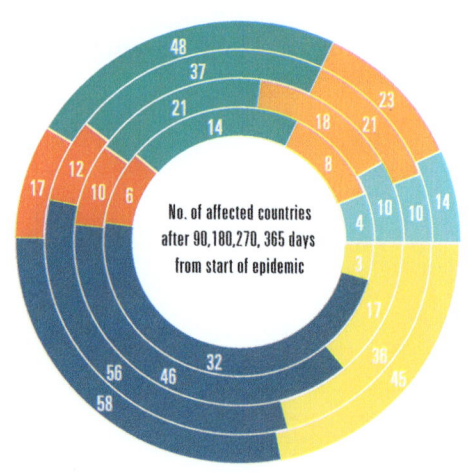

No. of affected countries
after 90,180,270, 365 days
from start of epidemic

▼ EPIDEMIC ACTIVITY

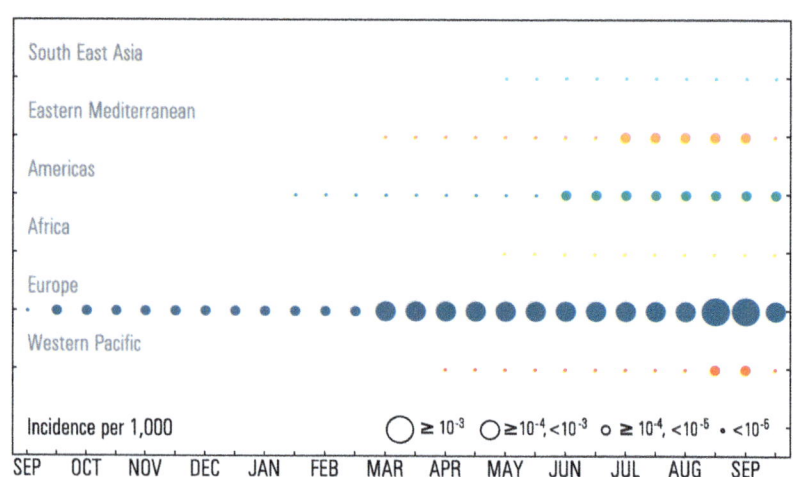

South East Asia

Eastern Mediterranean

Americas

Africa

Europe

Western Pacific

Incidence per 1,000   ○ ≥ 10⁻³   ○ ≥ 10⁻⁴, <10⁻³   ○ ≥ 10⁻⁴, <10⁻⁵   · <10⁻⁵

SEP OCT NOV DEC JAN FEB MAR APR MAY JUN JUL AUG SEP

WHO REGIONS

■ Americas   ■ E. Mediterranean
■ Africa   ■ South-East Asia
■ Europe   ■ Western Pacific

DECREASE IN R₀ | 30% AFTER 4 WEEKS
50% AFTER 7 WEEKS

INITIAL CONDITIONS

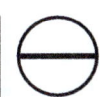

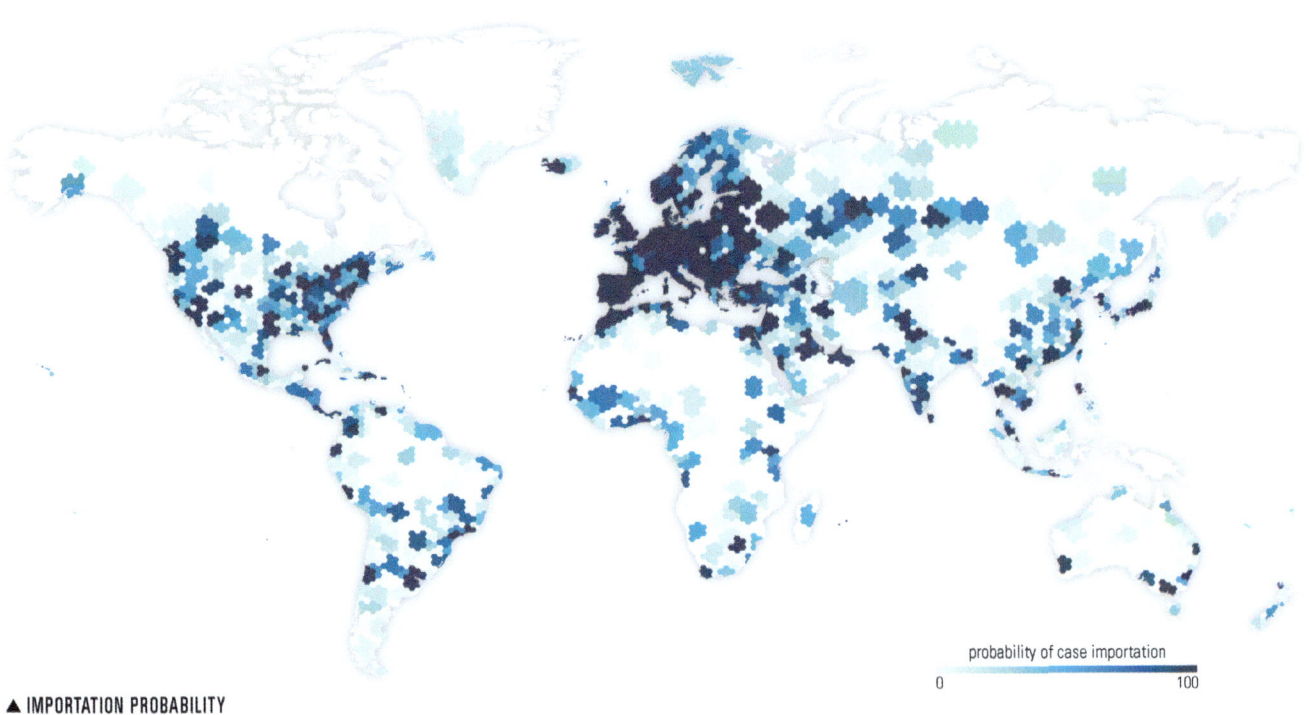

▲ IMPORTATION PROBABILITY

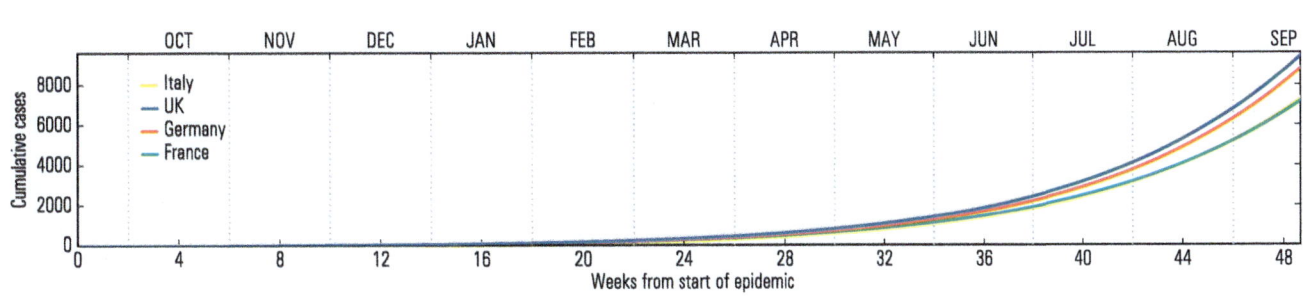

▲ COUNTRY LEVEL INCIDENCE

▼ INFECTION TREE GRAPH

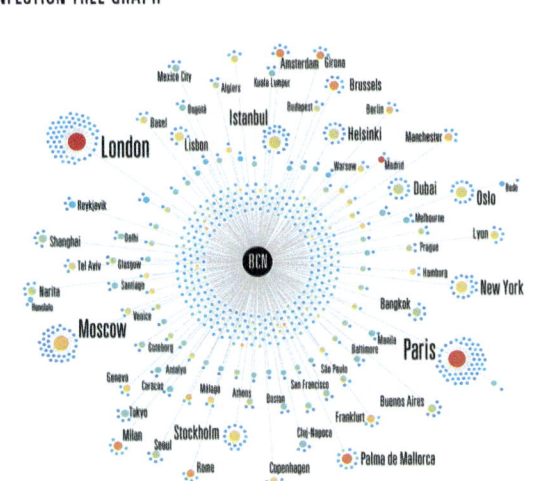

▼ ATTACK RATE COMPOSITION (ONE YEAR)

BARCELONA, SPAIN

▲ INFECTION TREE

time (days)
32   64   128   256

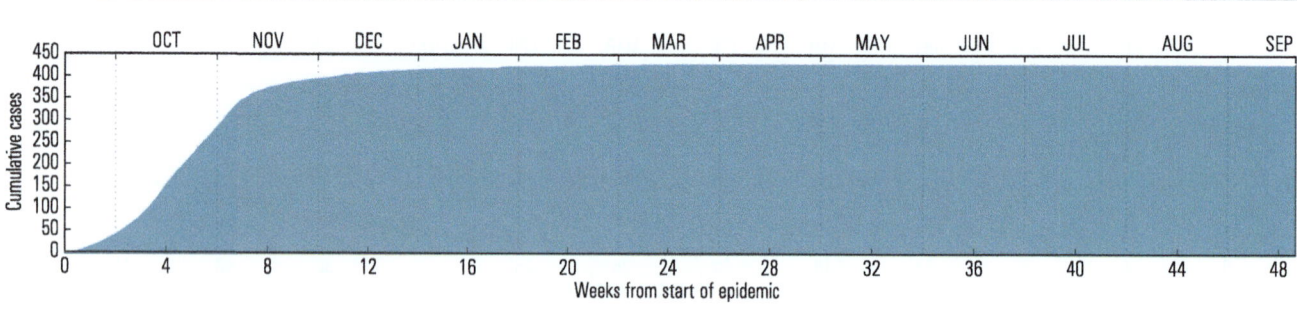

▲ CUMULATIVE INCIDENCE AND EPIDEMIC PEAKS

▼ AFFECTED COUNTRIES

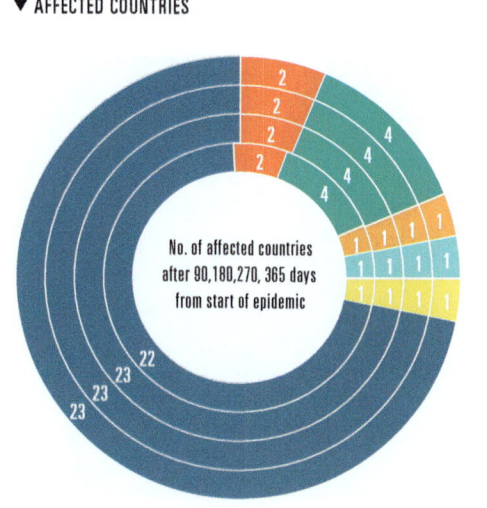

No. of affected countries after 90,180,270, 365 days from start of epidemic

▼ EPIDEMIC ACTIVITY

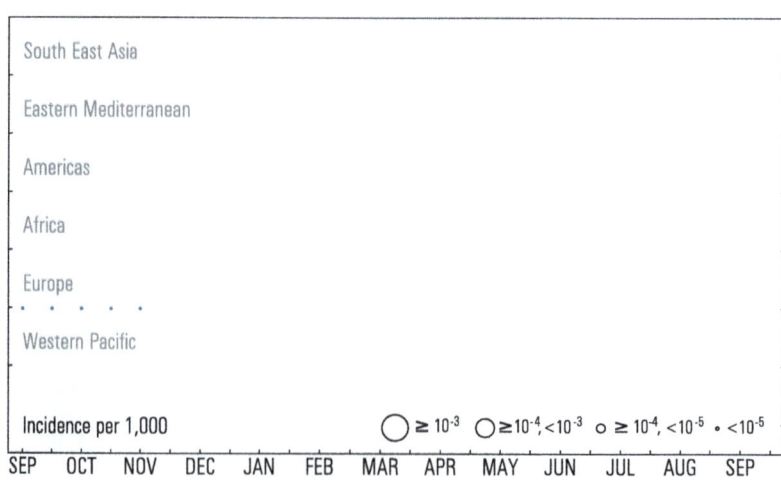

WHO REGIONS

- ■ Americas
- ■ Africa
- ■ Europe
- ■ E. Mediterranean
- ■ South-East Asia
- ■ Western Pacific

DECREASE IN R_0

50 % AFTER 4 WEEKS
90 % AFTER 7 WEEKS

INITIAL CONDITIONS

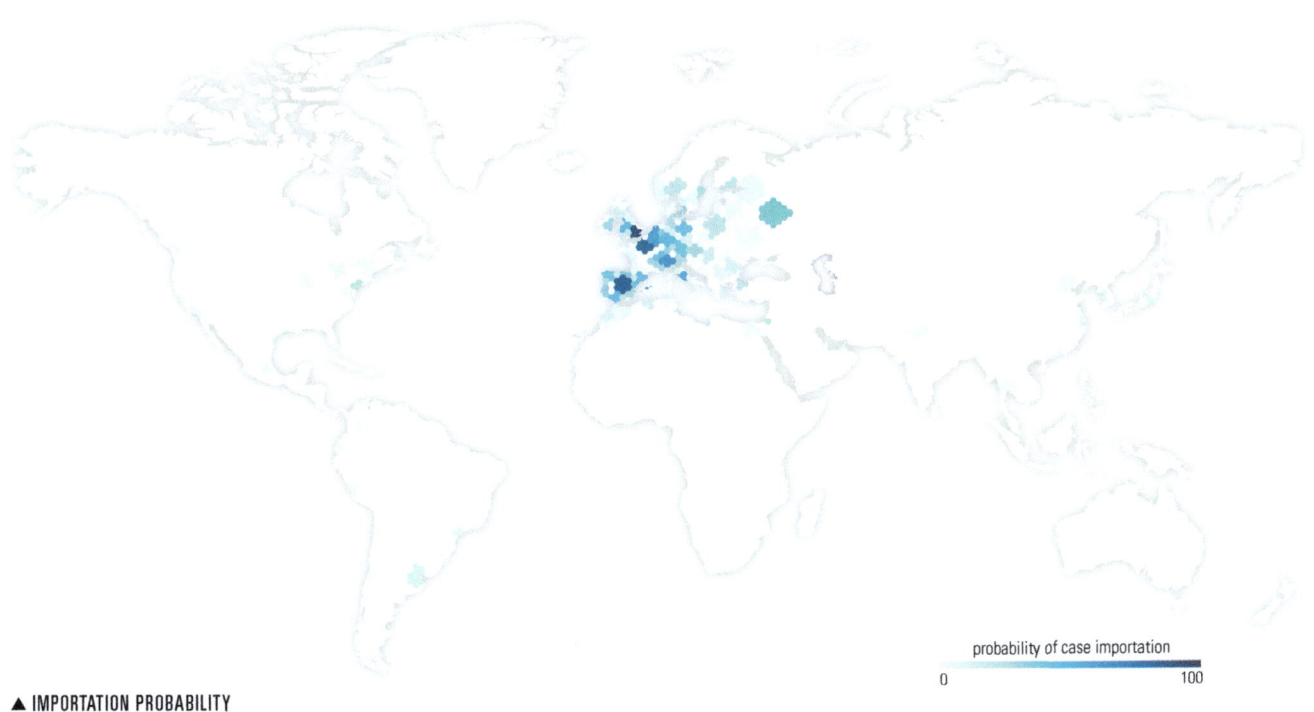

probability of case importation

0   100

▲ IMPORTATION PROBABILITY

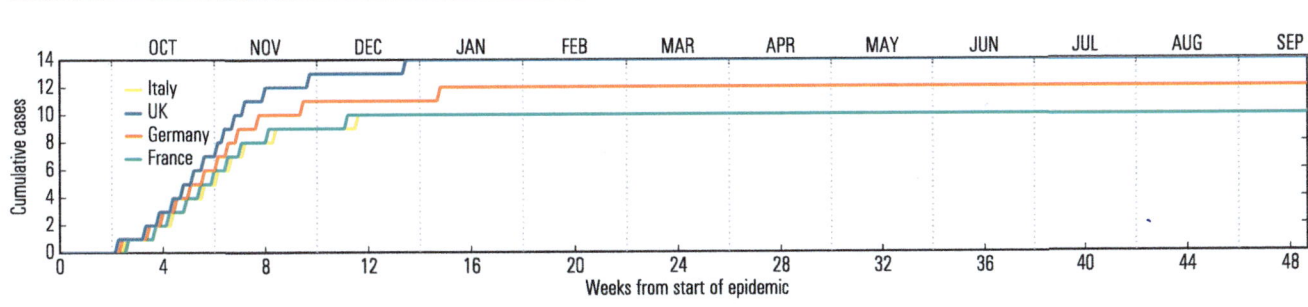

Italy
UK
Germany
France

Cumulative cases

Weeks from start of epidemic

OCT   NOV   DEC   JAN   FEB   MAR   APR   MAY   JUN   JUL   AUG   SEP

▲ COUNTRY LEVEL INCIDENCE

▼ INFECTION TREE GRAPH

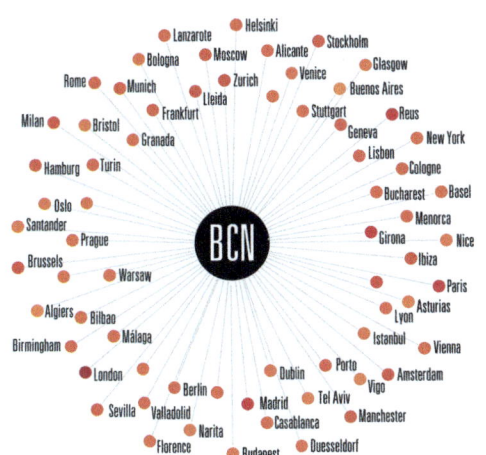

▼ ATTACK RATE COMPOSITION (ONE YEAR)

Europe
99%

GUANGZHOU, CHINA

▲ INFECTION TREE

time (days)

| 32 | 64 | 128 | 256 |

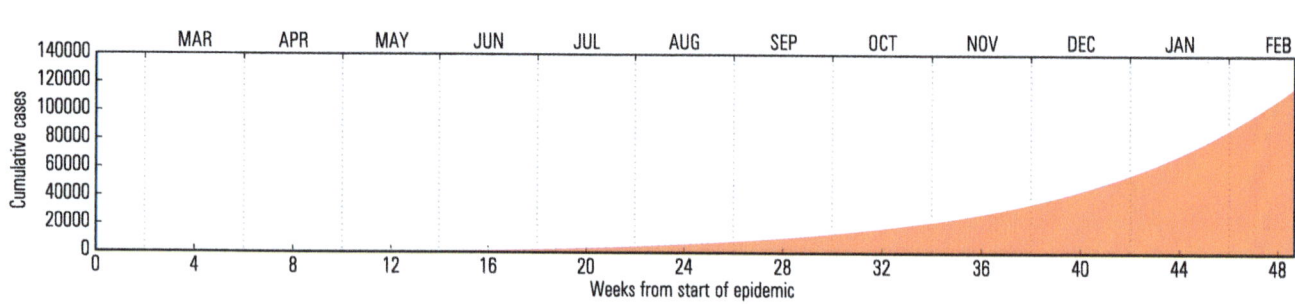

▲ CUMULATIVE INCIDENCE AND EPIDEMIC PEAKS

▼ AFFECTED COUNTRIES

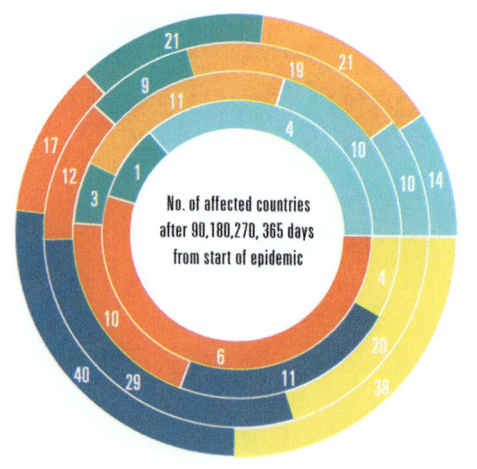

No. of affected countries after 90, 180, 270, 365 days from start of epidemic

▼ EPIDEMIC ACTIVITY

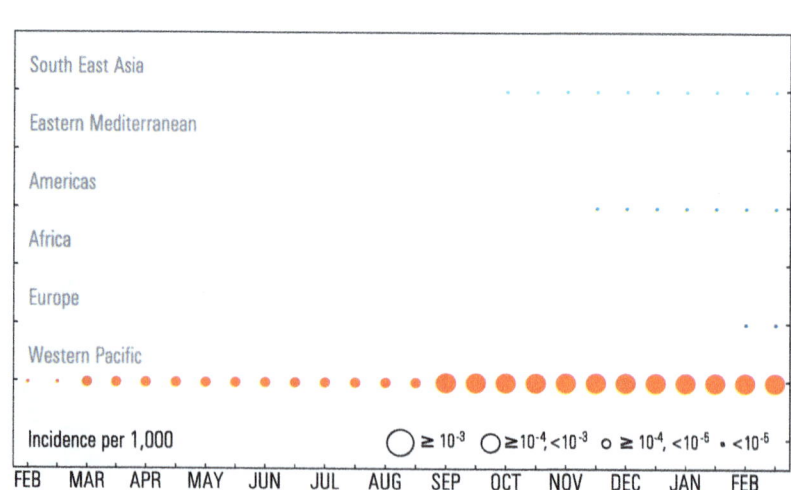

South East Asia

Eastern Mediterranean

Americas

Africa

Europe

Western Pacific

Incidence per 1,000   ○ ≥ 10⁻³ ○ ≥ 10⁻⁴, < 10⁻³ ○ ≥ 10⁻⁴, < 10⁻⁵ · < 10⁻⁵

FEB MAR APR MAY JUN JUL AUG SEP OCT NOV DEC JAN FEB

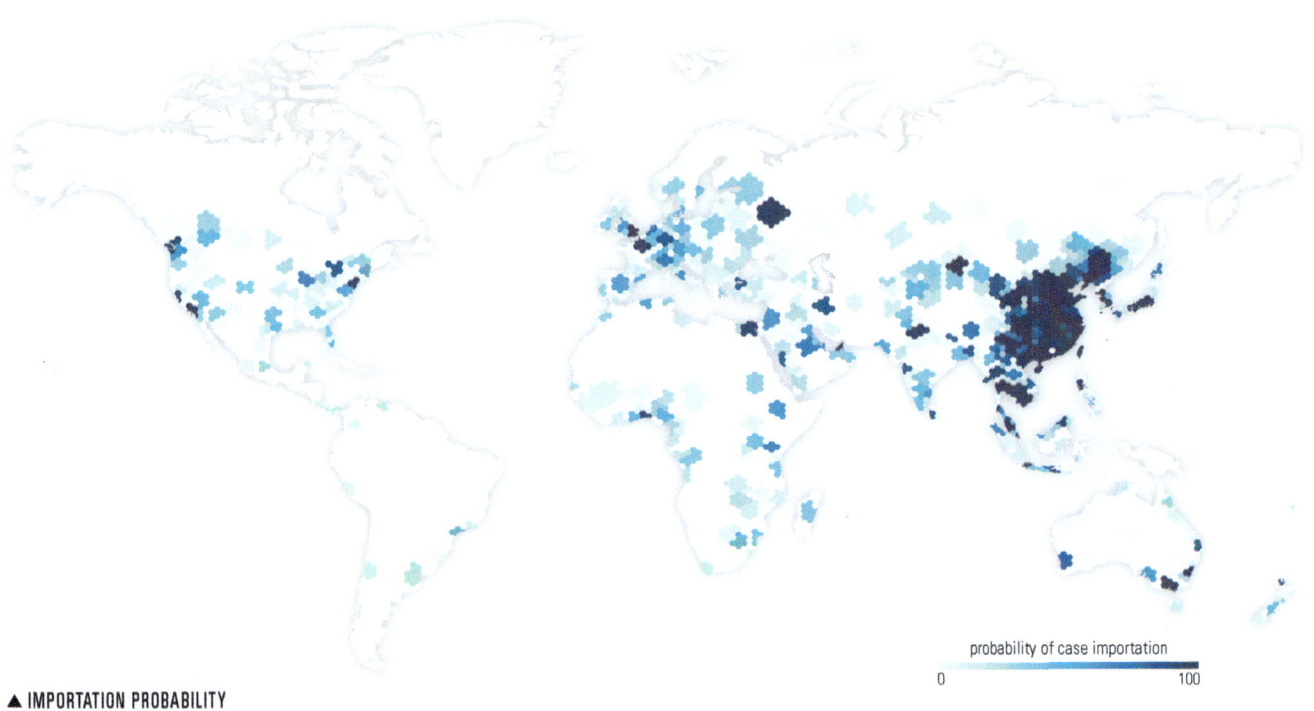

probability of case importation
0   100

▲ IMPORTATION PROBABILITY

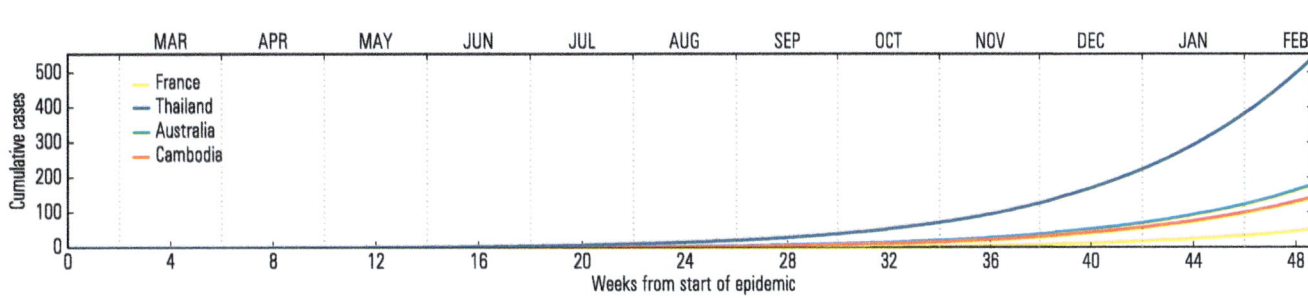

▲ COUNTRY LEVEL INCIDENCE

▼ INFECTION TREE GRAPH

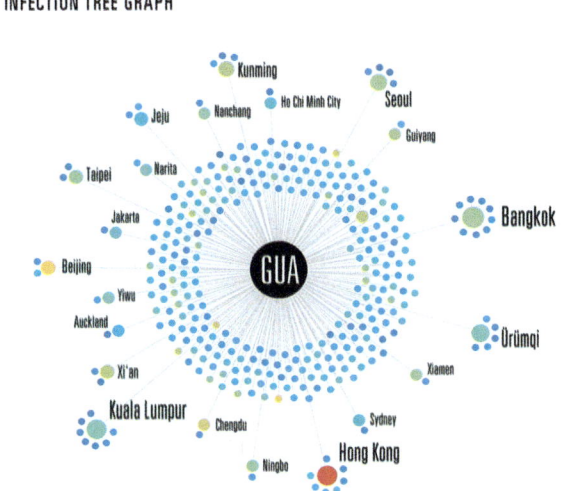

▼ ATTACK RATE COMPOSITION (ONE YEAR)

GUANGZHOU, CHINA

time (days)
32   64   128   256

▲ INFECTION TREE

▲ CUMULATIVE INCIDENCE AND EPIDEMIC PEAKS

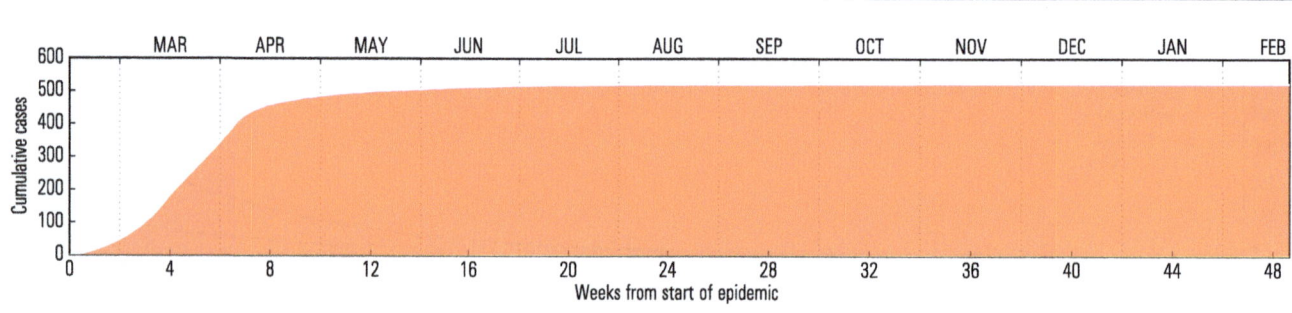

▼ AFFECTED COUNTRIES

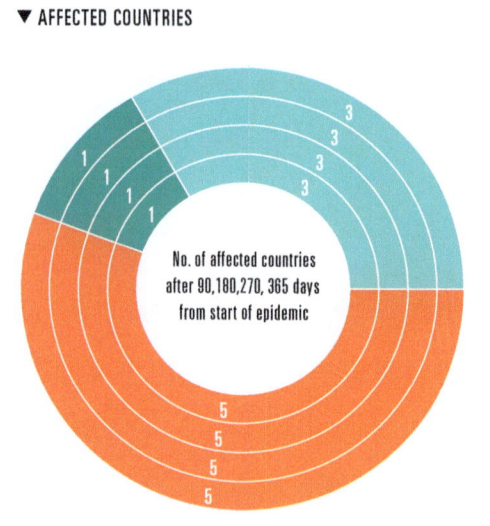

No. of affected countries
after 90,180,270, 365 days
from start of epidemic

▼ EPIDEMIC ACTIVITY

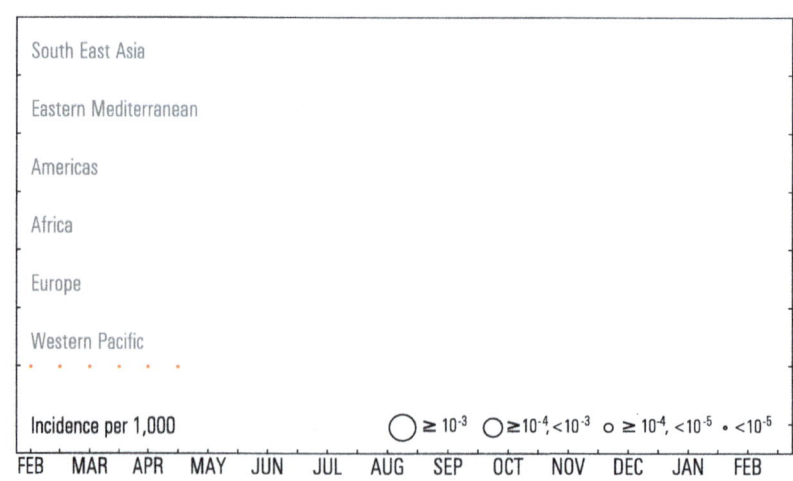

South East Asia

Eastern Mediterranean

Americas

Africa

Europe

Western Pacific

Incidence per 1,000   ◯ ≥ 10⁻³   ◯ ≥ 10⁻⁴, < 10⁻³   ○ ≥ 10⁻⁴, < 10⁻⁵   • < 10⁻⁵

FEB   MAR   APR   MAY   JUN   JUL   AUG   SEP   OCT   NOV   DEC   JAN   FEB

WHO REGIONS
■ Americas   ■ E. Mediterranean
■ Africa   ■ South-East Asia
■ Europe   ■ Western Pacific

DECREASE IN R₀   50% AFTER 4 WEEKS   90% AFTER 7 WEEKS

INITIAL CONDITIONS

probability of case importation
0   100

▲ IMPORTATION PROBABILITY

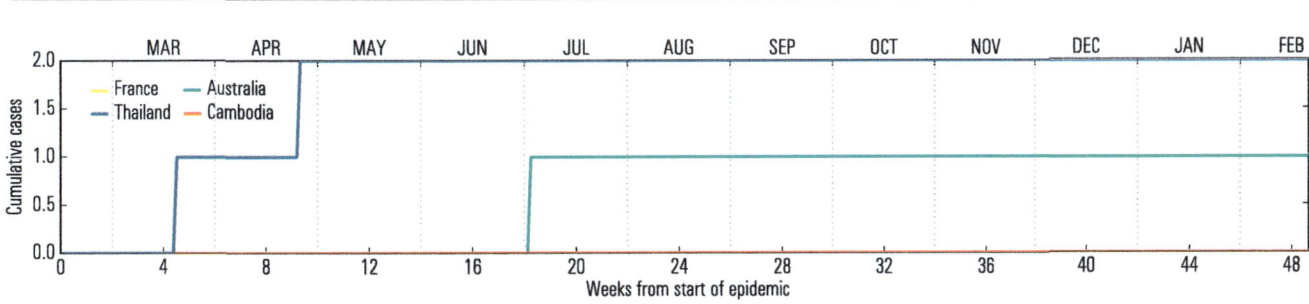

▲ COUNTRY LEVEL INCIDENCE

▼ INFECTION TREE GRAPH

▼ ATTACK RATE COMPOSITION (ONE YEAR)

JEDDAH, SAUDI ARABIA

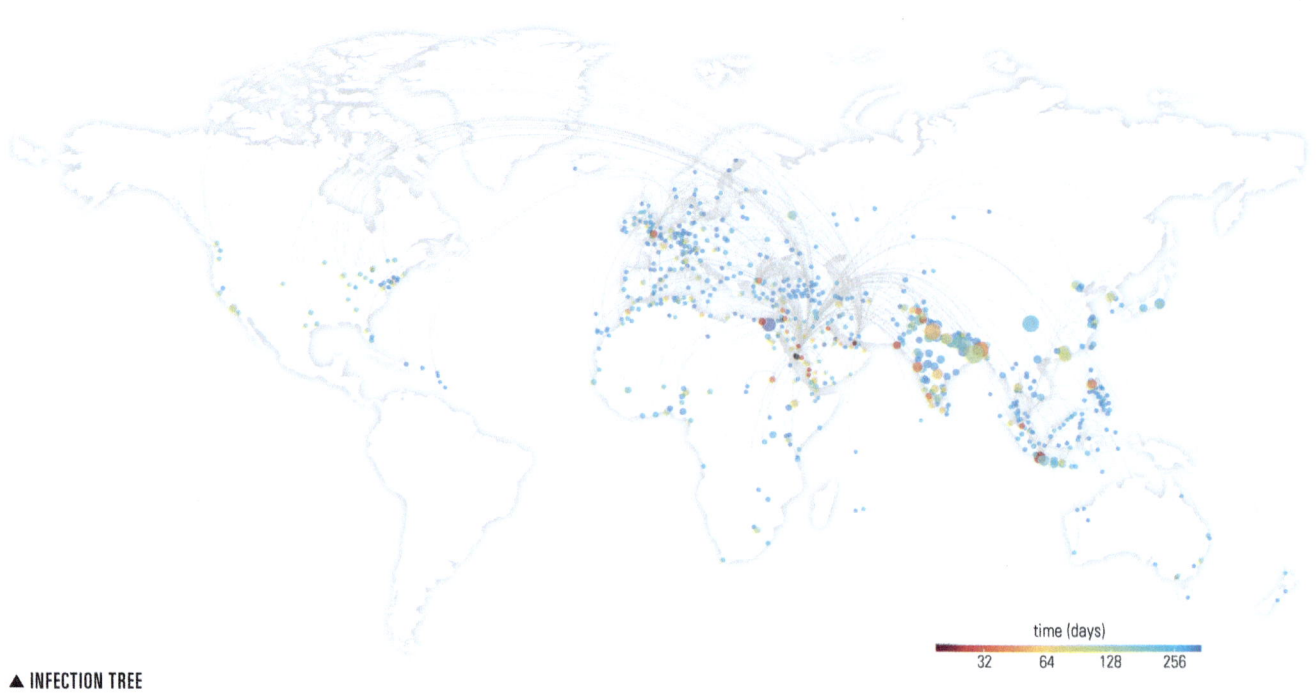

time (days)

32   64   128   256

▲ INFECTION TREE

▲ CUMULATIVE INCIDENCE AND EPIDEMIC PEAKS

▼ AFFECTED COUNTRIES

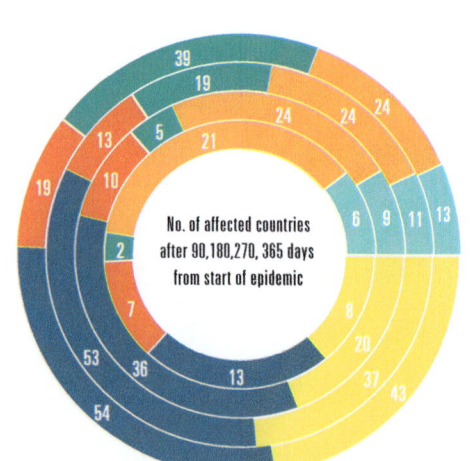

No. of affected countries after 90,180,270, 365 days from start of epidemic

▼ EPIDEMIC ACTIVITY

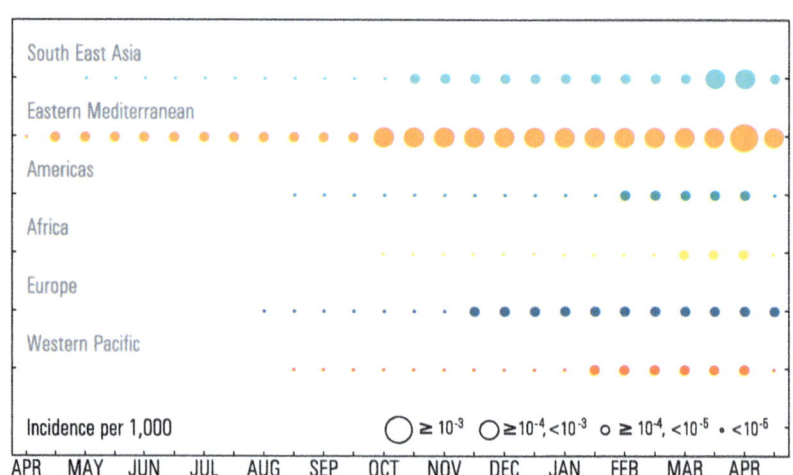

South East Asia

Eastern Mediterranean

Americas

Africa

Europe

Western Pacific

Incidence per 1,000   ◯ ≥ 10⁻³ ◯ ≥ 10⁻⁴, <10⁻³ ○ ≥ 10⁻⁴, <10⁻⁵ · <10⁻⁵

APR MAY JUN JUL AUG SEP OCT NOV DEC JAN FEB MAR APR

WHO REGIONS
- Americas
- Africa
- Europe
- E. Mediterranean
- South-East Asia
- Western Pacific

DECREASE IN R₀ | 30% AFTER 4 WEEKS | 50% AFTER 7 WEEKS

INITIAL CONDITIONS

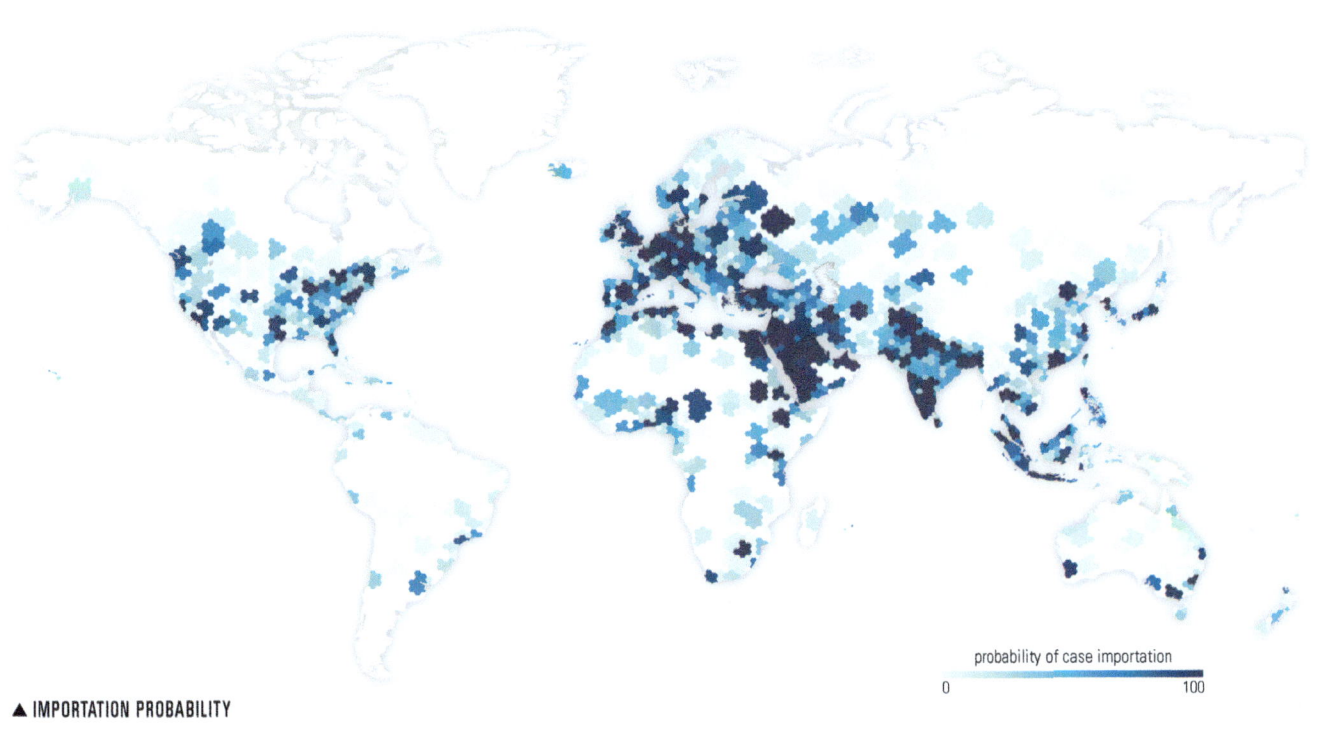

probability of case importation
0   100

▲ IMPORTATION PROBABILITY

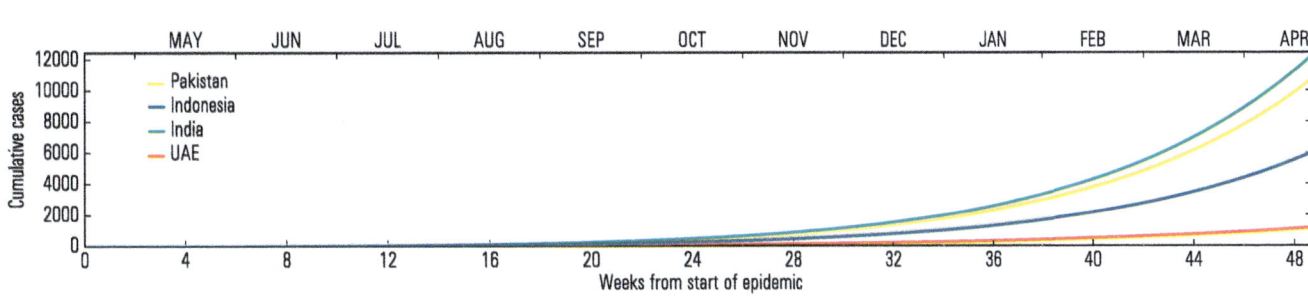

▲ COUNTRY LEVEL INCIDENCE

▼ INFECTION TREE GRAPH

▼ ATTACK RATE COMPOSITION (ONE YEAR)

Eastern Mediterranean 70%

South-East Asia 18%

Europe 6%

Western Pacific 4%

1%   1%

JEDDAH, SAUDI ARABIA

time (days)

32   64   128   256

▲ INFECTION TREE

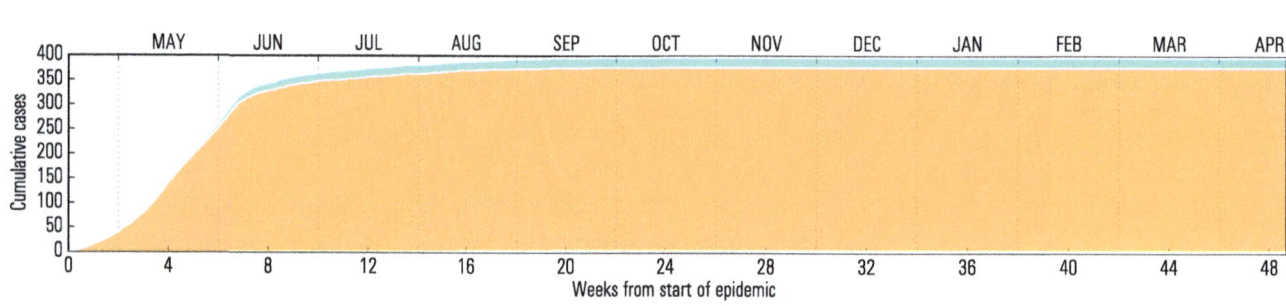

Weeks from start of epidemic

▲ CUMULATIVE INCIDENCE AND EPIDEMIC PEAKS

▼ AFFECTED COUNTRIES

No. of affected countries
after 90,180,270, 365 days
from start of epidemic

18
18
18
18

4 4 4
3
2 2 2
2
4
4
4
5 5 5
5
6
6
6
6

▼ EPIDEMIC ACTIVITY

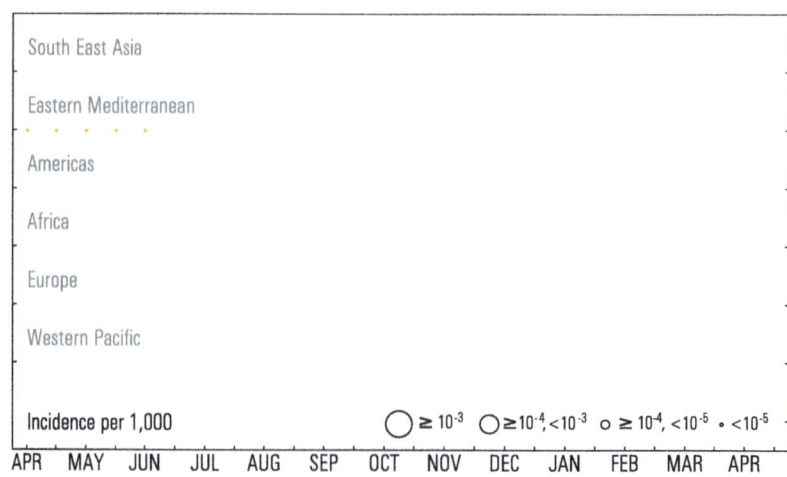

South East Asia

Eastern Mediterranean

Americas

Africa

Europe

Western Pacific

Incidence per 1,000   ◯ ≥ 10⁻³   ◯ ≥ 10⁻⁴, <10⁻³   ○ ≥ 10⁻⁴, <10⁻⁵   • <10⁻⁵

APR MAY JUN JUL AUG SEP OCT NOV DEC JAN FEB MAR APR

■ Americas   ■ E. Mediterranean
■ Africa   ■ South-East Asia
■ Europe   ■ Western Pacific

DECREASE IN R₀ : 50% AFTER 4 WEEKS
               : 90% AFTER 7 WEEKS

INITIAL CONDITIONS

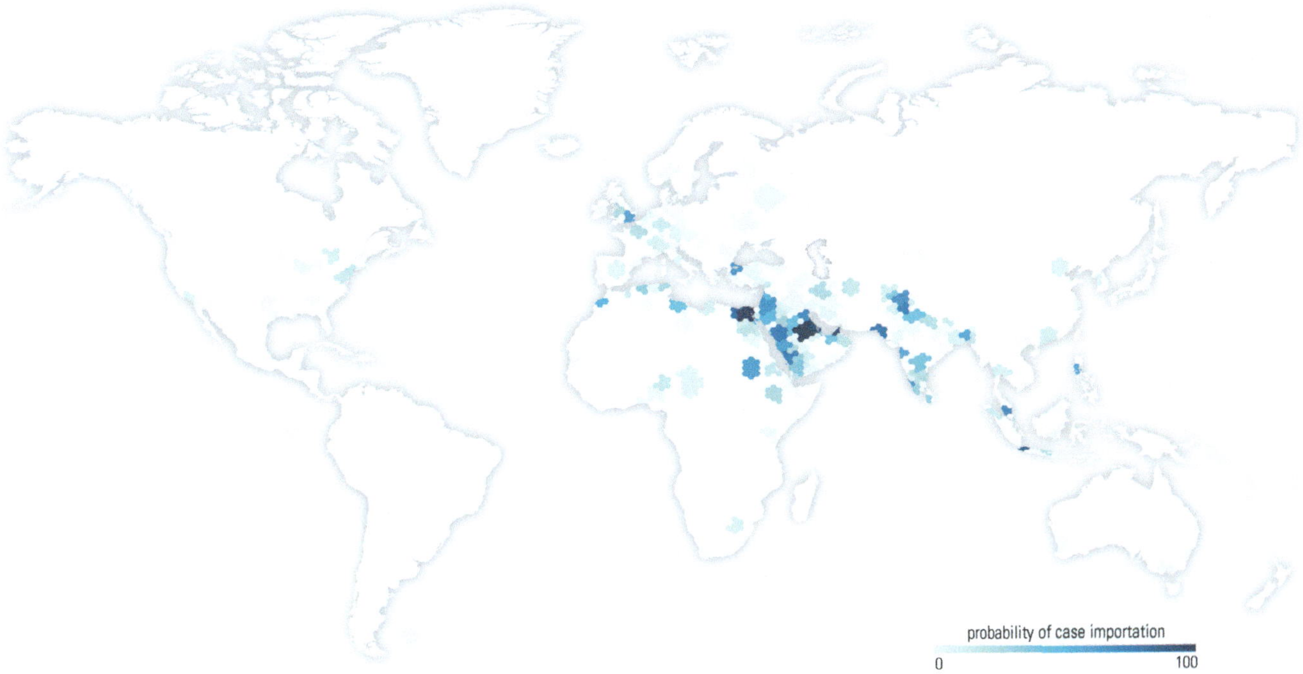

▲ IMPORTATION PROBABILITY

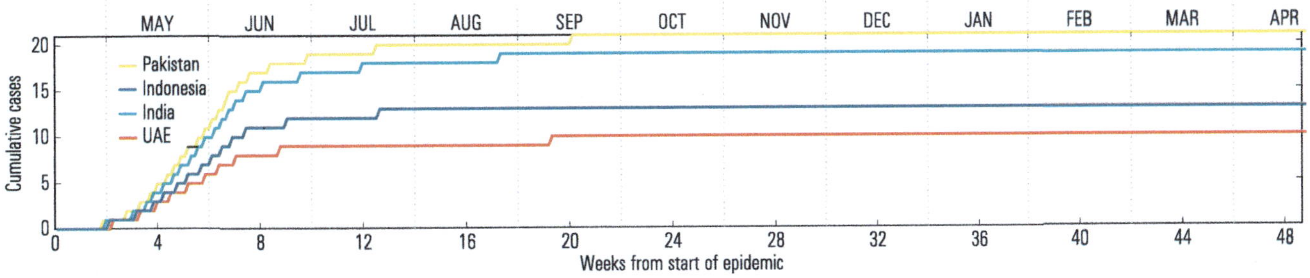

probability of case importation
0   100

▲ COUNTRY LEVEL INCIDENCE

Pakistan
Indonesia
India
UAE

Cumulative cases

Weeks from start of epidemic

▼ INFECTION TREE GRAPH

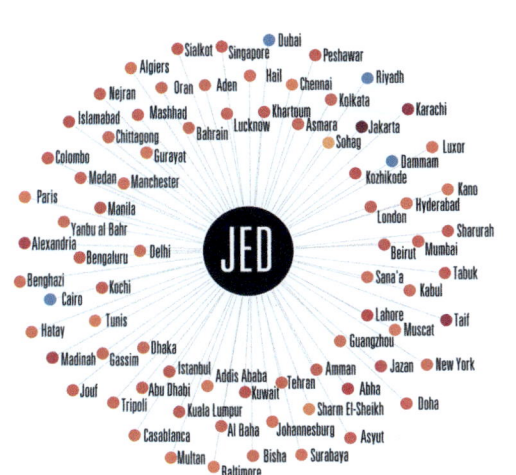

▼ ATTACK RATE COMPOSITION (ONE YEAR)

Eastern Mediterranean 95%

S-E Asia 5%

<1%

<1%

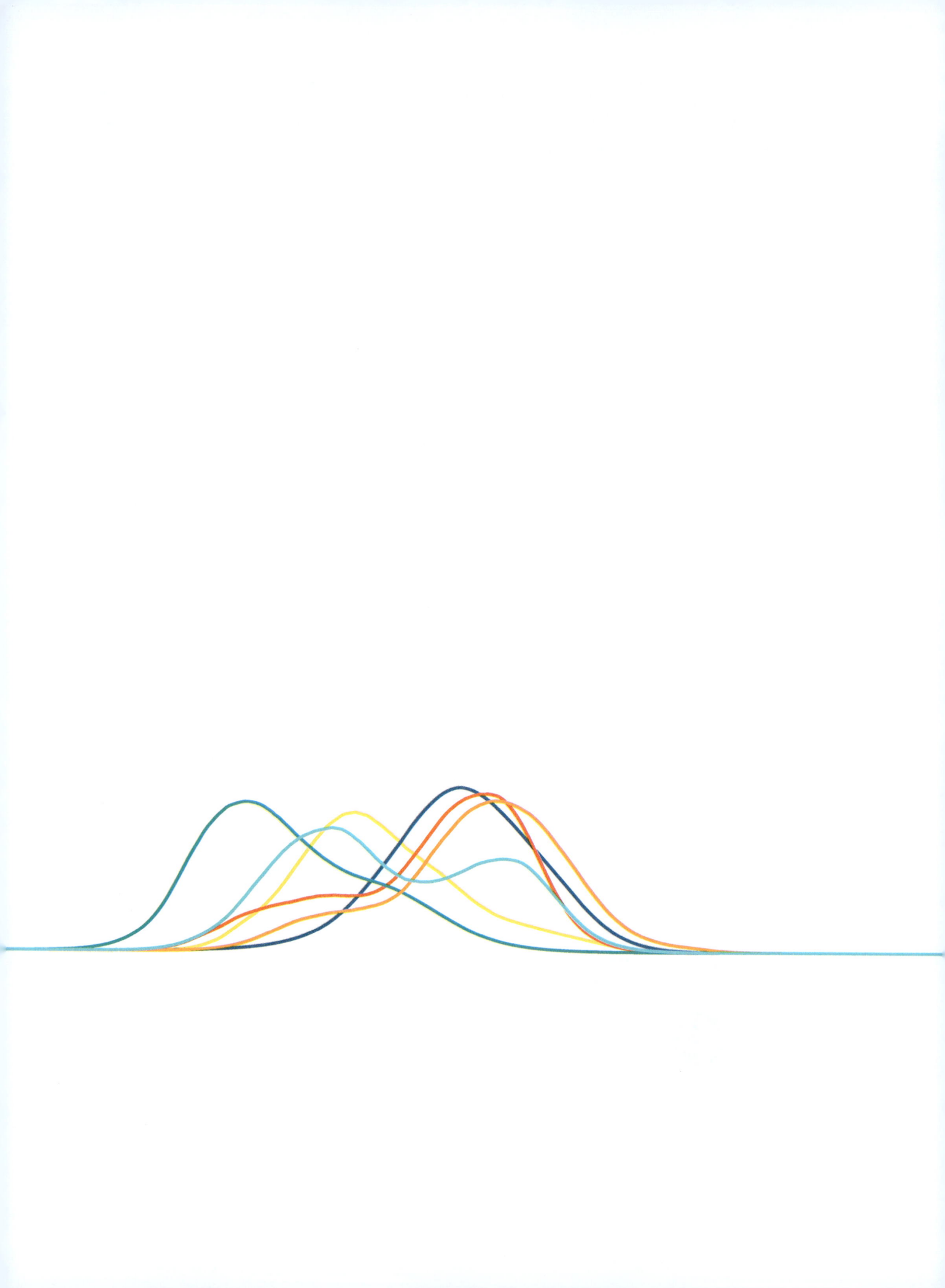

EBOLA VIRUS

In the case of Ebola outbreaks, we selected hypothetical scenarios considering a transmission rate and natural history of the disease similar to the 2014 EVD West African outbreak. As such, we are considering that transmission could also take place in the hospital setting and that even after dying, individuals could still infect the susceptible population.

As the transmission of this disease starts with highly symptomatic individuals, the isolation of infectious individuals makes the outbreak possible to contain. Our hypothetical scenarios consider different intervention times where the containment is put in place after 3 and 6 months, depending on the origin city.

Once again, contrary to influenza, we do not consider seasonal variations, as seasons have no impact on the spread of Ebola. Similarly to coronavirus, we only consider one starting date during the calendar year for each geographical location. In total we provide five scenarios, whose "how to read it" guide can be found on the following pages.

ORIGIN CITIES

Arua, Uganda
2,500
passengers/year

Kisangani, Dem. Rep. of the Congo
37,000
passengers/year

Lagos, Nigeria
2.9 million
passengers/year

STARTING CONDITIONS

- $R_0 = 1.8$
- **Starting date:** April 15

SCENARIOS

Reduction of transmissibility:

- $R_0 < 1$ after 24 weeks (Kisangani and Lagos)
- $R_0 < 1$ after 12 weeks (all cities)

CHART LAYOUT: EBOLA VIRUS

Location of pandemic origin

Infection tree

Each link describes an infection path along which the epidemic moved from one population to the other. The color of the nodes refers to the time lag with respect to the initial spreading.

Cumulative incidence and epidemic peaks

The colored regions describe the cumulative incidence at the regional level. The solid lines represent the regional epidemic profiles.

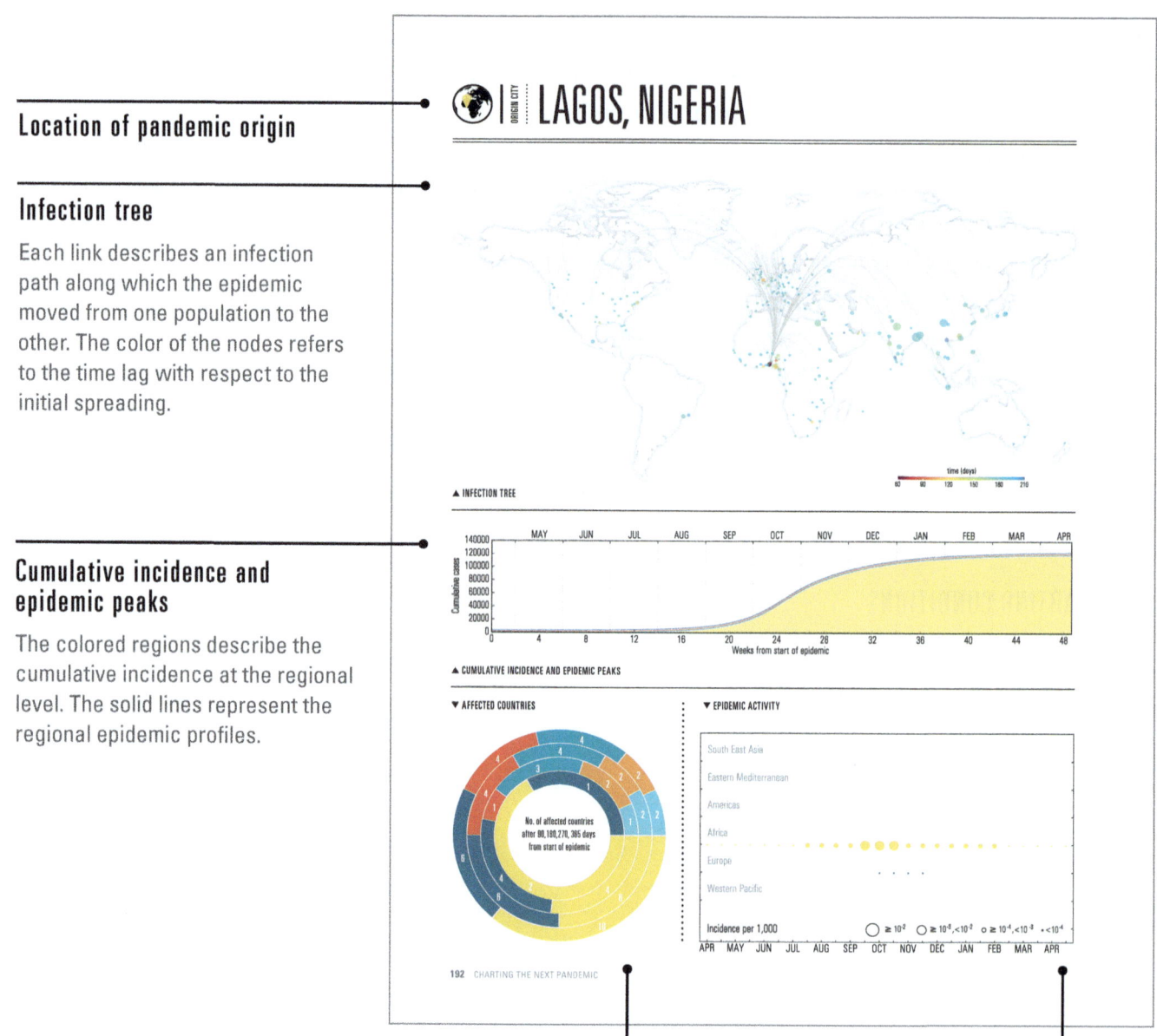

🌍| ORIGIN CITY | LAGOS, NIGERIA

▲ INFECTION TREE

▲ CUMULATIVE INCIDENCE AND EPIDEMIC PEAKS

▼ AFFECTED COUNTRIES

No. of affected countries after 90,180,270, 365 days from start of epidemic

▼ EPIDEMIC ACTIVITY

South East Asia
Eastern Mediterranean
Americas
Africa
Europe
Western Pacific

Incidence per 1,000

192 CHARTING THE NEXT PANDEMIC

Affected countries as a function of time

The concentric circles refer to 90, 180, 270, and 365 days from the start of the spreading, respectively. In each circle we report the number of countries affected by the outbreak in each region.

Epidemic activity

Epidemic activity at the regional level. The size of the circles indicates the level of activity, that is proportional to the number of cases.

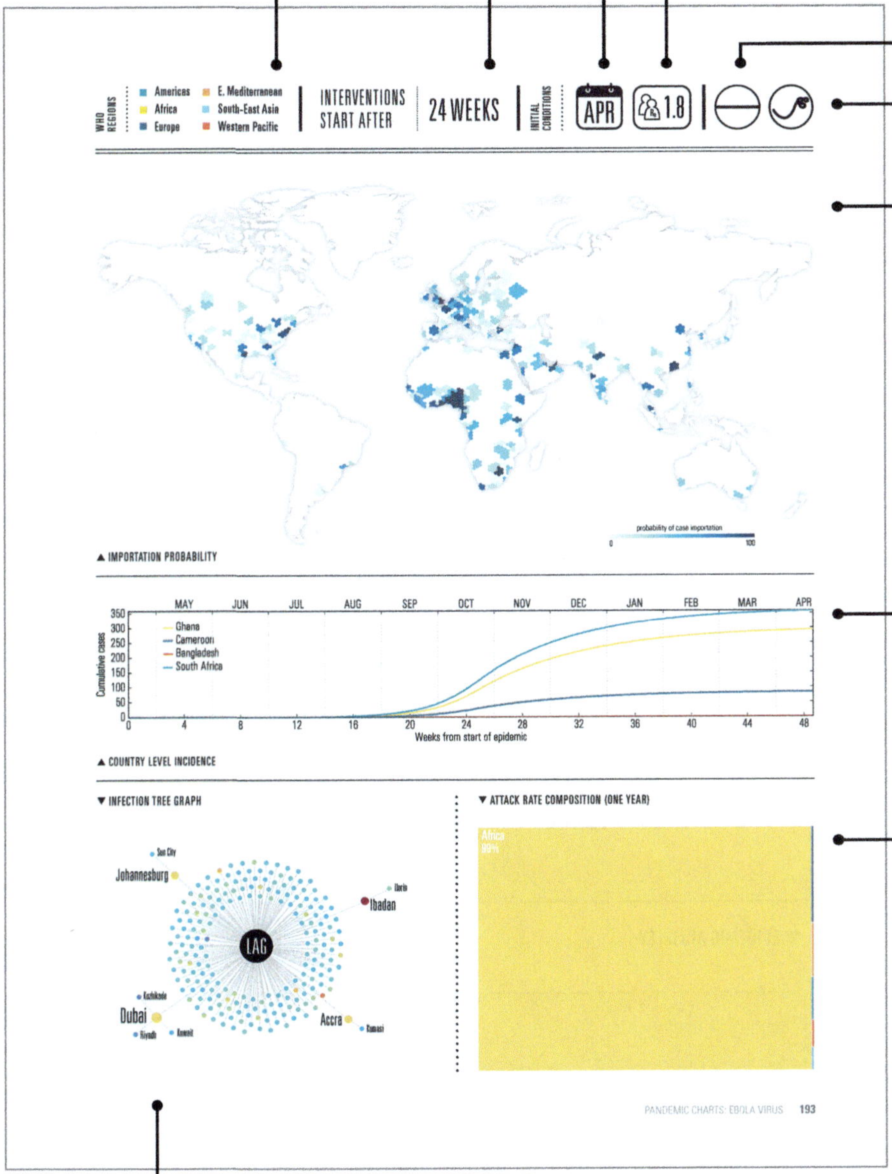

WHO regions

Interventions

Starting date

R_0 value

Seasonality

Disease type

Importation probability map
Probability of observing
an imported case.

Country level incidence
Daily incidence for
selected countries.

Attack rate composition
We illustrate the geographical
impact of the disease after one year
considering the share of clinical
cases in each WHO region.

Infection tree graph
Each link describes an infection path along which
the epidemic has moved from one population to the
other. The color refers to the time lag with respect
to the start of the epidemic.

WHO REGIONS
- Americas
- Africa
- Europe
- E. Mediterranean
- South-East Asia
- Western Pacific

INTERVENTIONS START AFTER

24 WEEKS

INITIAL CONDITIONS

APR 1.8

▲ IMPORTATION PROBABILITY

probability of case importation
0   100

Cumulative cases — MAY JUN JUL AUG SEP OCT NOV DEC JAN FEB MAR APR
- Ghana
- Cameroon
- Bangladesh
- South Africa

350
300
250
200
150
100
50
0
0 4 8 12 16 20 24 28 32 36 40 44 48
Weeks from start of epidemic

▲ COUNTRY LEVEL INCIDENCE

▼ INFECTION TREE GRAPH

Sun City
Johannesburg
Ilorin
Ibadan
LAG
Kishkede
Dubai
Accra
Kumasi
Riyadh
Kuwait

▼ ATTACK RATE COMPOSITION (ONE YEAR)

Africa
89%

ORIGIN CITY : **ARUA, UGANDA**

time (days)

60   90   120   150   180   210

▲ INFECTION TREE

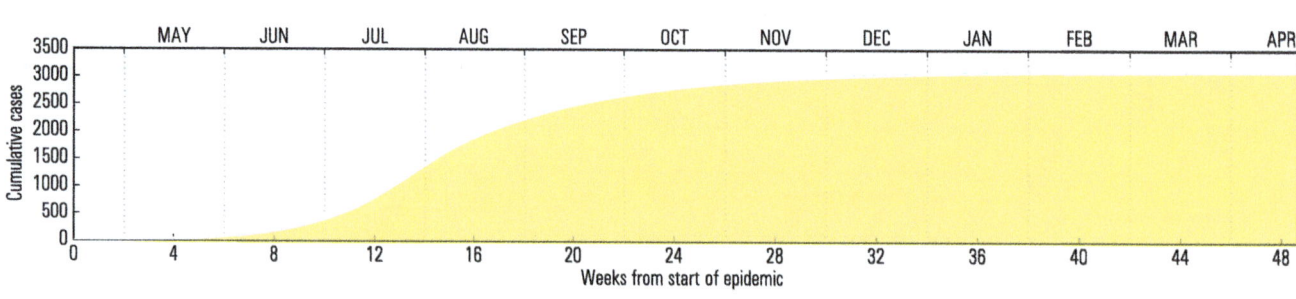

MAY   JUN   JUL   AUG   SEP   OCT   NOV   DEC   JAN   FEB   MAR   APR

Cumulative cases

Weeks from start of epidemic

▲ CUMULATIVE INCIDENCE AND EPIDEMIC PEAKS

▼ AFFECTED COUNTRIES

▼ EPIDEMIC ACTIVITY

No. of affected countries
after 90,180,270, 365 days
from start of epidemic

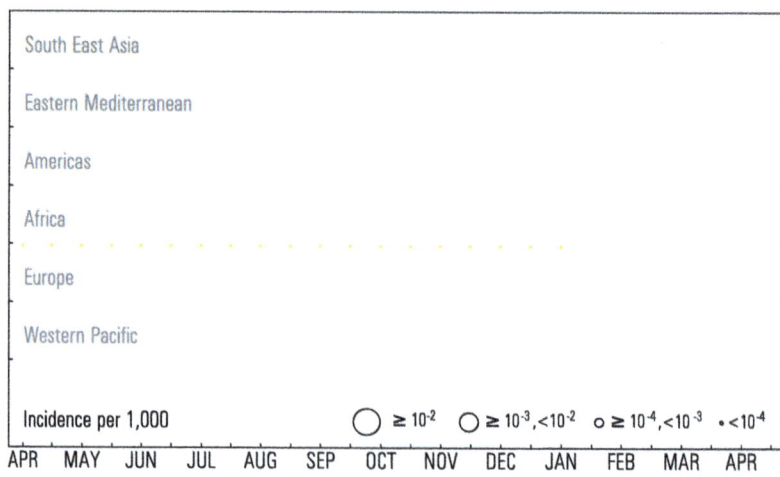

South East Asia

Eastern Mediterranean

Americas

Africa

Europe

Western Pacific

Incidence per 1,000   ◯ ≥ 10⁻²   ◯ ≥ 10⁻³,<10⁻²   o ≥ 10⁻⁴,<10⁻³   • <10⁻⁴

APR   MAY   JUN   JUL   AUG   SEP   OCT   NOV   DEC   JAN   FEB   MAR   APR

probability of case importation
0   100

▲ IMPORTATION PROBABILITY

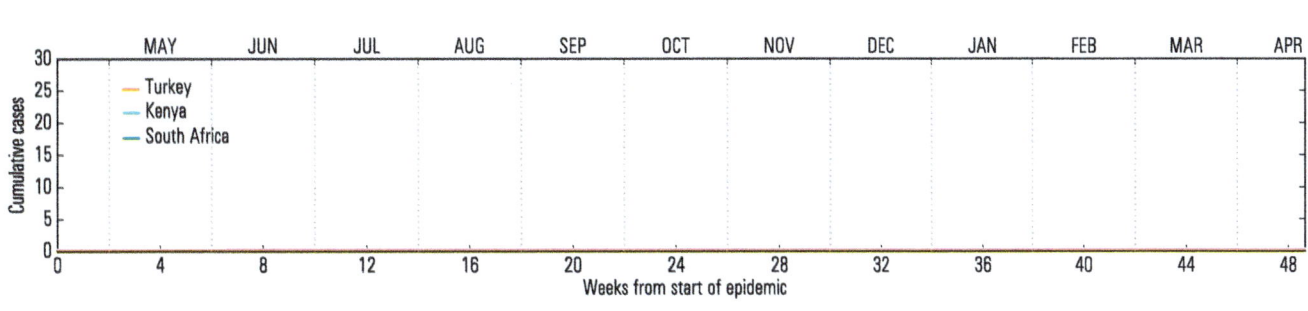

▲ COUNTRY LEVEL INCIDENCE

▼ INFECTION TREE GRAPH

▼ ATTACK RATE COMPOSITION (ONE YEAR)

ORIGIN CITY

KISANGANI, DEM. REP. OF THE CONGO

time (days)
60 90 120 150 180 210

▲ INFECTION TREE

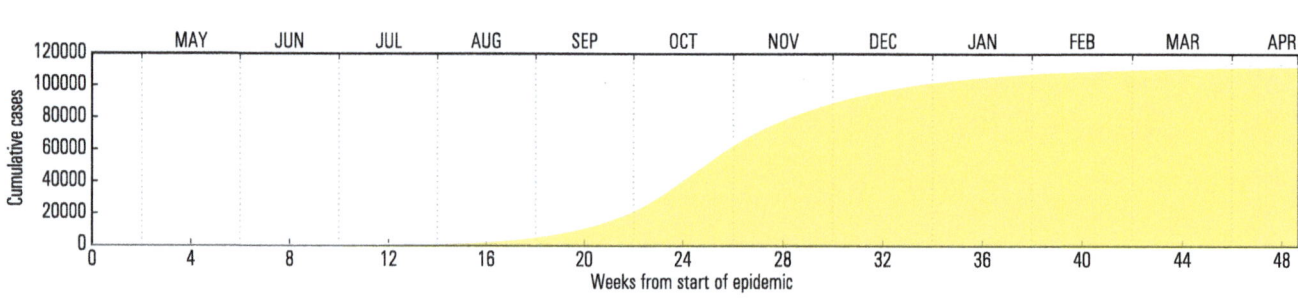

Cumulative cases

MAY   JUN   JUL   AUG   SEP   OCT   NOV   DEC   JAN   FEB   MAR   APR

120000
100000
80000
60000
40000
20000
0

0   4   8   12   16   20   24   28   32   36   40   44   48

Weeks from start of epidemic

▲ CUMULATIVE INCIDENCE AND EPIDEMIC PEAKS

▼ AFFECTED COUNTRIES

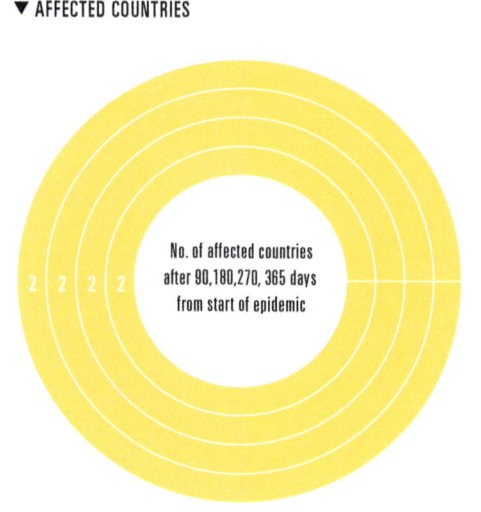

No. of affected countries
after 90,180,270, 365 days
from start of epidemic

▼ EPIDEMIC ACTIVITY

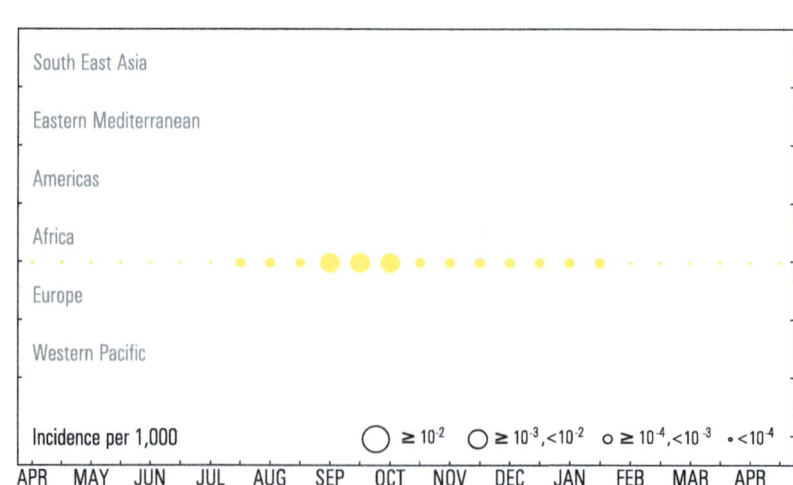

South East Asia

Eastern Mediterranean

Americas

Africa

Europe

Western Pacific

Incidence per 1,000   ◯ ≥ 10⁻²   ◯ ≥ 10⁻³, <10⁻²   o ≥ 10⁻⁴, <10⁻³   • <10⁻⁴

APR   MAY   JUN   JUL   AUG   SEP   OCT   NOV   DEC   JAN   FEB   MAR   APR

WHO REGIONS

- ■ Americas
- ■ Africa
- ■ Europe
- ■ E. Mediterranean
- ■ South-East Asia
- ■ Western Pacific

INTERVENTIONS START AFTER | 24 WEEKS | INITIAL CONDITIONS

probability of case importation
0 — 100

▲ IMPORTATION PROBABILITY

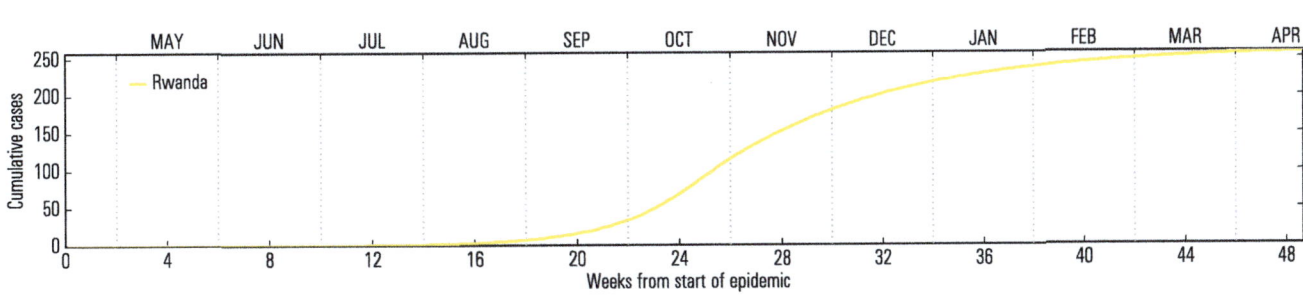

▲ COUNTRY LEVEL INCIDENCE

▼ INFECTION TREE GRAPH

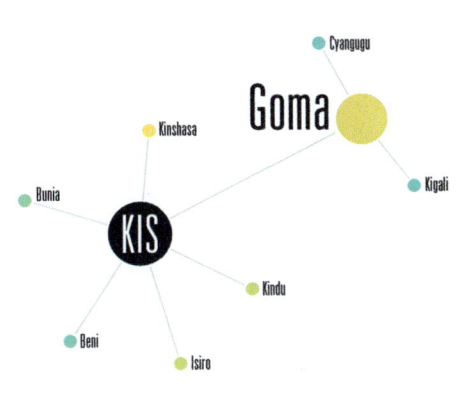

▼ ATTACK RATE COMPOSITION (ONE YEAR)

KISANGANI, DEM. REP. OF THE CONGO

ORIGIN CITY

time (days)

60   90   120   150   180   210

▲ INFECTION TREE

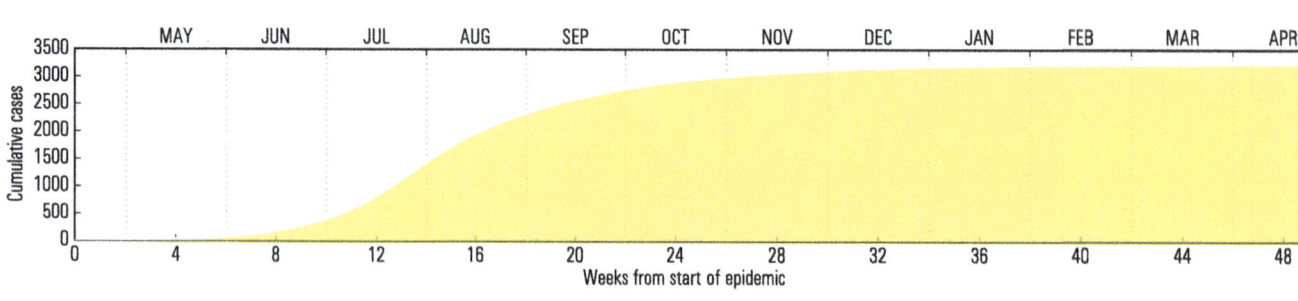

▲ CUMULATIVE INCIDENCE AND EPIDEMIC PEAKS

▼ AFFECTED COUNTRIES

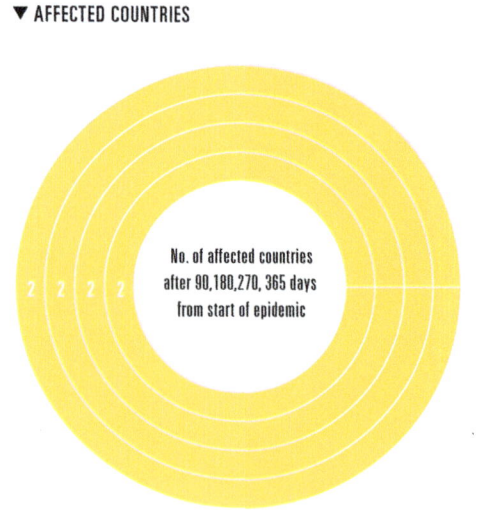

No. of affected countries
after 90,180,270, 365 days
from start of epidemic

▼ EPIDEMIC ACTIVITY

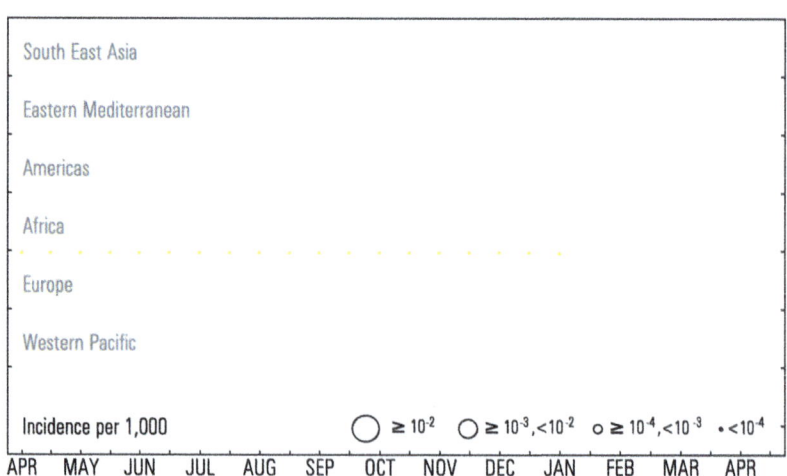

INTERVENTIONS START AFTER | **12 WEEKS** | INITIAL CONDITIONS

probability of case importation

0   100

▲ IMPORTATION PROBABILITY

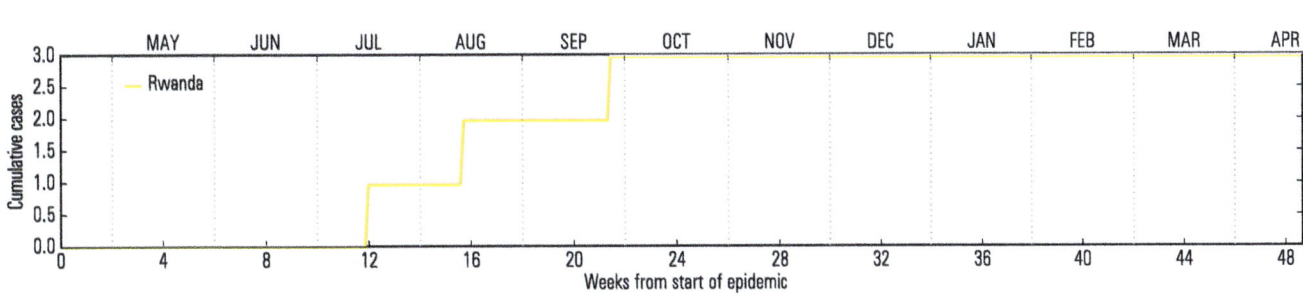

▲ COUNTRY LEVEL INCIDENCE

▼ INFECTION TREE GRAPH

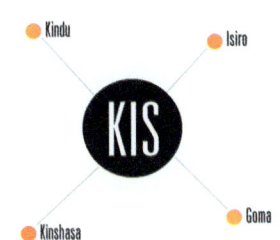

▼ ATTACK RATE COMPOSITION (ONE YEAR)

Africa
100%

LAGOS, NIGERIA

▲ INFECTION TREE

time (days)
60 90 120 150 180 210

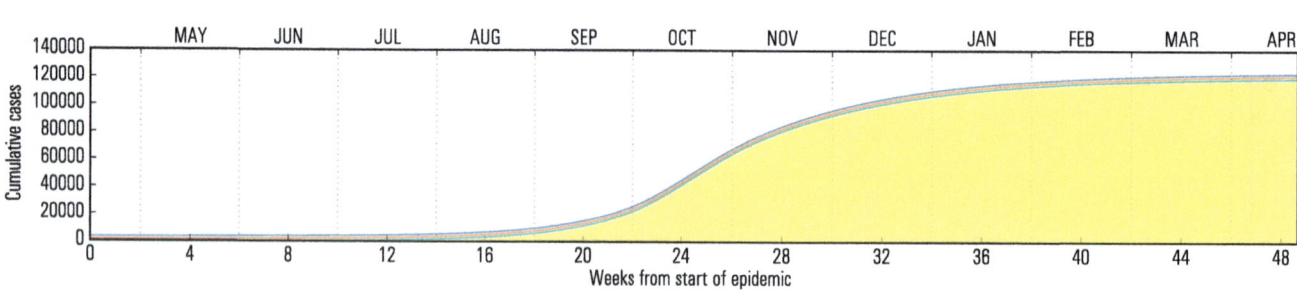

▲ CUMULATIVE INCIDENCE AND EPIDEMIC PEAKS

▼ AFFECTED COUNTRIES

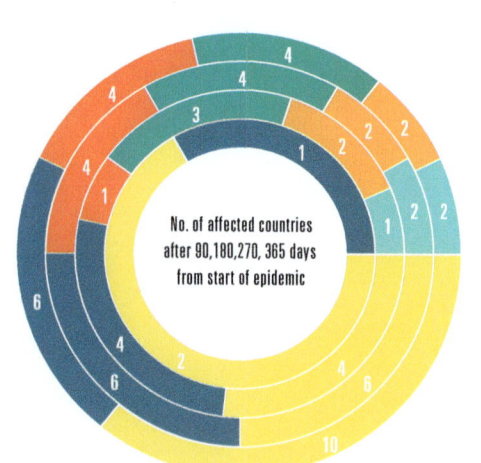

No. of affected countries after 90,180,270, 365 days from start of epidemic

▼ EPIDEMIC ACTIVITY

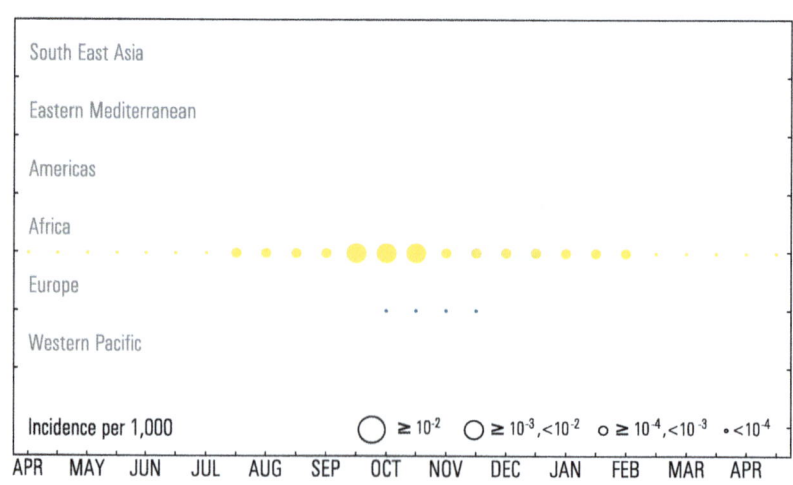

WHO REGIONS

- 🟩 Americas
- 🟨 Africa
- 🟦 Europe
- 🟧 E. Mediterranean
- 🟦 South-East Asia
- 🟧 Western Pacific

| INTERVENTIONS START AFTER | 24 WEEKS | INITIAL CONDITIONS

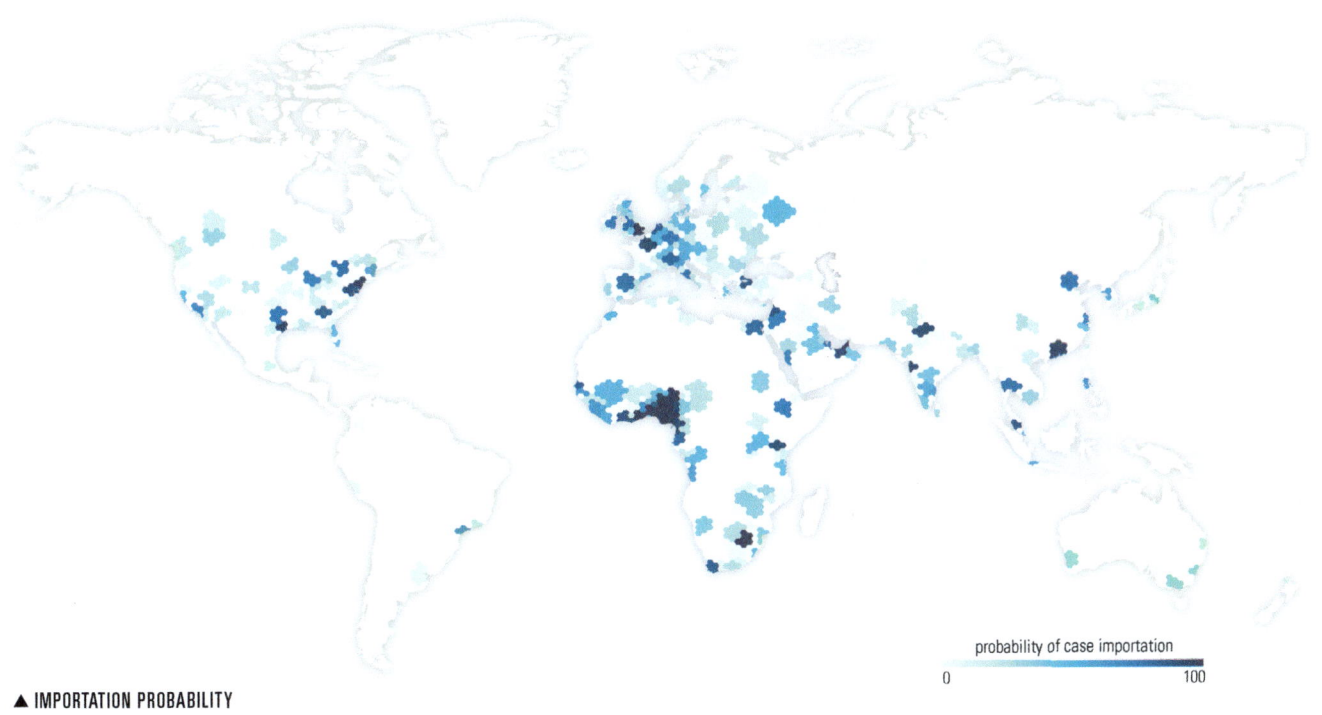

probability of case importation

0   100

▲ IMPORTATION PROBABILITY

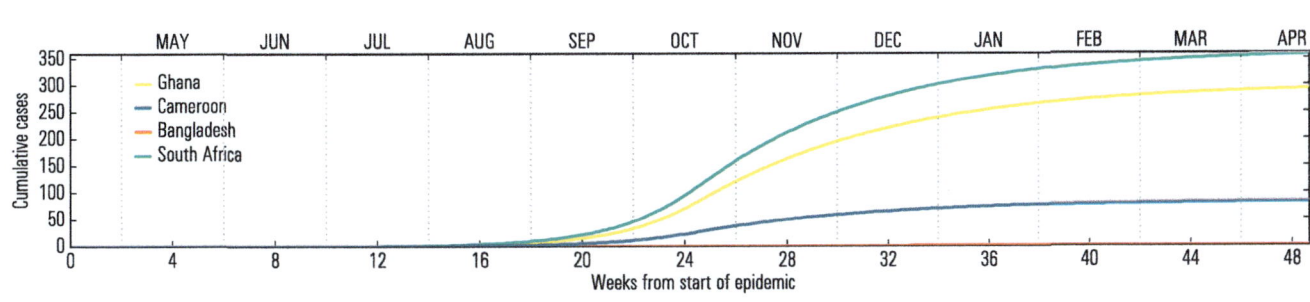

- Ghana
- Cameroon
- Bangladesh
- South Africa

Cumulative cases

Weeks from start of epidemic

▲ COUNTRY LEVEL INCIDENCE

▼ INFECTION TREE GRAPH

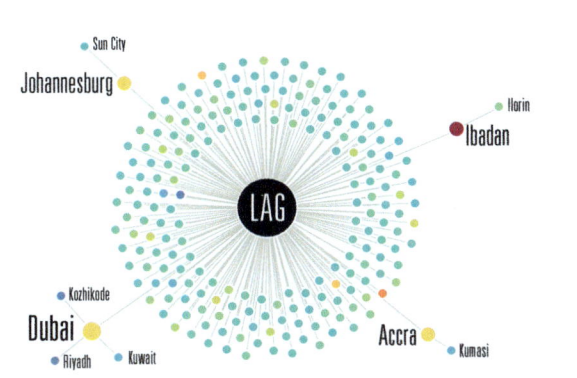

▼ ATTACK RATE COMPOSITION (ONE YEAR)

Africa
99%

LAGOS, NIGERIA

time (days)

60 90 120 150 180 210

▲ INFECTION TREE

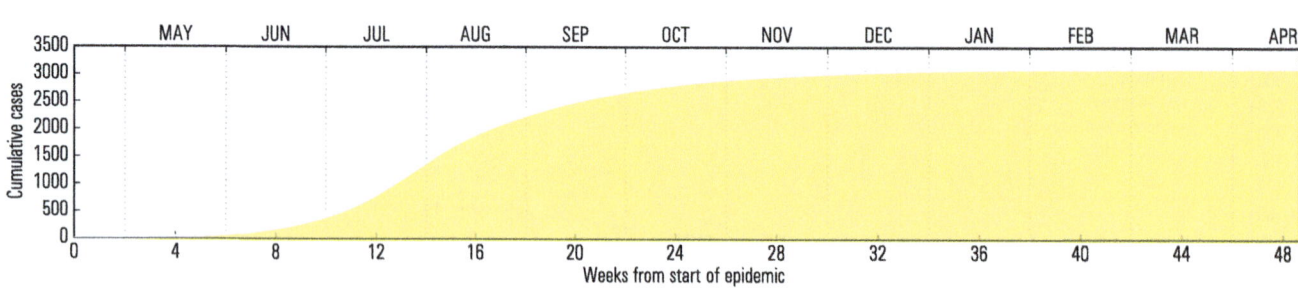

MAY   JUN   JUL   AUG   SEP   OCT   NOV   DEC   JAN   FEB   MAR   APR

Cumulative cases

Weeks from start of epidemic

▲ CUMULATIVE INCIDENCE AND EPIDEMIC PEAKS

▼ AFFECTED COUNTRIES

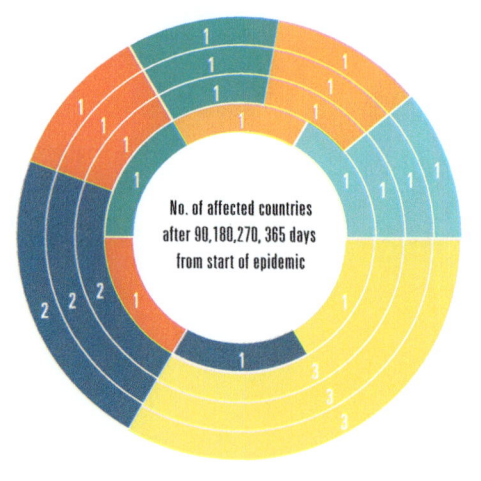

No. of affected countries after 90,180,270, 365 days from start of epidemic

▼ EPIDEMIC ACTIVITY

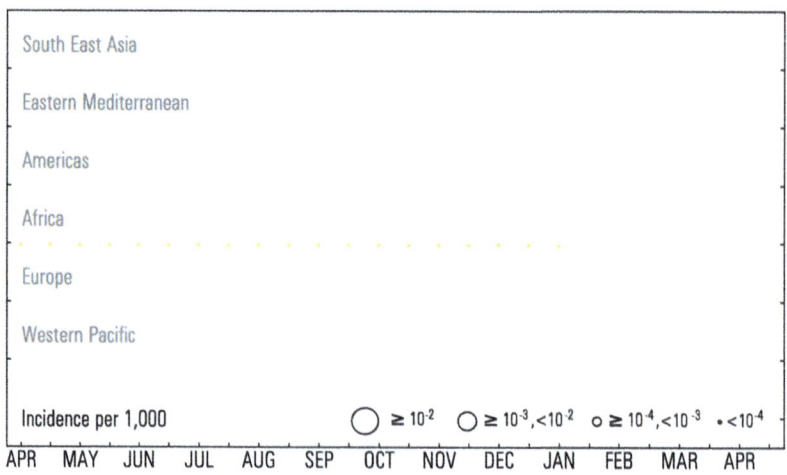

South East Asia

Eastern Mediterranean

Americas

Africa

Europe

Western Pacific

Incidence per 1,000   ◯ ≥ 10⁻² ◯ ≥ 10⁻³,<10⁻² ○ ≥ 10⁻⁴,<10⁻³ • <10⁻⁴

APR MAY JUN JUL AUG SEP OCT NOV DEC JAN FEB MAR APR

WHO REGIONS

■ Americas   ■ E. Mediterranean
■ Africa   ■ South-East Asia
■ Europe   ■ Western Pacific

INTERVENTIONS START AFTER | 12 WEEKS | INITIAL CONDITIONS

probability of case importation

0   100

▲ IMPORTATION PROBABILITY

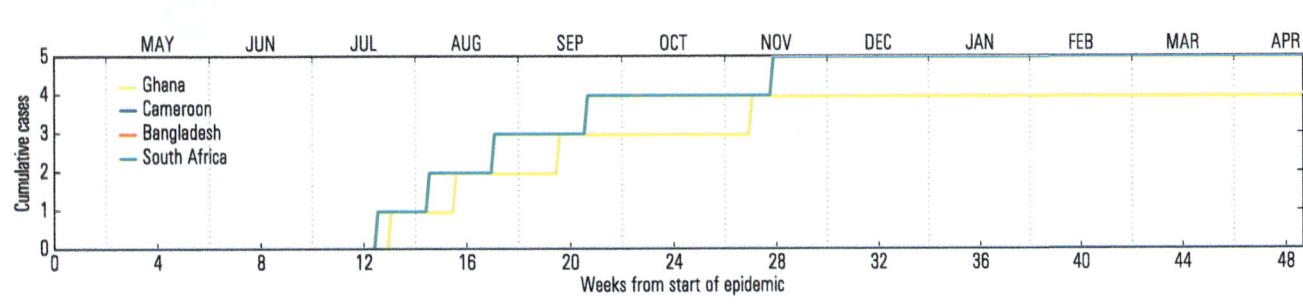

▲ COUNTRY LEVEL INCIDENCE

▼ INFECTION TREE GRAPH

▼ ATTACK RATE COMPOSITION (ONE YEAR)

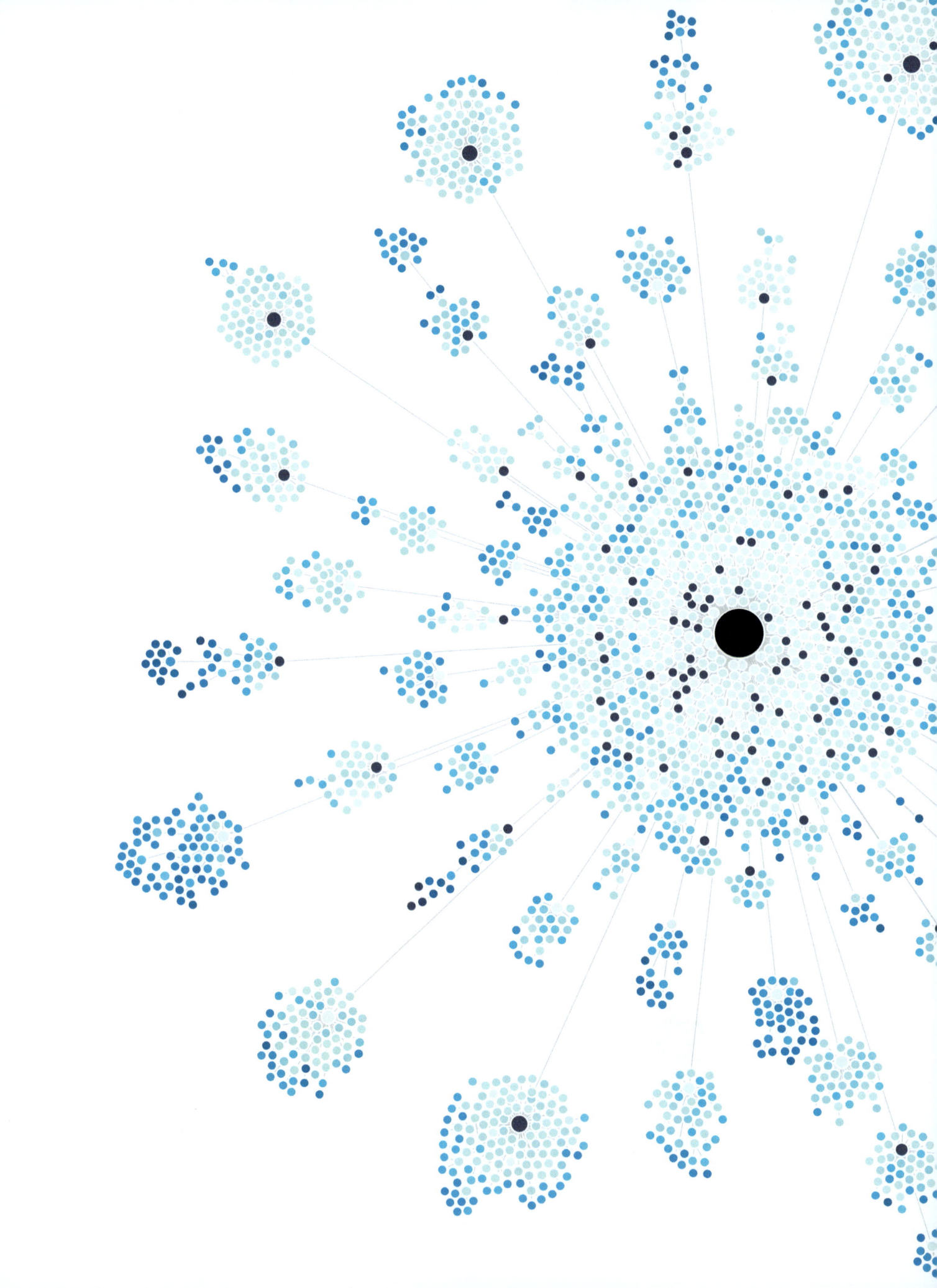

CHAPTER 6

COMPUTATIONAL MODELING OF "DISEASE X"

N THE PREVIOUS CHAPTERS, we have discussed only a few of the pathogens that pose a public health risk because of their epidemic potential. This list is in fact very long and also includes mosquito and other vector-borne diseases such as Zika and yellow fever, which require the modeling of the vector population along with humans. Furthermore, most priority lists for epidemic threats consider the so-called *Disease X*, i.e., the emergence of a future unknown pathogen with novel etiological characteristics and transmission mechanisms. The scenarios we have provided are therefore only a small fraction of the virtually infinite epidemic evolutions that might account for most known disease dynamics and the broad *Disease X* template.

In principle it is possible to imagine the creation of a database of potential epidemic evolutions that scans the parameter space defining general *Disease X* features. Besides the disease parameters, such a database should contain different initial conditions, both geographically and temporally. This database could then be used to implement machine-learning and artificial intelligence techniques aimed at matching the pattern of early stage outbreaks and automatically provide plausible epidemic scenarios. Such a plan, however, requires billions of simulations and presents us with computational, storage, and data mining challenges entailing an enormous amount of resources.

Another approach to the analysis of a generic *Disease X* is represented by the development of computational platforms able to accommodate the design and simulation of generic disease transmission mechanisms in large-scale information systems ingesting the vast amount of socio demographic data needed to simulate the disease evolution at the global level. The work presented in this book is based on the GLEAM framework that is publicly available as a computational platform for the simulation and exploration of general pandemic scenarios. This platform is a scientific application designed for researchers and students in the field of epidemiology who are interested in performing simulations of the spreading of infectious diseases on a global scale.

Figure 6.1 | GLEAM simulation platform

Illustration of the GLEAM simulation platform components.

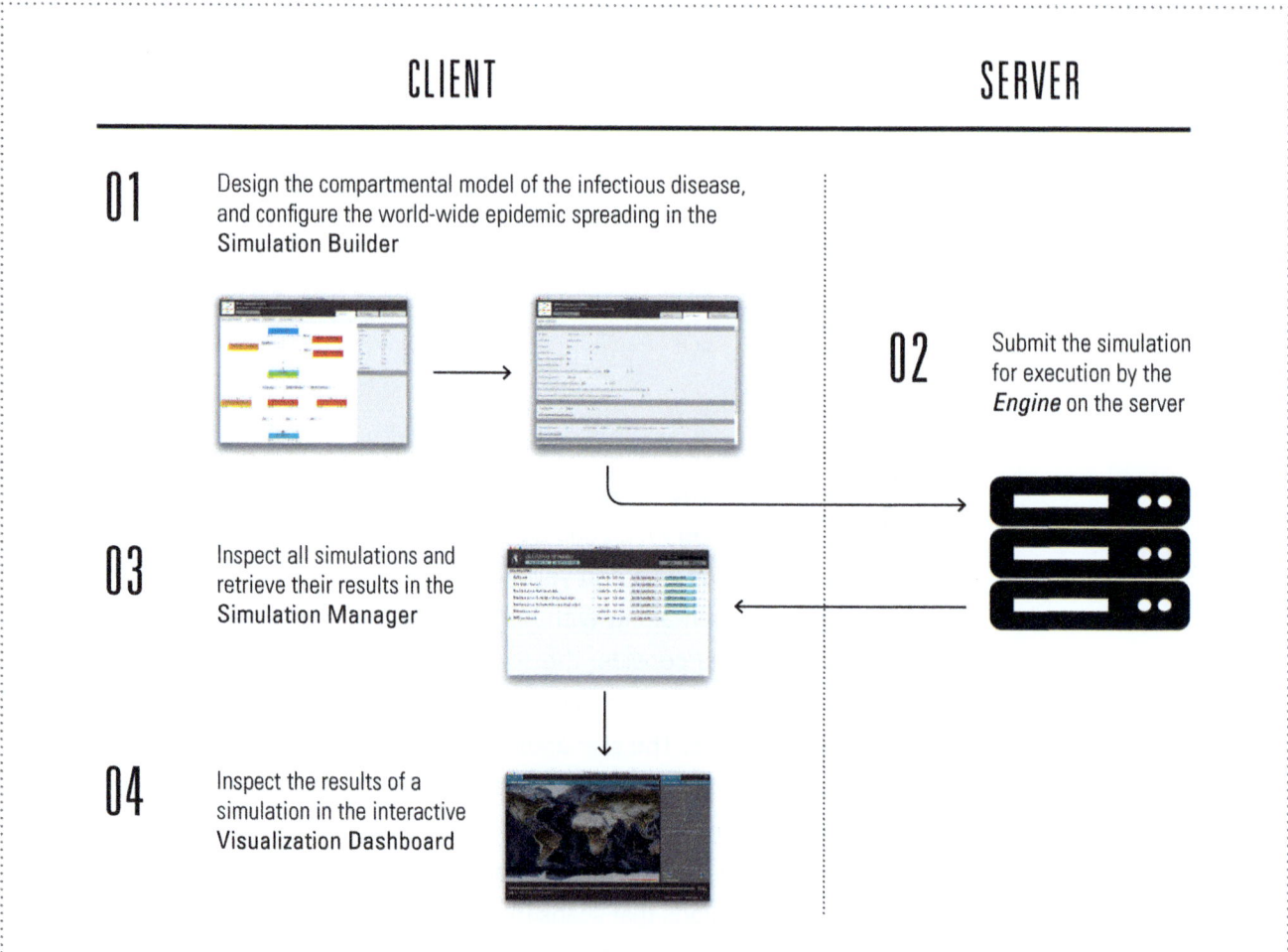

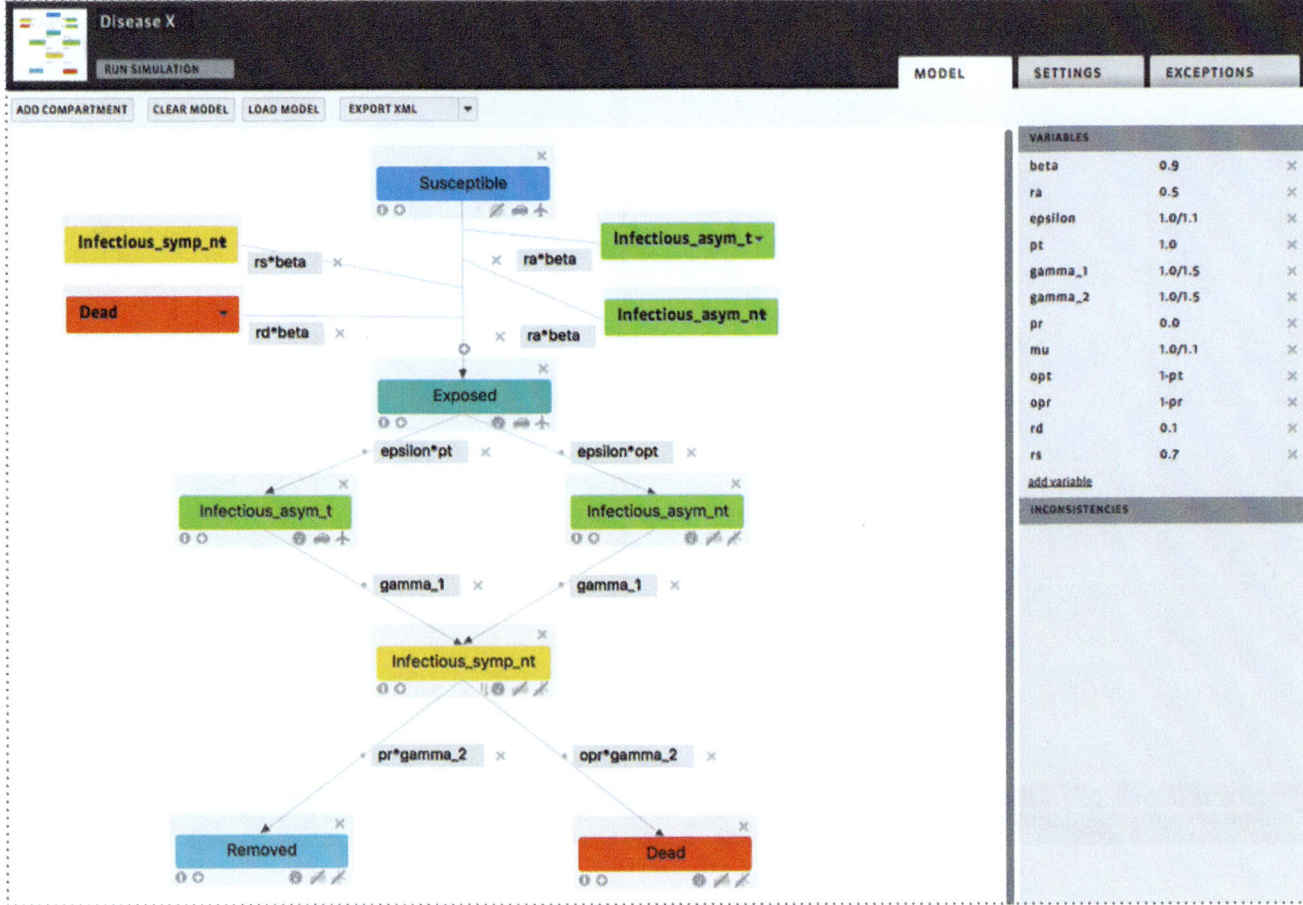

Figure 6.2 | The Model Builder

The Model Builder of the GLEAMviz tool allows the design of any model of infection dynamics. The tool is based on an editable diagram of the compartmental classification of the stages of the disease: it allows the user to add, move, or remove compartments, configure compartment settings, add or remove spontaneous transitions and infection transitions, set transition rates as constant values or functions of variables, and add, change, or remove infection source compartments.

The GLEAM computational platform consists of the GLEAM Server and the GLEAMviz Client application. The computational system architecture is illustrated in **FIGURE 6.1**. The GLEAM Server uses the GLEAM simulator as the numerical engine to perform the simulations on high-performance computers. The GLEAMviz Client is a desktop application through which users interact with the GLEAM Server. The client application is used to configure the simulations, to submit them for execution by the simulation engine on the server, and to retrieve and visualize the numerical results. The client consists of a few major components and is the main interface through which users interact with the GLEAM simulator. The simulation builder is the key component of the platform that allows the definition of the epidemic model and configuration of the simulation scenarios. It allows the definition of any natural history of the disease, including the setting of transmission mechanisms, as well as pharmaceutical and non-pharmaceutical interventions. The programming of the model is performed by means of a visual interface and a parameter panel (**FIGURE 6.2**), thus accommodating in principle a broad range of *Disease X* modeling. The model builder also allows the setting of initial conditions for the outbreak through the selection of more than 3,000 starting locations and the start date of the outbreak.

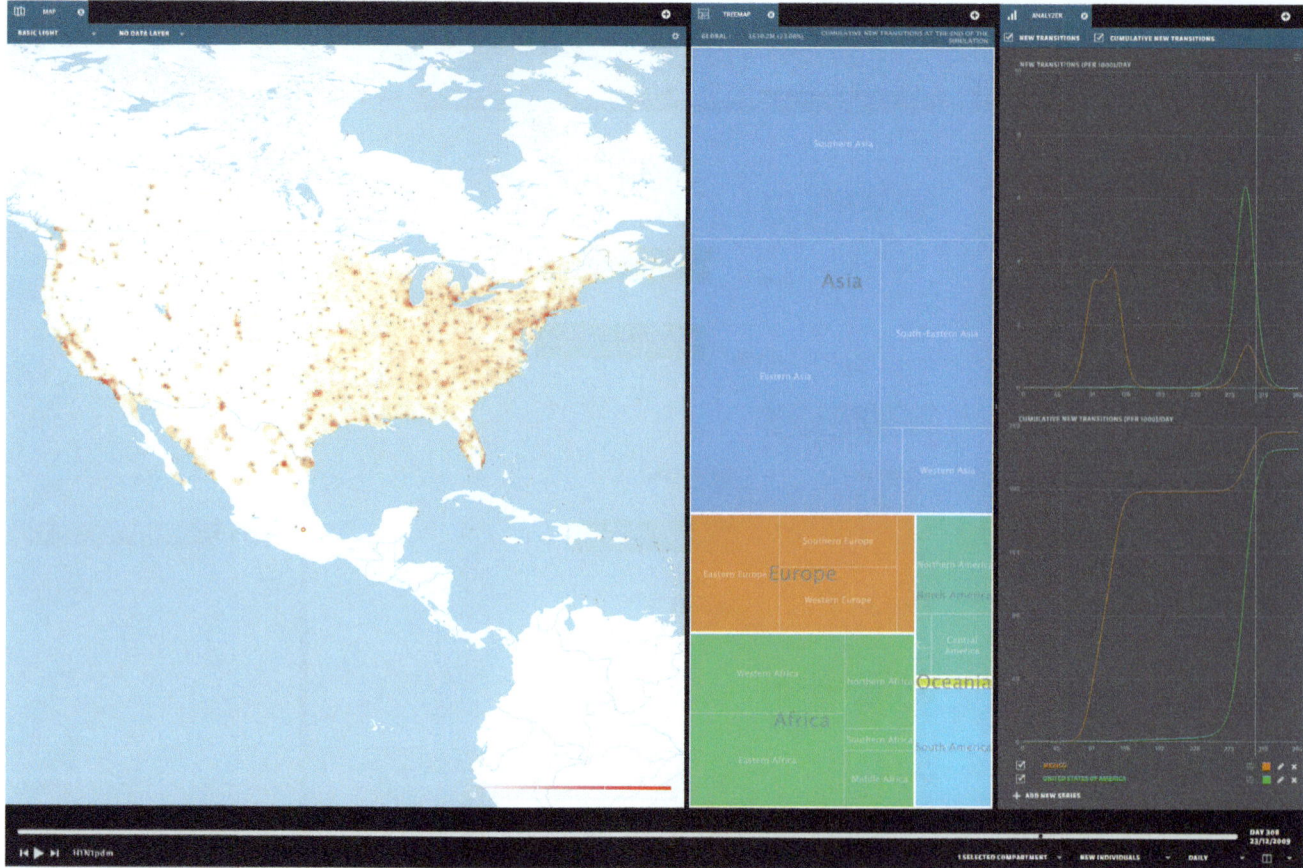

Figure 6.3 | The Visualization Dashboard

The Visualization Dashboard window is the main component of the GLEAMviz Client that allows the user to see the results of a simulation. The window opens when the user selects Show Dashboard in the Simulation manager.

THE COMPUTATIONAL ANALYSIS OF DISEASE X REQUIRES THE DESIGN AND SIMULATIONS OF A WIDE RANGE OF POTENTIAL DISEASE TRANSMISSION MECHANISMS.

The GLEAMviz client includes also a Visualization Dashboard that allows the user to visualize and explore the results of a simulation. The Visualization Dashboard consists of a configurable set of visualization widgets that can be opened and arranged at will by the user according to the size of the display. Currently the public edition of the GLEAMviz client supports many of the visualizations presented in the previous chapters, including geographical mapping, treemap and invasion tree visualizations, and the possibility of analyzing incidence and prevalence profiles (FIGURE 6.3).

The GLEAM computational platform provides a visual way to set up simulations, develop disease models, and evaluate simulation results using a variety of maps, charts, and data analysis tools. The GLEAMviz simulator system is a software platform in continuous evolution: new tools and functionalities will be added in the future, and all the basic features, including the core data upon which the model is based, are updated and upgraded regularly. In other words it allows the definition of ad hoc scenarios and possible disease evolutions in a virtually infinite number of combinations of infectious agents, initial conditions, and intervention planning. The public version of GLEAMviz is not meant for heavy-duty research, but it allows any readers willing to invest some time in familiarizing with the platform to recover the results of this book and design other possible scenarios, eventually exploring the next *Disease X*.

OUTLOOK

....

WE HOPE THAT WITH THIS BOOK WE HAVE BEEN ABLE TO PROVIDE A GLIMPSE INTO THE EMERGING SCIENCE OF COMPUTATIONAL EPIDEMIC MODELING, WHICH PROMISES TO QUANTITATIVELY PROJECT THE "WHEN," "WHERE," AND "HOW" OF EPIDEMICS.

ADVANCES IN MEDICAL TREATMENT, improved healthcare systems, and knowledge of infectious disease agents have provided humankind with an armory of weapons in the fight against infectious diseases. The introduction of antibiotics. The culling of common diseases such as measles, mumps, rubella, and polio. The eradication of smallpox. Each new milestone led practitioners and the public at large to believe that in a few decades we could defeat infectious diseases once and for all. Unfortunately, infectious diseases have had quite a comeback in recent years, with several outbreaks of coronaviruses (more commonly known as SARS and MERS), a major flu pandemic, and an Ebola epidemic of unprecedented proportions. Vector-borne diseases are also on the rise, as exemplified by the elusive Zika epidemic which marched through the Americas in 2015 and 2016. We also realized that our societies are now far more interconnected than in the past. Diseases are able to take advantage of our transportation infrastructures, hopping flights from city to city and country to country, spreading with unprecedented ease. More troubling, the vastly interdependent social, technological, and economic systems characterizing our modern world would be extremely fragile with respect to the disruptive mortality rates of past epidemics like the infamous influenza pandemic of 1918.

The war against infectious diseases is, and will always be, fought in the hospital trenches by medical doctors, healthcare workers, and public health responders. We hope, however, that with this book we have been able to provide a glimpse into the emerging science of computational epidemic modeling, which promises to quantitatively project the "when," "where," and "how" of epidemics.

Computing and data sciences are the keys to predictive modeling, accessing and harnessing also a wealth of novel "intel" such as electronic health records, social media, Internet, mobile phones, and remote sensors. From mining Twitter posts in order to analyze the flu season to using cell-phone data and satellite imagery in order to understand the population movements driving the dissemination of epidemics, a computational approach would strengthen the usual disease surveillance system and provide public health institutions with new intelligence to fight infectious diseases.[1]

In view of the huge potential of predictive modeling, we must also be aware of the many foundational and technical challenges, as well as limitations that have yet to be overcome. While hurricanes do not care about our forecasts, people change their behavior according to the awareness and knowledge of the risks poised by an epidemic. The real-time modeling of the feedback loop between the disease progression and the behavioral adaptation of social systems is still a major issue affecting the predictive power of mathematical and computational models. Another issue is the lack of systematic understanding of the predictive performance of different types of models. How does the accuracy of prediction scale with the complexity of the model? How does data quality affect the reliability of epidemic predictive models? Can we quantify the impact of the intrinsic noisy data-gathering process on the predictions? Which ensemble modeling approach provides the best predictive power? These and many other research questions are still objects of research, often through post hoc assessment[2] or ad hoc "forecasting challenges,"[3] that compare predictive modeling results from different teams and modeling approaches against real-world outbreak data. This is a slow learning process

DATA-DRIVEN MODELS DO MORE THAN JUST FORECAST. MORE, IN GENERAL, DESCRIPTIVE AND PREDICTIVE MODELS PROVIDE RATIONALES AND ANALYSIS TO QUANTITATIVELY SUPPORT PUBLIC HEALTH DECISIONS AND INTERVENTION PLANS.

1 Shweta Bansal et al., "Big data for infectious disease surveillance and modeling," The Journal of Infectious Diseases 214, S375–S379 (2016)

2 Jean-Paul Chretien et al., "Mathematical modeling of the West Africa Ebola epidemic," Elife 4, e09186 (2015)

3 Matthew Biggerstaff et al., "Results from the Centers for Disease Control and Prevention's Predict the 2013–2014 Influenza Season Challenge," BMC Infectious Diseases 16, 357 (2016); "CHIKV Challenge Announces Winners, Progress toward Forecasting the Spread of Infectious Diseases", www.darpa.mil/news-events/2015-05-27

because, luckily enough, epidemic outbreaks are not presenting us with a continuous and abundant flow of events to predict as in meteorology science. For this reason, challenges featuring synthetic datasets of simulated outbreaks have now been introduced to assess predictive modeling performance in a controlled environment.[4]

Data-driven models, however, do more than just forecast. More in general, descriptive and predictive models provide rationales and analysis to quantitatively support public health decisions and intervention plans. They are used to increase situational awareness, for intervention planning, and to find epidemiological explanations. Models also provide ways to create counterfactual scenarios otherwise impossible to explore and to look into causal arguments that can be checked against real-world data. More importantly, they also provide means to fill the gaps when data are scarce or not available. For example, the 2015 Zika epidemic in the Americas remained under the radar of the medical and research communities until early 2016, but it likely emerged in Brazil during the last months of 2013. Computational models are, in this case, used to recreate the evolution of the disease and potentially shed light on the many puzzles concerning the past impact of the epidemic in various parts of Latin America.[5]

The field of predictive epidemic modeling is rapidly evolving, and several more policies and interdisciplinary scientific challenges need to be addressed, from rigorously validating new methodologies to defining appropriate data sharing policies to be adopted during health crises. There is an urgent need to develop the collaborations and scientific initiatives needed to advance the field: it is not a matter of "if" but "when" we will fight the next battle against a pandemic pathogen, and this is a call to arms that should bring the entire research community together, as well as the rest of society.

---

4 The RAPIDD Ebola challenge, Comparison of disease forecasting models, www.ebola-challenge.org
5 T Alex Perkins et al., "Model-based projections of zika virus infections in childbearing women in the Americas," Nature Microbiology 1, 16126 (2016)

Suggested Reading

This book is not intended as an academic review of the field of computational and data-driven epidemic modeling. As such, we did not provide references within the text to the huge body of work in the area, as well as to the many seminal works upon which the field is built. Here we propose a number of further readings related to some of the topics discussed throughout the book. This list is far from being exhaustive, and it has obvious biases based on what could be the natural follow-up to dive deeper into what we presented in the book. We apologize in advance to all the outstanding colleagues and fantastic scientists who feel that their contributions are missing from the following list.

EPIDEMIC MODELING TEXTBOOKS

Roy M Anderson, Robert M May, and B Anderson, *Infectious diseases of humans: dynamics and control*, Vol. 28 (Wiley Online Library, 1992).

Odo Diekmann and Johan Andre Peter Heesterbeek, *Mathematical epidemiology of infectious diseases: model building, analysis and interpretation*, Vol. 5 (John Wiley & Sons, 2000).

Lisa Sattenspiel, *The Geographic Spread of Infectious Diseases: Models and Applications: Models and Applications* (Princeton University Press, 2009).

Emilia Vynnycky and Richard White, *An introduction to infectious disease modelling* (Oxford University Press, 2010).

Matt J Keeling and Pejman Rohani, *Modeling infectious diseases in humans and animals* (Princeton University Press, 2011).

PERSPECTIVES AND COMMENTARIES

Stephen B Thacker and Donna F Stroup, "Future directions for comprehensive public health surveillance and health information systems in the United States," American Journal of Epidemiology **140**, 383–397 (1994).

Maria D Van Kerkhove, Tommi Asikainen, Niels G Becker, Steven Bjorge, Jean-Claude Desenclos, Thais dos Santos, Christophe Fraser, Gabriel M Leung, Marc Lipsitch, and Ira M Longini Jr, "Studies needed to address public health challenges of the 2009 H1N1 influenza pandemic: insights from modeling," PLoS Medicine **7**, e1000275 (2010).

Gilles Hejblum, Michel Setbon, Laura Temime, Sophie Lesieur, and Alain-Jacques Valleron, "Modelers' perception of mathematical modeling in epidemiology: a web-based survey," PloS One **6**, e16531 (2011).

Marcel Salathe, Linus Bengtsson, Todd J Bodnar, Devon D Brewer, John S Brownstein, Caroline Buckee, Ellsworth M Campbell, Ciro Cattuto, Shashank Khandelwal, and Patricia L Mabry, "Digital epidemiology," PLoS Computational Biology **8**, e1002616 (2012).

Madhav Marathe and Anil Kumar S Vullikanti, "Computational epidemiology," Communications of the ACM **56**, 88–96 (2013).

Eric T Lofgren, M Elizabeth Halloran, Caitlin M Rivers, John M Drake, Travis C Porco, Bryan Lewis, Wan Yang, Alessandro Vespignani, Jeffrey Shaman, and Joseph NS Eisenberg, "Opinion: Mathematical models: A key tool for outbreak response," Proceedings of the National Academy of Sciences **111**, 18095–18096 (2014).

Benjamin M Althouse, Samuel V Scarpino, Lauren Ancel Meyers, John W Ayers, Marisa Bargsten, Joan Baumbach, John S Brownstein, Lauren Castro, Hannah Clapham, and Derek AT Cummings, "Enhancing disease surveillance with novel data streams: challenges and opportunities," EPJ Data Science **4**, 17 (2015).

Hans Heesterbeek, Roy M Anderson, Viggo Andreasen, Shweta Bansal, Daniela De Angelis, Chris Dye, Ken TD Eames, W John Edmunds, Simon DW Frost, and Sebastian Funk, "Modeling infectious disease dynamics in the complex landscape of global health," Science **347**, aaa4339 (2015).

Jean-Paul Chretien, David Swedlow, Irene Eckstrand, Dylan George, Michael Johansson, Robert Huffman, and Andrew Hebbeler, "Advancing epidemic prediction and forecasting: A new us government initiative," Online Journal of Public Health Informatics **7** (2015).

Shweta Bansal, Gerardo Chowell, Lone Simonsen, Alessandro Vespignani, and Cécile Viboud, "Big data for infectious disease surveillance and modeling," The Journal of Infectious Diseases **214**, S375–S379 (2016).

National Science and Technology Council Pandemic Prediction and Forecasting Science and Technology Working Group, "Towards epidemic prediction: Federal efforts and opportunities in outbreak modeling," (2016).

M Mitchell Waldrop, "News feature: Special agents offer modeling upgrade," Proceedings of the National Academy of Sciences **114**, 7176–7179 (2017).

METAPOPULATION MODELS

Lisa Sattenspiel and Klaus Dietz, "A structured epidemic model incorporating geographic mobility among regions," Mathematical Biosciences **128**, 71–91 (1995).

Alun L Lloyd and Robert M May, "Spatial heterogeneity in epidemic models," Journal of Theoretical Biology **179**, 1–11 (1996).

1943 Michael E. Gilpin, Ilkka Hanski, and ScienceDirect (Online service), *Metapopulation biology: ecology, genetics, and evolution* (San Diego, CA : Academic Press, 1997).

Bryan Grenfell and John Harwood, "(Meta) population dynamics of infectious diseases," Trends in ecology & evolution **12**, 395–399 (1997).

Matt J Keeling and Pejman Rohani, "Estimating spatial coupling in epidemiological systems: a mechanistic approach," Ecology Letters **5**, 20–29 (2002).

Matt J Keeling, Ottar N Bjørnstad, and Bryan T Grenfell, "Metapopulation dynamics of infectious diseases," in *Ecology, genetics and evolution of metapopulations* (Elsevier, 2004) pp. 415–445.

Duncan J Watts, Roby Muhamad, Daniel C Medina, and Peter S Dodds, "Multiscale, resurgent epidemics in a hierarchical metapopulation model," Proceedings of the National Academy of Sciences of the United States of America **102**, 11157–11162 (2005).

Steven Riley, "Large-scale spatial-transmission models of infectious disease," Science **316**, 1298–1301 (2007).

Vittoria Colizza and Alessandro Vespignani, "Epidemic modeling in metapopulation systems with heterogeneous coupling pattern: Theory and simulations," Journal of Theoretical Biology **251**, 450–467 (2008).

Marco Ajelli, Bruno Gonçalves, Duygu Balcan, Vittoria Colizza, Hao Hu, José J Ramasco, Stefano Merler, and Alessandro Vespignani, "Comparing large-scale computational approaches to epidemic modeling: agent-based versus structured metapopulation models," BMC Infectious Diseases **10**, 190 (2010).

Matt J Keeling, Leon Danon, Matthew C Vernon, and Thomas A House, "Individual identity and movement networks for disease metapopulations," Proceedings of the National Academy of Sciences **107**, 8866–8870 (2010).

Frank Ball, Tom Britton, Thomas House, Valerie Isham, Denis Mollison, Lorenzo Pellis, and Gianpaolo Scalia Tomba, "Seven challenges for metapopulation models of epidemics, including households models," Epidemics **10**, 63–67 (2015).

GLOBAL TRANSPORTATION NETWORKS AND EPIDEMICS

Leonid A Rvachev and IM Longini, "A mathematical model for the global spread of influenza," Math Biosci **75**, 1–22 (1985).

Rebecca F Grais, J Hugh Ellis, and Gregory E Glass, "Assessing the impact of airline travel on the geographic spread of pandemic influenza," European Journal of Epidemiology **18**, 1065–1072 (2003).

Lars Hufnagel, Dirk Brockmann, and Theo Geisel, "Forecast and control of epidemics in a globalized world," Proceedings of the National Academy of Sciences of the United States of America **101**, 15124–15129 (2004).

Vittoria Colizza, Alain Barrat, Marc Barthelemy, and Alessandro Vespignani, "The role of the airline transportation network in the prediction and predictability of global epidemics," Proceedings of the National Academy of Sciences of the United States of America **103**, 2015–2020 (2006).

Andrew J Tatem, Simon I Hay, and David J Rogers, "Global traffic and disease vector dispersal," Proceedings of the National Academy of Sciences **103**, 6242–6247 (2006).

Cécile Viboud, Mark A Miller, Bryan T Grenfell, Ottar N Bjørnstad, and Lone Simonsen, "Air travel and the spread of influenza: important caveats," PLoS Medicine **3**, e503 (2006).

T Déirdre Hollingsworth, Neil M Ferguson, and Roy M Anderson, "Will travel restrictions control the international spread of pandemic influenza?" Nature Medicine **12**, 497–499 (2006).

Ben S Cooper, Richard J Pitman, W John Edmunds, and Nigel J Gay, "Delaying the international spread of pandemic influenza," PLoS Medicine **3**, e212 (2006).

Joshua M Epstein, D Michael Goedecke, Feng Yu, Robert J Morris, Diane K Wagener, and Georgiy V Bobashev, "Controlling pandemic flu: the value of international air travel restrictions," PloS One **2**, e401 (2007).

Duygu Balcan, Bruno Gonçalves, Hao Hu, José J Ramasco, Vittoria Colizza, and Alessandro Vespignani, "Modeling the spatial spread of infectious diseases: The global epidemic and mobility computational model," Journal of Computational Science **1**, 132–145 (2010).

Michael A Johansson, Neysari Arana-Vizcarrondo, Brad J Biggerstaff, J Erin Staples, Nancy Gallagher, and Nina Marano, "On the treatment of airline travelers in mathematical models," PLoS One **6**, e22151 (2011).

Liang Mao, Xiao Wu, Zhuojie Huang, and Andrew J Tatem, "Modeling monthly flows of global air travel passengers: An open-access data resource," Journal of Transport Geography **48**, 52–60 (2015).

HUMAN MOBILITY AND EPIDEMIC DIFFUSION

Cécile Viboud, Ottar N Bjørnstad, David L Smith, Lone Simonsen, Mark A Miller, and Bryan T Grenfell, "Synchrony, waves, and spatial hierarchies in the spread of influenza," Science **312**, 447–451 (2006).

Duygu Balcan, Vittoria Colizza, Bruno Gonçalves, Hao Hu, José J Ramasco, and Alessandro Vespignani, "Multiscale mobility networks and the spatial spreading of infectious diseases," Proceedings of the National Academy of Sciences **106**, 21484–21489 (2009).

Filippo Simini, Marta C Gonzalez, Amos Maritan, and Albert-Laszlo Barabasi, "A universal model for mobility and migration patterns," Nature **484**, 96–100 (2012).

James Truscott and Neil M Ferguson, "Evaluating the adequacy of gravity models as a description of human mobility for epidemic modelling," PLoS Computational Biology **8**, e1002699 (2012).

Segolene Charaudeau, Khashayar Pakdaman, and Pierre-Yves Boëlle, "Commuter mobility and the spread of infectious diseases: application to influenza in France," PloS One **9**, e83002 (2014).

Michele Tizzoni, Paolo Bajardi, Adeline Decuyper, Guillaume Kon Kam King, Christian M Schneider, Vincent Blondel, Zbigniew Smoreda, Marta C Gonzalez, and Vittoria Colizza, "On the use of human mobility proxies for modeling epidemics," PLoS Computational Biology **10**, e1003716 (2014).

Amy Wesolowski, Taimur Qureshi, Maciej F Boni, Pål Roe Sundsøy, Michael A Johansson, Syed Basit Rasheed, Kenth Engø-Monsen, and Caroline O Buckee, "Impact of human mobility on the emergence of dengue epidemics in Pakistan," Proceedings of the National Academy of Sciences **112**, 11887–11892 (2015).

Dirk Brockmann and Dirk Helbing, "The hidden geometry of complex, network-driven contagion phenomena," Science **342**, 1337–1342 (2013).

AGENT-BASED MODELING APPROACHES

Stephen Eubank, Hasan Guclu, VS Anil Kumar, Madhav V Marathe, Aravind Srinivasan, Zoltan Toroczkai, and Nan Wang, "Modelling disease outbreaks in realistic urban social networks," Nature **429**, 180–184 (2004).

Neil M Ferguson, Derek AT Cummings, Simon Cauchemez, Christophe Fraser, Steven Riley, Aronrag Meeyai, Sopon Iamsirithaworn, and Donald S Burke, "Strategies for containing an emerging influenza pandemic in southeast Asia," Nature **437**, 209–214 (2005).

Ira M Longini, Azhar Nizam, Shufu Xu, Kumnuan Ungchusak, Wanna Hanshaoworakul, Derek AT Cummings, and M Elizabeth Halloran, "Containing pandemic influenza at the source," Science **309**, 1083–1087 (2005).

Timothy C Germann, Kai Kadau, Ira M Longini, and Catherine A Macken, "Mitigation strategies for pandemic influenza in the United States," Proceedings of the National Academy of Sciences **103**, 5935–5940 (2006).

M Elizabeth Halloran, Neil M Ferguson, Stephen Eubank, Ira M Longini, Derek AT Cummings, Bryan Lewis, Shufu Xu, Christophe Fraser, Anil Vullikanti, and Timothy C Germann, "Modeling targeted layered containment of an influenza pandemic in the United States," Proceedings of the National Academy of Sciences **105**, 4639–4644 (2008).

Stefano Merler and Marco Ajelli, "The role of population heterogeneity and human mobility in the spread of pandemic influenza," Proceedings of the Royal Society of London B: Biological Sciences **277**, 557–565 (2010).

Dennis L Chao, M Elizabeth Halloran, Valerie J Obenchain, and Ira M Longini Jr, "Flute, a publicly available stochastic influenza epidemic simulation model," PLoS Computational Biology **6**, e1000656 (2010).

John J Grefenstette, Shawn T Brown, Roni Rosenfeld, Jay DePasse, Nathan TB Stone, Phillip C Cooley, William D Wheaton, Alona Fyshe, David D Galloway, and Anuroop Sriram, "Fred (a framework for reconstructing epidemic dynamics): an open-source software system for modeling infectious diseases and control strategies using census-based populations," BMC Public Health **13**, 940 (2013).

Stefano Merler, Marco Ajelli, Laura Fumanelli, Marcelo FC Gomes, Ana Pastore y Piontti, Luca Rossi, Dennis L Chao, Ira M Longini, M Elizabeth Halloran, and Alessandro Vespignani, "Spatiotemporal spread of the 2014 outbreak of Ebola virus disease in Liberia and the effectiveness of non-pharmaceutical interventions: a computational modelling analysis," The Lancet Infectious Diseases **15**, 204–211 (2015).

Lander Willem, Frederik Verelst, Joke Bilcke, Niel Hens, and Philippe Beutels, "Lessons from a decade of individual-based models for infectious disease transmission: a systematic review (2006-2015)," BMC Infectious Diseases **17**, 612 (2017).

REAL-TIME PREDICTIVE MODELING

Hiroshi Nishiura, "Prediction of pandemic influenza," European Journal of Epidemiology **26**, 583–584 (2011).

Michele Tizzoni, Paolo Bajardi, Chiara Poletto, José J Ramasco, Duygu Balcan, Bruno Gonçalves, Nicola Perra, Vittoria Colizza, and Alessandro Vespignani, "Real-time numerical forecast of global epidemic spreading: case study of 2009 A/H1N1pdm," BMC Medicine **10**, 165 (2012).

Jeffrey Shaman, Alicia Karspeck, Wan Yang, James Tamerius, and Marc Lipsitch, "Real-time influenza forecasts during the 2012–2013 season," Nature Communications **4**, 2837 (2013).

Marcelo FC Gomes, Ana Pastore y Piontti, Luca Rossi, Dennis Chao, Ira Longini, M Elizabeth Halloran, and Alessandro Vespignani, "Assessing the international spreading risk associated with the 2014 West African Ebola outbreak," PLoS Currents **6** (2014).

Jean-Paul Chretien, Steven Riley, and Dylan B George, "Mathematical modeling of the West Africa Ebola epidemic," Elife **4**, e09186 (2015).

Gabriel Rainisch, Jason Asher, Dylan George, Matt Clay, Theresa L Smith, Christine Kosmos, Manjunath Shankar, Michael L Washington, Manoj Gambhir, and Charisma Atkins, "Estimating Ebola treatment needs, United States," Emerging Infectious Diseases **21**, 1273 (2015).

Shihao Yang, Mauricio Santillana, and Samuel C Kou, "Accurate estimation of influenza epidemics using Google search data via argo," Proceedings of the National Academy of Sciences **112**, 14473–14478 (2015).

Matthew Biggerstaff, David Alper, Mark Dredze, Spencer Fox, Isaac Chun-Hai Fung, Kyle S Hickmann, Bryan Lewis, Roni Rosenfeld, Jeffrey Shaman, and Ming-Hsiang Tsou, "Results from the Centers for Disease Control and Prevention's Predict the 2013–2014 Influenza Season Challenge," BMC Infectious Diseases **16**, 357 (2016).

Michael A Johansson, Nicholas G Reich, Aditi Hota, John S Brownstein, and Mauricio Santillana, "Evaluating the performance of infectious disease forecasts: A comparison of climate-driven and seasonal dengue forecasts for Mexico," Scientific Reports **6** (2016).

Kelly R Moran, Geoffrey Fairchild, Nicholas Generous, Kyle Hickmann, Dave Osthus, Reid Priedhorsky, James Hyman, and Sara Y Del Valle, "Epidemic forecasting is messier than weather forecasting: The role of human behavior and internet data streams in epidemic forecast," The Journal of Infectious Diseases **214**, S404–S408 (2016).

Cécile Viboud, Kaiyuan Sun, Robert Gaffey, Marco Ajelli, Laura Fumanelli, Stefano Merler, Qian Zhang, Gerardo Chowell, Lone Simonsen, and Alessandro Vespignani, "The RAPIDD ebola forecasting challenge: Synthesis and lessons learnt," Epidemics **22**, 13–21 (2018), https://doi.org/10.1016/j.epidem.2017.08.002.

James Manyika, Michael Chui, Brad Brown, Jacques Bughin, Richard Dobbs, Charles Roxburgh, and Angela H Byers, "Big data: The next frontier for innovation, competition, and productivity," (2011).

Duygu Balcan, Hao Hu, Bruno Gonçalves, Paolo Bajardi, Chiara Poletto, José J Ramasco, Daniela Paolotti, Nicola Perra, Michele Tizzoni, Wouter Van den Broeck, Vittoria Colizza, and Alessandro Vespignani, "Seasonal transmission potential and activity peaks of the new influenza A (H1N1): a Monte Carlo likelihood analysis based on human mobility," BMC medicine **7**, 45 (2009).

T Alex Perkins, Amir S Siraj, Corrine W Ruktanonchai, Moritz UG Kraemer, and Andrew J Tatem, "Model-based projections of Zika virus infections in childbearing women in the Americas," Nature Microbiology **1**, 16126 (2016).

Qian Zhang, Kaiyuan Sun, Matteo Chinazzi, Ana Pastore y Piontti, Natalie E. Dean, Diana Patricia Rojas, Stefano Merler, Dina Mistry, Piero Poletti, Luca Rossi, Margaret Bray, M. Elizabeth Halloran, Ira M. Longini, and Alessandro Vespignani, "Spread of Zika virus in the Americas," Proceedings of the National Academy of Sciences, **114**, 22 (2017).

The manufacturer's authorised representative in the EU is Springer
Nature Customer Service Centre GmbH, Europaplatz 3, 69115 Heidelberg,
Germany. If you have any concerns regarding our products, please
contact ProductSafety@springernature.com

Printed and bound by CPI Group (UK) Ltd, Croydon, CR0 4YY

27/04/2026

02097588-0001